ROUTLEDGE LIBRARY EDITIONS:
INTERNATIONAL BUSINESS

THE LARGE INTERNATIONAL FIRM IN DEVELOPING COUNTRIES

THE LARGE INTERNATIONAL FIRM IN DEVELOPING COUNTRIES

The International Petroleum Industry

EDITH T. PENROSE

With a Chapter on the Oil Industry in Latin America by
P. R. ODELL

Volume 30

Routledge
Taylor & Francis Group
LONDON AND NEW YORK

First published in 1968

This edition first published in 2013
by Routledge
2 Park Square, Milton Park, Abingdon, Oxon, OX14 4RN

Simultaneously published in the USA and Canada
by Routledge
605 Third Avenue, New York, NY 10017

Routledge is an imprint of the Taylor & Francis Group, an informa business

British Library Cataloguing in Publication Data
A catalogue record for this book is available from the British Library

ISBN: 978-0-415-63009-2 (Set)
ISBN: 978-0-415-65843-0 (Volume 30) (hbk)

Publisher's Note
The publisher has gone to great lengths to ensure the quality of this reprint but points out that some imperfections in the original copies may be apparent.

Disclaimer
The publisher has made every effort to trace copyright holders and would welcome correspondence from those they have been unable to trace.

THE LARGE INTERNATIONAL FIRM IN DEVELOPING COUNTRIES

THE INTERNATIONAL PETROLEUM INDUSTRY

BY

EDITH T. PENROSE

With a Chapter on the Oil Industry in Latin America by

P. R. ODELL

London

GEORGE ALLEN AND UNWIN LTD

RUSKIN HOUSE MUSEUM STREET

FIRST PUBLISHED IN 1968

PRINTED IN GREAT BRITAIN
in 10 *on* 11 *point Times Roman*
BY SIMSON SHAND LTD
LONDON, HERTFORD AND HARLOW

For E.F.P.

NOTES ON UNITS AND TERMINOLOGY

1. UNITS

The petroleum industry uses an unfortunate variety of units in published statistical tables. Crude-oil production and reserves and refinery throughput may be expressed in barrels (42 US gallons) per year, barrels per day, long tons or metric tons per year. (In Japan usually kilolitres.)

Metric tons per year are obviously the scientifically preferable unit, but all of the American companies consistently use barrels per day, and I have followed suit, with the only excuse that it saved the labour of conversion.

Metric tons per year have been converted to barrels per day by multiplying by 0·02, which reflects an average world gravity. Barrels per day can be converted to tons per year by multiplying by 50.

I have used an exchange rate of £1=$2·80. For other currencies I have used an average rate for the relevant period.

2. TERMINOLOGY

In the oil industry the terms '*producing*' and '*production*' refer to crude-oil production only, which is called an '*upstream*' activity. Refining and marketing are called '*downstream*' activities.

Gross crude-oil production includes royalty oil and oil the producer uses in operating the oilfield. *Net crude production* excludes this royalty. Most figures are given gross because the producer normally retains the royalty oil, but most of the American independent companies (i.e. non-'majors') give net figures only in their published reports. Gross crude includes natural gas liquids contained in the oil (condensate).

Two types of companies are referred to in this study. The first is the internationally consolidated group, which includes the parent companies and all subsidiaries and affiliates so far as information is available to include them. For this type I use the terms *firm*, *group*, or *Company* (with a capital C) interchangeably. The second type of company is the subsidiary or affiliate, which may or may not be locally incorporated. For this type, I use the term *company* (with a small c).

The seven largest international oil Companies are referred to as the *majors*. They are Standard Oil Company (New Jersey), Royal Dutch/Shell, British Petroleum, Gulf Oil Corporation, Texaco, Standard Oil Company of California, and Mobil Oil Company. Compagnie Française des Pétroles is often referred to as a major company, but is not so considered here. Other companies are referred to as 'independents' and 'state' or 'government' companies. Here the term *independent* will be used to include all the non-major companies as well as the national oil companies set up in the oil-producing countries where no distinction between these two groups is required.

ACKNOWLEDGMENTS

This book could not have been written without the help of large numbers of people to whom I want publicly to express my deepest gratitude.

Perhaps my most pervasive debt is to J. E. Hartshorn (which will undoubtedly surprise him) for his valuable book on *Oil Companies and Governments*, for thought-provoking discussions, and for a careful and detailed criticism of an earlier draft of the manuscript which forced me to do much re-thinking and re-writing. He alone will recognize the extent of his influence if he reads this book again. But, as with all others mentioned here, he cannot be held responsible for anything I have written, nor can it be assumed he is even in agreement.

Several other economists associated with oil in one way or another also read all or parts of the manuscript and gave me valuable criticism: Dr P. H. Frankel of Petroleum Economics Ltd, Mr W. J. Levy, Professor Zuhayr Mikdashi of AUB, and Mr Thomas Stauffer of Harvard. They, too, will find changes made in consequence. I should also like to thank Mr W. L. Newton of Petroleum Economics Ltd, and Mr Edward Symonds of the National City Bank of New York for discussions and information. I have learned much from numerous conversations and seminar sessions with Mr Francisco Parra of the Organization of Petroleum Exporting Countries, and from other members of the staff of OPEC. Professor M. A. Adelman of MIT, Professor Peter Bauer of the London School of Economics, and Professor Fritz Machlup of Princeton University (who are not concerned in any way with a study of the oil industry but who are interested in the economic problems) also read parts of the manuscript and forced me to clarify my arguments.

Many of the ideas have been tried out in seminars on the economics of the international petroleum industry organized over four years in the London School of Economics and the School of Oriental and African Studies by myself and my colleague, Dr P. R. Odell, from whom I have learned much; the criticisms of my students and other members of the seminar have had more influence than they realized. In particular, I have drawn on the work of Dr B. Dasgupta, who assisted me at times in my research, Mr S. Madujibeya and Mr J. Fernandez. I appreciate as well the criticism of Mr D. Waddell and Dr E. Ahiram. Seminars held by OPEC in Geneva and Vienna, by the Kuwait Institute of Economic and Social Planning in the Middle East, by the American University in Beirut, and by Harvard have been most valuable.

I would like to express my appreciation to the Ministry of Mines and Hydrocarbons in Caracas for an invitation to visit Venezuela, and to Dr Alireo Parra, Chief of the Economic Section of the Ministry and now doing his turn as chairman of the Board of Governors of OPEC. From him and others in Venezuela I learned much about the industry in that country.

The London School of Economics generously financed a trip to the Far East in the course of my research on the industry. Without this much-appreciated help I would not have had the opportunities of first-hand discussions with officials of the governments and the oil Companies in Tehran, Pakistan, India, Ceylon, Hong Kong, and Tokyo. I must express my thanks for the help and information given me by friends in Iraq and Beirut, by the National

Iranian Oil Company, the Pakistan Institute for Economic Development, the oil Companies and government officials in India (including Mr Malaviya, Mr S. S. Khera, and Mr K. R. Damle), the Ceylon Petroleum Corporation, the Union Oil Company in Hong Kong, and, in Tokyo, by the Arabian Oil Company. The Arabian Oil Company provided an outstanding example of the famed Japanese hospitality. In all these countries, the Western-owned oil Companies—notably Esso, Shell and Burmah Shell—were most co-operative.

The oil Companies without exception, but in varying degrees, have responded helpfully to my requests for information. In London, both British Petroleum and Shell International have been generous with information and criticism. Their officials and statisticians went over the tables and an earlier draft of the entire manuscript and commented freely (and often caustically). I am particularly grateful for help on historical material to Mrs Rose Greaves of British Petroleum, and to Sir Maurice Bridgeman for his discussions and comment. My work has benefited especially from the criticism and wise comment of Mr Brandon Grove of Mobil Oil in London, who has been liberal in giving time to read and discuss a draft of the manuscript. All of the major American Companies, as well as Burmah Oil, reviewed the statistical material and the parts of the manuscript dealing with the description of their characteristics and operations in Chapter IV. Jersey Standard's comments were especially helpful. Needless to say, none of the Companies or individuals takes any responsibility for the way in which I have dealt with the material or indeed for its accuracy. It will be clear to the reader that the Companies will not necessarily have agreed with my analysis or conclusions; they may think the result a poor return for the help they have given. Nevertheless, I am indeed grateful.

And finally, there are hardly words left to express my debt to those who lived through the book with me. Mrs Judith Mabro, my secretary, took care of a myriad of details over the years. Miss Valerie Lowenhoff typed and retyped the manuscript with patience, good humour and great skill—but above all, with intelligence. Her contribution cannot be adequately acknowledged in a line of text. Because of my husband's criticism, the arguments are clearer and the expression of them more polished than they would otherwise have been. In addition, he richly deserves, and here receives, my profound gratitude not only for enduring without complaint the inevitable 'long hours of silence', but even more for suffering so many hours of talk on one subject!

Edith T. Penrose

School of Oriental and African Studies
University of London
August 1967

CONTENTS

LIST OF TABLES AND CHARTS

Chapter I

INTRODUCTION

This study is concerned with the economics of large international firms, but it is at the same time a study of one of the world's greatest industries in which a few large firms are of overwhelming importance. In 1966 the world's petroleum production exceeded 34 million barrels a day, or nearly one and three-quarter thousand million metric tons a year, which is over half a ton for each of the world's estimated 3,000,000,000 people, and output has been rising about three times as fast as population. The US was still by far the single largest petroleum-producing country in the world, as it has been, except for a few years around the turn of the century, for over 100 years. Production in the Western Hemisphere was surpassed only in 1964 by that of the Eastern Hemisphere. The Middle East accounted for about half of the output of the Eastern Hemisphere (71 per cent if we exclude the Communist countries), while African output rose at an average rate of 41 per cent a year between 1961 and 1966.

Outside North America and the Communist countries, the greater part of this oil is produced, refined into products and marketed by seven Western oil Companies (the so-called 'international majors'). Most of the crude oil produced enters into international trade, although some is refined in the crude-oil producing countries, and the products are exported; except for North America and the USSR, very little petroleum is consumed in the major countries producing it. Some is exported to countries in Africa, Asia, notably Japan, and Latin America, but most of it goes to the industrialized countries of Western Europe.

Thus, crude oil is one of the major commodities in world trade, yet most of the crude crossing international frontiers is transferred between the affiliates of a small group of international firms. The way in which prices, quantities, and the distribution of investment and of the income originating in the industry are determined bears only a faint resemblance to the processes traditionally analysed in the theory of international trade and investment. The economics of the large international firm, including the significance of the fact that international economic relations in a number of important industries are very largely conducted within the administrative framework of a few such firms, is a subject worthy of economic analysis in its own right.[1]

[1] Maurice Byé's comment that 'The mutual relations of large firms within one and the same industry . . . have an impact on the world economy which is at least as important as that which the theory of the market assigns to factor proportions', applies with full force to this industry, as we shall see. 'Self-Financed Multiterritorial Units and their Time Horizon', *International Economic Papers*, No. 8, translated from Professor Byé's article published in *Revue d'économie politique*, June 1957.

The competitive relationships among the oil Companies, and their policies with respect to supply, prices (especially inter-affiliate transfer prices) and investment have very important effects on the national income and balance of payments of developed and underdeveloped, exporting and importing countries alike. But in this study we shall be concerned only with the significance of the organization and operations of the international firms for the economies of the underdeveloped countries of Asia, Africa, and Latin America, including exporting and importing countries, and countries that are also minor producers. We shall not deal with their activities within the industrialized countries of the world, including the USSR (and Eastern Europe) and Japan. The study is primarily an attempt to analyse the structure of the industry, the development of the firms in it and the economic influence both have had. We shall examine the way in which supply has been controlled, the significance and functions of the prices of crude oil and of products, the relationship among the oil Companies, and the role of governments. For the most part these subjects do not lend themselves easily to measurement.

In Chapter II the general issues relating to the operations of large international firms are analysed and the groundwork is laid for a study of the international petroleum firms. A brief history of the industry follows in Chapter III. This is fairly well known to anyone who has been concerned with the subject at all, but it is essential background for those who are not familiar with it. The history of the petroleum firms in Chapter III is, with some exceptions, less well known, but I have tried to avoid recapitulating in any great detail those aspects of the historical record that have been widely described elsewhere. Unfortunately, we do not have very complete and detailed historical information about the international oil firms. In spite of their enormous economic importance for national economies, many of them have not been generous with published information, especially financial information, and notably information about their affiliates in the underdeveloped countries.[1] We are often ignorant of why important policies were adopted in the past, and frequently even of what policies were adopted: the terms of agreements arrived at many years ago are still secret; and the particulars of operations and actions in many parts of the world are not publicly known. It is perhaps partly for this reason that no comprehensive general history of the international industry has yet been written, and only a few reasonably well-documented and balanced studies of individual firms, periods, or areas are available for use in a study of this kind.[2]

[1] There are, however, very great differences between the Companies in this respect. The annual reports of most Companies are steadily becoming more informative. Shell, Jersey Standard and Mobil Oil now provide a great deal of information (but not about their non-consolidated affiliates). British Petroleum has been re-casting its own reports to the advantage of the reader seeking information, but still gives somewhat less than the most informative of the American Companies, with one very important exception, which is worth more than much of that given by other Companies—BP is the only Company to include in its financial statements the Group proportion of capital expenditure, retained income, depreciation, etc., of its associated companies.

[2] Notable among these is the two-volume history of the Standard Oil Company (New Jersey) by G. S. Gibb and F. H. Knowlton and R. W. Hidy and M. E. Hidy (see bibliography). The four-volume study of the early history of the Royal Dutch by F. C. Gerretson is also a notable contribution.

Thus, attempts to appraise the impact of the international oil firms on the economies in which they operate have given rise to much controversy in a number of countries, developed and underdeveloped alike, controversy that has been carried on in a fog of ignorance. Even governments have taken action against the firms or refrained from taking action, with very little of the relevant information at their disposal. Judgments have often been faulty for lack of information; they have also been clouded (and not only in the underdeveloped countries) by the emotive use of words. For example, if one wants to minimize misunderstanding and obtain a measure of agreement, at least on the nature of the issues, the word 'monopoly' is better avoided so far as possible. From very ancient times monopoly has, with few exceptions (e.g. a patent monopoly), been associated with practices that enrich sellers in ways universally considered to be socially undesirable and that certainly do not commend themselves to consumers. It is a word fraught with emotional overtones, and unless used as a technical term of art, it is more likely to confuse than to clarify issues.

THE PROBLEMS OF APPRAISAL

When approaching the appraisal of large firms, economists almost instinctively think in terms of monopolistic pricing and market behaviour, and of restrictions on output with a consequent distortion in the use of resources. This is a very narrow way of looking at the issues involved, as has been pointed out by Schumpeter and many others, who have stressed particularly the contributions that large firms have made to technological progress—contributions which can to some extent be attributed to the existence of a degree of monopoly and to 'product competition' instead of price competition. However, this simply illustrates the more general proposition that to appraise the activities of large firms it is necessary to look not only at the way in which they raise their money, that is, largely at their pricing policies, but also at how they use their money. Refusal to do this can be defended only on the general ground that price competition alone can produce an 'optimum' allocation of resources and that, except in rare and carefully defined instances, monopolistic elements necessarily produce a malallocation of resources in almost all circumstances.

In the modern world of 'research-based' industries, patent protection, brand allegiance, and a large variety of financial and other advantages of sheer size, almost any large firm has considerable 'monopoly power'. A large firm is essentially a 'planned economy'; its planning not only includes the setting of both internal and external prices but also the long-run programming of production in the light of its own resources as well as of market demand. Such a firm works intensively to develop particular fields which are deemed profitable, moving into greener ones at the margin as new opportunities emerge, but devoting much of its energy and its finance to creating these new opportunities for itself. Even if higher profits are obtained from a restriction of output in one direction or in one country, they may be ploughed back to expand output or create entirely new products in some other, often unpredictable, direction, or in another country. It is extremely difficult to insist *a priori* with any conviction that gains from this sort of thing are *inevitably* outweighed by the losses that economic theory traditionally attaches to monopolistic pricing and the consequent failure to expand the output of existing goods to the point of lowest average cost.

Nevertheless, with all its weaknesses when used as a basis for normative judgments, the traditional theory, well supported by empirical evidence, does demonstrate that in a private-enterprise economy the efficacy of the profit motive as a spur to enterprise and good management, and the significance of profits as an indication of managerial efficiency and as a meaningful guide to the use of resources, are intimately bound up with the degree of price competition to which firms are subject. Both of these functions of profit are less likely to be fulfilled satisfactorily as competition becomes less pressing. Moreover, it is clear that when the 'market' is not permitted, or is not able, to make the economic allocations, some authority must, and there is certainly no presumption that powerful private interests will be the most appropriate arbiters of the public interest.

From this point of view international firms create special difficulties. It is hard enough to define the 'public interest' within a single country, but at least there exists an authority specifically charged with responsibility for it; internationally there is as yet no such authority except in special cases. Moreover, in a world in which there are great inequalities in the international distribution of population, natural resources and skills, coupled with severe barriers to the movement of goods and services between the several economies, economic theory provides no 'objective' general criteria with respect to which the performance of a particular industry can be judged.

All this, however, does not mean that we cannot come to conclusions; it means only that we must face up to the fact that the acceptability of the conclusions depends upon the acceptability of the 'value judgments' on which they rest. From the outset, therefore, it is important to distinguish two purposes that can be served by any economic discussion. First, a straightforward analysis of the economic and institutional factors at work, and of their results in the past and present, can provide an explanation of what happened or is happening, or a prediction of what is likely to happen. This does not involve a judgment on the desirability of the results in the past or the future; the purpose is to explain, not to appraise, or to analyse consequences, not to recommend courses of action. For this purpose, traditional economic theory used to explain the direction of change in an industry as a whole is an indispensable tool, even when the assumptions are highly abstract.

Second, an evaluation of the desirability of alternative outcomes can provide criteria for the formulation of policies in so far as outcomes can be influenced by the actions of particular groups, be they Companies or governments. Here the purpose is to make recommendations about appropriate choices in the light of the criteria chosen. It is often said that the second kind of problem does not come within the scope of economics and consequently is not within the competence of an economist *qua* economist, since account must be taken of considerations which have nothing to do with the 'science' of economics. I would not quarrel with this provided it is not interpreted to mean that a student of a particular problem is precluded from passing judgment on any aspect of the problem simply because he happens to be trained in economics. All that matters is that the judgments made be appraised in the light of the arguments and evidence put forward, and that no one is induced to accept them under the illusion that their validity is enhanced by the fact that an economist made them!

Thus, it is important that as clear a distinction as possible should be drawn between the kind of analysis that is essentially of cause and effect, and the kind that involves judgments about the desirability of particular conditions or outcomes. This being done, there is much to be said in a study such as this one for presenting conclusions about appropriate policy; otherwise, the reader whose chief interest may be in the practical question of what ought to be done may be left with a finely balanced statement of the pros and cons and a feeling of intense frustration and extreme irritation with the author for not making up his mind! But if any reader should be under the illusion that economic analysis alone can arrive at objectively demonstrable and unequivocably 'right' policies, I want unequivocably to disillusion him.

In the petroleum world the men responsible for policy are grappling with seemingly insoluble problems as they struggle to defend the interests of their respective groups in an industry where political relationships and institutional forms alike are being severely strained by changing economic and political realities. The great international oil Companies, long the dominant force in the industry, are by no means of common mind, but so far they present a reasonably united front, putting their case before the publics in many countries, listing their achievements, recounting the benefits they have showered equally on producers and consumers, defending their innocence against charges of exploitation and deception, and occasionally warning the world of dire results should their financial position appreciably worsen.

The interests of governments are of many kinds: oil is often literally the life-blood of their economies and vital to their military security; it plays a role in external political alignments and responsibilities; it influences, sometimes powerfully, their balance of payments; it is a significant item for their tax receipts in domestic currency; and it is responsible for the creation of vested interests which often exercise considerable economic and political influence. Each government is faced with a different set of circumstances, and the relative importance of the individual ingredients varies, necessitating different weighting in the determination of the general policies to be adopted.

In spite of divergent interests, the great oil-producing countries, conscious of their strategic position and openly endeavouring to consolidate it by the type of collective action that they are the first to condemn when they suspect either Companies or consumers of practising it, demand greater and greater revenues as their 'just right', pleading their poverty and the need for funds for their economic development, and charging the Companies with despoliation and deprivation in the past. They have, in fact, done very well out of the industry, for the oil Companies, in the last fifteen years at least, have paid handsomely for their petroleum rights (and for peace and quiet); most of the producing countries obtain from profits and royalties alone foreign exchange receipts per head of their populations which are many times those received by a large number of the other underdeveloped countries of Asia and Africa from all exports. The fact that crude-oil producing countries are still economically and socially very little developed cannot be blamed on the oil Companies.

The rich industrialized importing countries, united in their need for secure long-term supplies of energy but divided on almost all other issues that face the industry, go their separate ways in defending what each conceives to be its particular interest. The poor underdeveloped importing countries, struggling with intransigent balance of payments problems, alternate between grateful acceptance of capital investment for exploration, production or refining from the oil Companies and bewilderment over and suspicion of their price, distribution, and financial policies. They, too, plead their poverty and their problems of economic development, and on occasion have protested forcefully at the bland insistence that prices are set by 'market forces', before which the powerful and the weak alike must give way.

The ultimate consumers—industrialists, households, and consuming agencies of governments—are but occasionally heard from. These are often affected as much by government tax policies as by the actions of the industry itself. Motorists, the chief consumers of gasoline, periodically complain of high taxes, for gasoline taxes exceed 70 per cent of the price in some countries; industrialists and households continue steadily to substitute oil or gas for solid fuel in spite of widespread government measures designed to protect the coal industry.

The oil Companies have endeavoured to sell their oil at prices below those of other energy equivalents; the industry is technologically progressive and pays its workers well; in addition to the revenues paid directly to governments it has spent millions of dollars and pounds sterling, spread among almost all the countries of Asia and Africa and Latin America, searching for oil, building refineries, establishing distribution centres, rendering technical aid, improving agriculture and water supplies, and in numerous other ways directly contributing to economic development.

And yet the industry is in trouble: the great Companies are looked on with suspicion nearly everywhere, their political relations with the producing countries are frequently strained, and the underdeveloped importing countries often regard them with very great suspicion occasionally even going so far as to expropriate their properties. Is this merely the resentment of the rich by the poor? Are the uneasy relations between the oil-producing countries and the Companies simply the result of the fact that large amounts of money are involved and that human greed is never satisfied? For one must not forget that the total amount of profit arising in the industry before taxes is very great indeed. Or are there other issues which arise not only from the structure and activities of the petroleum Companies, but have also deeper sources relating to the role and operations of international firms generally? This is the subject matter of our enquiry.

Chapter II

THE NATURE AND ECONOMIC SIGNIFICANCE OF THE LARGE INTERNATIONAL FIRM

Economic theory has never comfortably digested the large diversified firm. The theoretical system which provides the economic justification of a competitive economy takes grossly inadequate account of these great administrative organizations, and the theory of international trade and investment virtually ignores them, in spite of the fact that a very important part of both is carried on within the compass of their administrative framework. Because of the large international firm, the flow of international trade and investment, the structure of international prices, and the international distribution of income are subject to forces that are at times very different from those presumed in ruling economic theory. Yet it is to this theory that even the great international corporations openly appeal in demanding their independence from government 'intervention'.[1]

The study of the development, the structure, and the operations of the international petroleum industry is a study in the economics of large international firms. The development of modern capitalist enterprise is epitomized in the history of the petroleum firms; in the international sphere at least, the issues that their operations pose for both economic analysis and public policy are, in large part and on an enlarged scale, the fundamental issues associated with the general role of the large firm in the modern economy. To be sure, the international petroleum industry, and therefore the firms in it and the relationships among them, are in several ways unique, but this can be said of any industry and any firm; the uniqueness may at times obscure, but it does not blot out, the underlying similarity of the economic problems involved in the analysis of the operations of large international firms in this and other industries.[2]

[1] I do not imply here that the analytical tools of traditional economic theory are useless and inappropriate techniques of analysis, but only that inappropriate use is often made of them and inappropriate conclusions drawn. As I suggest later, the 'real world' cannot in any meaningful sense be measured against highly abstract economic 'models' of the type represented by the theory of perfect competition, but by the same token such models cannot be used in arguments against government intervention to restrict the freedom of action of private enterprises simply because they operate in a 'market' or 'competitive' system and regardless of the kind of power they exercise.

[2] The analysis in this chapter is based primarily on a study of British and American firms, for it is in these countries that almost all of the really large international firms have their headquarters, and it may not apply without some modification to Continental European

The technological changes in industry and agriculture that vastly increased the productive powers of men in the eighteenth and early nineteenth centuries began in Europe. Along with the new technology, increased capital formation and the break-up of older social attitudes and relationships, which laid the foundations for new patterns of working and living, came new developments in industrial organization. The modern corporation, or limited liability company, is a direct descendant of the older joint stock company, which was first used by merchant traders and later by various types of public utilities, and which in the early nineteenth century invaded industry on a small scale. In its modern form this creature of the law has become the dominant type of industrial and commercial organization. The intricate network of control over a large number of legal entities, with different functions, that can be erected has provided a magnificent instrument for extending the capacity of a few men, not only to organize and direct the exploitation of the opportunities for profitable investment opened up by technological changes in industry and transport and by rapidly expanding markets, but also to create new technology and new opportunities.

The nineteenth century saw changes in social outlook, most marked perhaps in the US, that gave great prestige to men possessed of money and of power in business, without inquiring too closely into just how these had been obtained. This, together with the invention of limited liability, gave rise to a situation—and again especially in the US—in which the aggressive 'go-getter' not only became the model of social as well as of business virtue, but also had the ideal instrument to hand for the moulding of an industrial empire. The stakes were high, but in the early days the risks were great; rivalry was often intense and various ways of mitigating its effects were commonplace—price rings, pools, trusts, cartels, monopolistic acquisition and merger.

The history of the petroleum industry in the US up to the First World War displays, sometimes in exaggerated form, the most colourful characteristics of the corporate business of that period: aggressive enterprise, ruthless competitive tactics often explicitly designed to eliminate competitors and obtain monopolistic control of markets, supplies or productive facilities, brilliant organizing ability, technological progressiveness, and a steadfast belief in the social and economic desirability of unregulated, uninhibited business enterprise. But some of the competitive devices adopted by business were so ruthlessly successful in achieving their objectives that great popular resentment was aroused, leading in 1890 to the Sherman Antitrust Act, which was followed in later years by further legislation and numerous court decisions, of which one of the most famous was the dissolution of the Standard Oil Company of New Jersey in 1911.

Attempts to legislate against abuse of market power or undue extension of monopolistic control have gone further in the US than in most other countries, but public supervision of the operations of large corporations has by and large been confined to those aspects of their financial activities that could lead to

firms. In particular, some Continental governments apparently supervise the operations of their international firms rather more closely than do the governments of the USA and Great Britain. Moreover, it is much more difficult to obtain information about them.

fraud, the illegal evasion of taxes, the milking of shareholders, or the grosser forms of monopolistic restrictive practices. Even the largest corporation is still looked on as a private unit in a competitive enterprise system and thus, in principle, is entitled to much the same rights as private individuals are. There are, of course, significant exceptions; publicly held corporations, for example, are usually required to publish audited balance sheets and income statements, but even here the governing principle seems to be the protection of shareholders. The firms are still allowed a very high degree of privacy in their business affairs and a very great measure of independence from any kind of public scrutiny and control in spite of the great significance of their operations for the national economies in which they do business. Subsidiaries of foreign firms are not even required to report publicly on their affairs if their shares are not listed for public trading.[1]

This attitude towards the right of firms to conduct activities with a minimum of public intervention or supervision rests on certain inherited assumptions about the nature of firms and their relationship to the economic system. Historically, the business enterprise was simply an organization set up and owned by, and therefore accountable to, private individuals. It was merely an instrument of its owners, and the limited liability company was only a device enabling individuals to use their initiative and enterprise and risk their funds in productive ventures, while at the same time limiting the extent of their personal liability. It seemed natural that firms thus conceived as instruments of, and accountable to, their private owners should in principle be entitled to the same type of privacy and freedom of action as the law accorded to private individuals.

As corporate capitalism developed, the need for a variety of special regulations was recognized, the regulations taking different forms at different times in different countries, but neither in theory nor in practice has much serious attempt been made to come to grips with the issues raised by the fact that very large areas of national and international economic activity are in the hands of great autonomous bureaucracies, whose resources may exceed those of many a medium-sized country. These firms—or 'groups' as they are sometimes called—possess extensive power for the exercise of which they are in practice directly accountable to no one, although there are many real and important restraints on their use of power. Their activities are largely guided by a search for profit, but their motive does not seem to be the simple human desire to make money in order to enrich either the particular individuals who own them or those who are responsible for their management, even though some enrichment of these individuals accompanies and is a necessary condition of their success.[2]

THE AUTONOMY AND MOTIVATION OF LARGE FIRMS

Firms are essentially administrative planning units. Their administrative structure usually permits an extensive devolution of authority to departments

[1] Again, however, the USA seems to be taking the lead in obtaining information from such subsidiaries if their shares are traded 'over the counter'. It was announced in August 1966, that some eighty foreign firms had for the first time to file information about their activities with the US Government. (*The Times* (London), August 12, 1966.)

[2] See E. T. Penrose, *The Theory of the Growth of the Firm*, Chapter II.

and subsidiary units, but even where there is a great deal of decentralization of responsibility, the activities of established firms are nevertheless conducted within an administrative planning framework with recognized lines of authority and responsibility. Boards of Directors, assisted by 'high level' committees, are responsible for the general structure of the organization, lay down general lines of policy and provide for financial discipline and the appointment of senior executives. Even though the administrative boundaries of a large firm may be blurred at the edges because of the network of subsidiaries and affiliates through which it operates, often involving joint ventures with other firms or even with governments, it is by and large an identifiable, recognizable planning organization.

Within the somewhat permissive framework set by the law, the large firm is a self-governing bureaucracy operating in accordance with a loosely defined code of business ethics. In the smaller corporations, and in those few large ones in which ownership is heavily concentrated in the hands of a few individuals or families, the actions of management may reflect decisions taken primarily by the owners; in large corporations where ownership is widely spread, individual shareholders rarely have any influence on managerial decisions, except for those who are also among the directors and managers. Indeed, the chain of decision may be reversed, for the decisions of owners voting in poorly attended shareholders' meetings, or by proxies in the hands of the management, are more likely to reflect previously worked out managerial decisions than the other way round. Typically, the large firms have spread over a wide variety of activities and have a large number of shareholders, but are effectively controlled by management. Accountable in principle and in law to shareholders, they are in fact autonomous bureaucracies.

The bureaucracy, or administrative organization, of a large firm is 'autonomous' in the sense that it does not execute the orders nor carry out the policy of any other body, including the body of its own shareholders, but makes its own policies and its own decisions. To be sure, such a firm is subject to a wide variety of external and internal constraints, including constraints imposed by competition, by the 'countervailing power' of other groups, by the press and other organs of public opinion, by governments, or by the attitudes of groups associated with the firm itself, such as its labour force or its shareholders, both of which are necessary providers of productive services to a firm, and as such can make their influence felt.[1] 'Autonomy' does not imply omnipotence, but it does imply the absence of an effective overall control by any outside body.

Managements often protest that they have very little autonomy, that they are 'responsible', or even subservient, to shareholders. This 'responsibility' may be a legal truth but it is more often than not an administrative fiction when the firm has a large number of shareholders. The appearance of accountability is provided in a firm's Annual Reports to shareholders. In these, firms not only meet their legal obligations (often minimally) in reporting their finances, but also give a great deal of (carefully selected) information not required by the law, which is designed to strengthen the appeal of the firm to

[1] In some firms a closely-knit group may hold a large block of shares and be represented on the Board of Directors because of this, but such groups could hardly be said to 'represent' the mass of the shareholders.

existing and potential investors, and more broadly, to serve important purposes of public relations. Annual Reports of large firms (especially American and British firms) are today not only more lavishly produced and illustrated than in earlier days, but they also give more information than ever before. This can hardly be attributed to an increase in the control of shareholders, although it may well reflect not only an increased appreciation by the firm of the value of its 'public image' but even, for some firms perhaps, a subliminal appreciation of a genuine responsibility to a wider public than its own shareholders.

The fact that existing shareholders have little control over large widely-held firms has important implications for the financial policy of such firms, and is of special significance for international firms. It is commonly assumed that firms want to maximize profits, but I find it more plausible to assume, for large firms at least, that they are more interested in their retained funds than in total profits; so far as they try to 'maximize' anything, they appear to try to maximize retained funds net of dividends and taxes.[1] This objective of firms is implied in their dual emphasis on growth on the one hand, and on 'self-financing' on the other, for the ability to finance expansion through internal resources seems always to be regarded as a mark of merit. Depreciation and similar increments to reserves which usually appear as costs contribute to retained funds and, together with retained profits, are sometimes the cheapest and usually the safest and most convenient sources of investible funds. The primary objective of large firms is to maximize their own retained funds for the purpose of reinvestment in order to ensure the survival and continued growth of the organization; in many ways this is equivalent to maximization of long-term profits. Shareholders are not thereby 'robbed', for they enjoy an appreciation of their assets—which may not be income for tax purposes but is an alternative gain often bearing a lower rate of tax or none at all—when firms retain and reinvest earnings profitably.

When directors of firms emphasize in speeches and Annual Reports that they are dealing with 'shareholders' money' they may in a sense be correct so far as legal ownership is concerned,[2] but they may be very misleading if the impression is given, as is apparently sometimes intended, that shareholders are the *source* of the firm's funds which have been voluntarily made available

[1] I do not want here to enter into the controversy over whether firms in fact attempt to maximize anything, or are instead content with a 'satisfactory' level of profit or activity. I merely want to shift the emphasis from total profit to retained earnings. This makes little difference for most of the theoretical purposes served by the hypothesis of profit maximization, but it does explicitly indicate that firms can be expected to try to minimize taxes and to pay out in dividends only the amounts necessary to maintain their long-run financial position in capital markets. Other assumptions, for example, the assumption that firms maximize sales subject to a profit constraint, are not relevant for present purposes, as will be seen.

[2] I use the phrase 'in a sense' advisedly. See Robin Marris, *The Economic Theory of Managerial Capitalism*, pp. 11 ff, for an illuminating discussion of the position of shareholders: 'Strictly, however, all that a shareholder owns is his bundle of rights. His shares are his property, the company is not. The shareholders are not the legal owners of the assets of the company, nor even, in many countries, of the current profits before distribution.' (p. 12). It should also be noted that in some Continental countries specific legal injunctions are placed on directors to protect the rights of minority shareholders.

for investment by the firm. The notion that large firms are dealing with their 'shareholders' money' has been used in a highly misleading way in a number of contexts.

THE FINANCIAL ROLE OF SHAREHOLDERS

Shareholders have a claim on the income of their firms, but the amount of payment against this claim is defined from period to period by the firm's board of directors,[1] who are in principle elected by the body of the shareholders, but can in practice be removed only with great difficulty. Now, if we assume that a firm attempts to maximize retained funds, then from its point of view, all outlays relating to current operations will appear as costs—not necessarily costs of production but simply costs of being in business. Thus, we do in fact sometimes find taxes, including income taxes, treated as costs, and were it not for traditional notions about the relationship between shareholders and firms, we might well have found dividends also recorded as costs, since a dividend payment is a loss of funds to the firm. There is no obvious reason why firms should pay out more than is necessary to maintain their financial standing, or occasionally to give it a special boost if fresh approaches to the capital market are intended in the near future.

Moreover, we do in fact often find that movements in the ratio of dividends payments to total profit are inversely related to the rate of profit, for by maintaining the level of payments when the rate of profit is down a firm may help to offset the effect of reduced profitability on its standing in financial markets, and when profits are buoyant the prospects of growth with its consequent appreciation of share values may well remove the need to raise the rate of dividend proportionately. The relationship of dividend payments to profitability for the major petroleum Companies is shown in Chart I for the period 1954–66. The indexes of profitability and per cent of income paid out moved in a similar way for all Companies, and a clear inverse relationship appears for all Companies except Texaco. The empirical evidence on the factors determining dividend payments does not give any support to the assumption that large firms are willing to permit shareholders themselves to determine the level of dividend payments.[2] Indeed, the judgment of a firm's directors about the amount of funds needed by the firm to finance its expansion plans may, together with expected market reactions, be decisive.[3]

[1] '. . . it is the directors who determine the dividend, and they have gradually acquired discretion to withhold considerable proportions of current profits, which then, either as fixed or liquid assets, become the property of the company. This capital accumulated from retained profits "belongs" to the shareholders only to the extent provided by their specific rights. . . . But the shareholders cannot in general directly initiate a capital distribution except by enforcing total liquidation, i.e. by causing the assets to be sold at break-up value.' *Ibid*, p. 13.

[2] For a useful survey of the empirical literature see D. M. Lamberton, *The Theory of Profit*. R. Marris has carefully analysed the theoretical factors affecting the determination of dividends.

[3] Compare, for example, the following statements to shareholders by the Chairmen of British Petroleum and Shell Transport respectively:

'In a company such as ours, the amount to be distributed as dividend must be related not only to the profit for the preceding year, but also to the magnitude of our future capital

It is essential for a big firm to be thought well of in financial markets, which are nowadays increasingly dominated by large institutional investors whose affairs are managed by men highly skilled in appraising investment prospects and who, like a radar screen, are continually scanning the market. Such institutions are interested in placing their funds profitably and reasonably safely; they are not interested in controlling the business decisions of the firms whose shares they hold, providing things are going well. In these circumstances, it is most misleading to look at the retained earnings of firms as the voluntary savings of existing shareholders, who can in normal circumstances neither force a greater distribution nor turn out management for holding back funds.

The purchase of outstanding shares as a means of individual saving is, of course, voluntary, but if the existing shareholders of a particular firm whose directors withhold a large proportion of its profits should prefer a greater current income over an appreciation of the value of assets, they can always sell their shares and buy those of a firm whose directors follow a different policy. If there were always willing buyers, this would be a matter of indifference to the firm; only if the financial markets were unwilling to absorb such shares at acceptable prices, and thus imperil the financial status of the firm, would the directors have to reconsider their policies.

It follows from all this that the amount of dividends paid out is determined largely by the financial markets in the light of the particular characteristics of a firm, its past habits, and the other opportunities open to investors.[1] In other words, dividends have to be paid, sometimes on a handsome scale, but these payments are the cost to the firm of maintaining its financial reputation as a good investment; they are the cost of ensuring access to new funds if necessary.[2] Indeed, if we forget the incidental and purely legal aspect of shareholding—the fact that it is an ownership interest—and look at it simply as a way in which individuals can hold their assets with a risk of loss or chance of gain directly correlated with the financial success of the firm whose shares they hold, then there is no particular reason why shareholders should be paid more than is necessary to induce them to hold the shares. Firms provide a service to the community by offering, among other things, a variety of profitable ways in which individuals can hold their assets; shareholders perform a service by holding industrial assets and permitting a private financial market to exist.

In any event, few large firms rely significantly on their shareholders for new funds, although of course they may occasionally raise some additional money from them. Many, if not most firms, prefer what is called 'self-financing',

---

commitments and to our ability to raise finance in the form of loan capital on reasonable terms.' British Petroleum, *Annual Report*, 1964, p. 6.

'. . . the principles governing the Company's dividend policy remain unchanged, namely that dividend contributions will represent the maximum amounts which we believe to be compatible with the continuing capital requirements of the business'. Chairman, Shell Transport and Trading Company at the Annual General Meeting, 1966.

[1] The widespread use of stock options and incentive bonus schemes for management may also influence, at the margin, the decisions of business executives, since they thus have a personal financial interest in the price of their firm's shares.

[2] If firms pay only the minimum dividends necessary, then this payment can be looked on as a kind of 'normal profit' (in the Marshallian sense), or the supply price of capital to the firm.

Chart I

INDEXES OF PROFITABILITY AND PAY-OUT OF DIVIDENDS. SEVEN MAJOR COMPANIES, 1954–66.

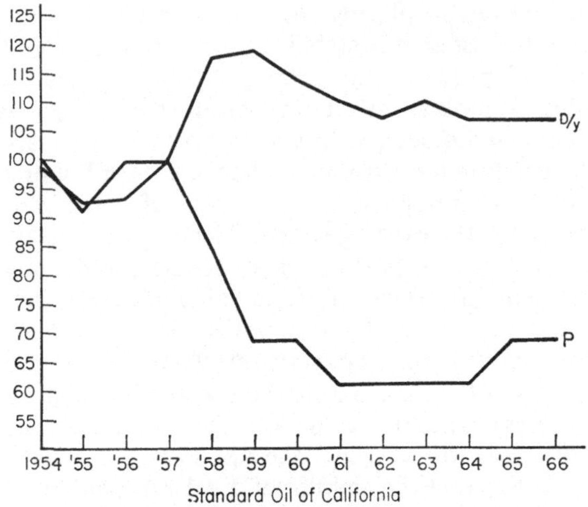

P=Ratio Income to Assets
D/y=Ratio of Dividends to Income

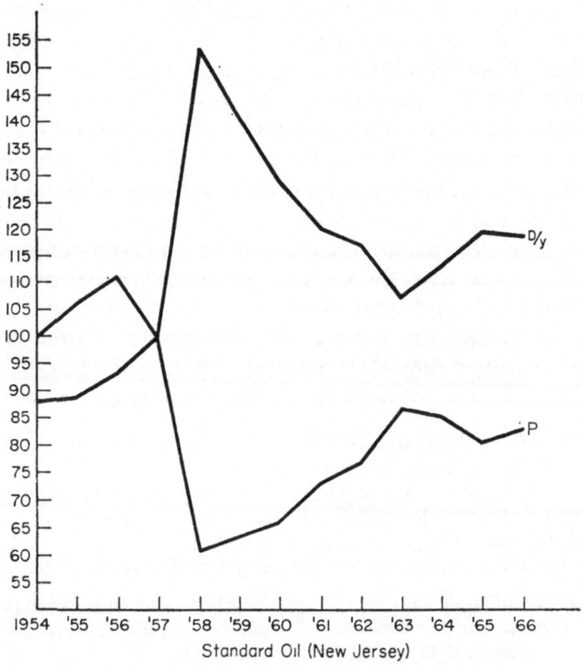

P=Ratio Net Income to Shareholders' Equity
D/y=Ratio of Dividends to Net Income

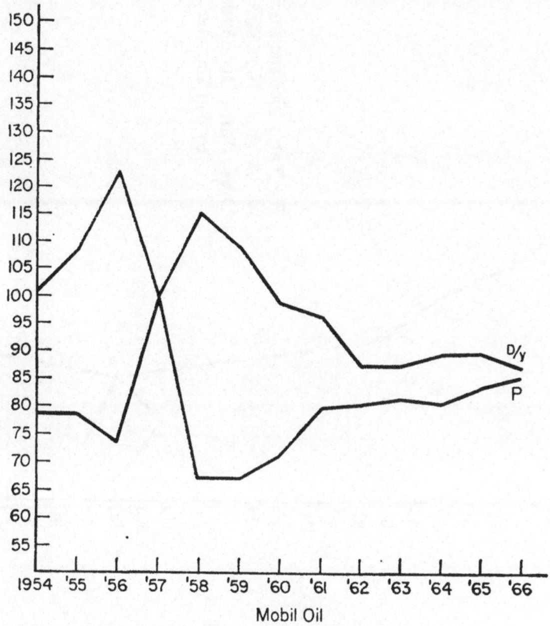

P=Ratio of Income to Average Total Assets
D/y=Ratio of Dividends to Income

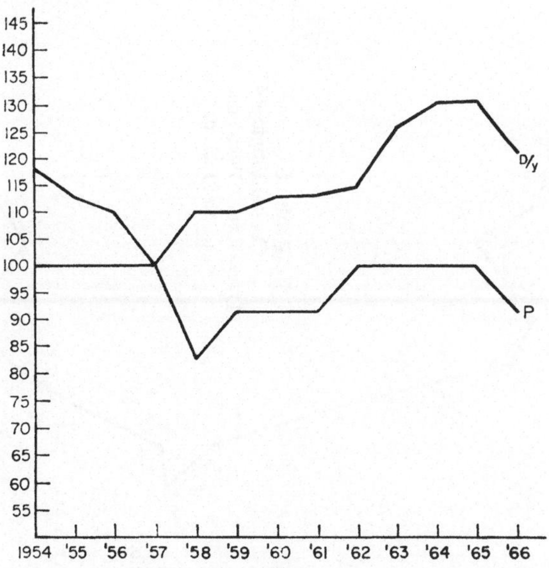

P=Ratio of Income to Total Assets
D/y=Ratio of Dividends to Income

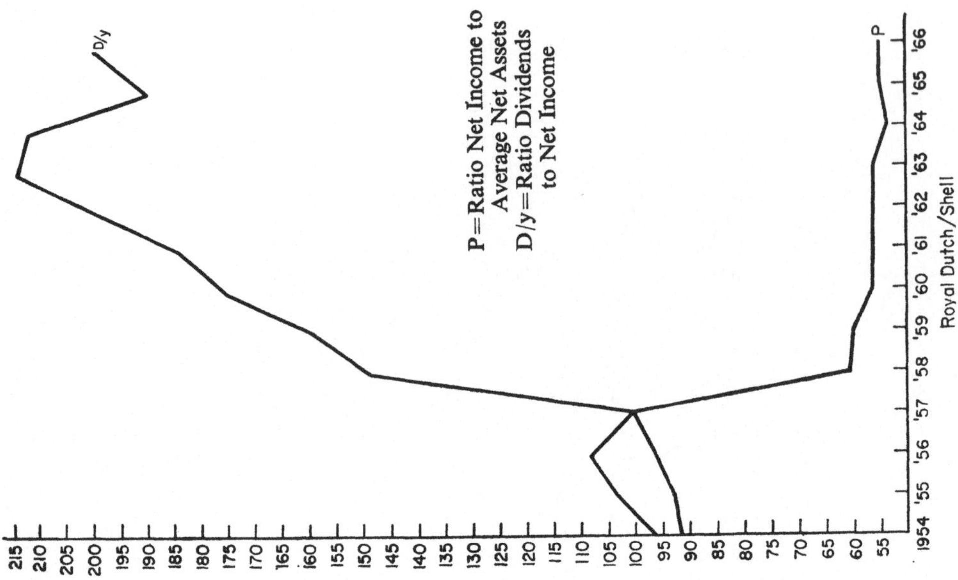
P=Ratio Net Income to Average Net Assets
D/y=Ratio Dividends to Net Income
D/y
P
215 210 205 200 195 190 185 180 175 170 165 160 155 150 145 140 135 130 125 120 115 110 105 100 95 90 85 80 75 70 65 60 55
1954 '55 '56 '57 '58 '59 '60 '61 '62 '63 '64 '65 '66
Royal Dutch/Shell

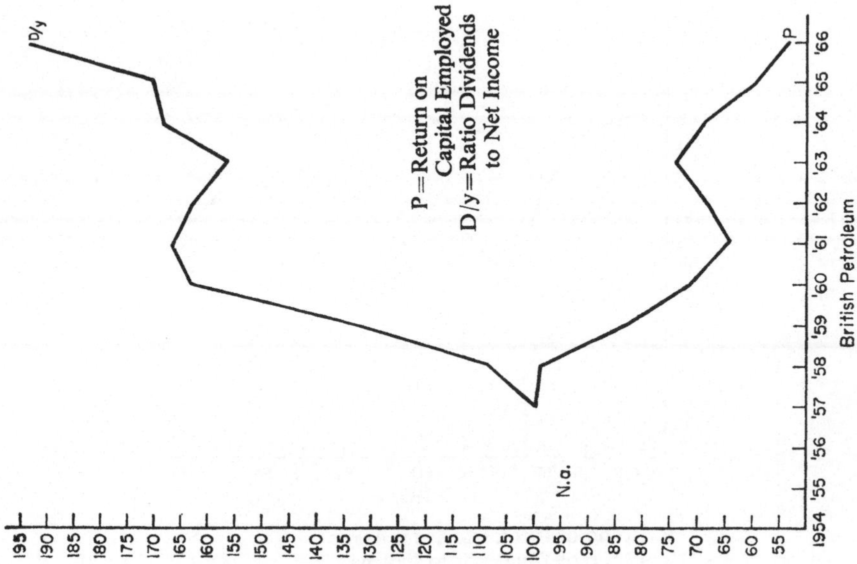
P=Return on Capital Employed
D/y=Ratio Dividends to Net Income
D/y
P
N.a.
195 190 185 180 175 170 165 160 155 150 145 140 135 130 125 120 115 110 105 100 95 90 85 80 75 70 65 60 55
1954 '55 '56 '57 '58 '59 '60 '61 '62 '63 '64 '65 '66
British Petroleum

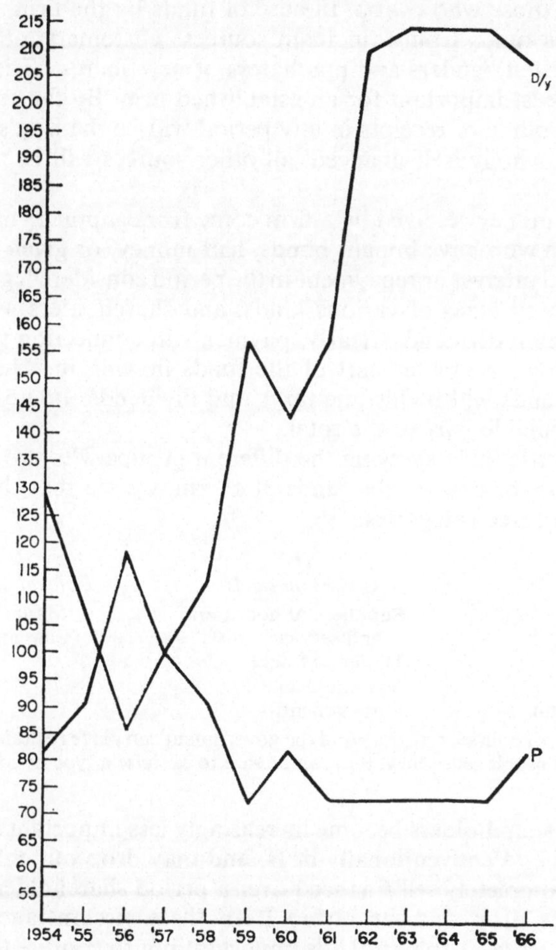

Gulf Oil Co.

P=Ratio Income to Assets
D/y=Ratio Dividends to Income

which means that they raise their money through the prices charged to their customers.

'Self-Financing'

In this respect it is useful to distinguish three broad groups: those who are the *source* of funds to the firm, those who have a *claim* on the funds received by the firm, and those who *control* the use of funds by the firm.

A firm receives funds from four main sources: customers, other firms in which it has invested, lenders and purchasers of new shares.[1] The last group is normally the least important for an established firm. By far and away the most important source of receipts in any period will be the firm's customers. In firms that are wholly 'self-financed' all other sources will be negligible at all times.

Claims on the money received by a firm come from suppliers of goods and services, creditors who have bought bonds, lent money, or given trade credit and are entitled to interest or repayment in the period considered, governments which have imposed taxes of various kinds, and shareholders who must be paid some minimum dividend. Clearly, payments in connection with current output will absorb the greater part of the funds flowing in. The other two chief claims on funds will be income taxes and dividends. Interest payments are usually a negligible part of the total.

If we make a little table showing the different groups who are the sources of, have claims on, or control the funds of a firm, we see that shareholders[2] can appear in all three categories.

I Sources of funds	*II* Claims on funds	*III* Control of funds*
Customers	Suppliers of goods and and services	Management
Lenders		Shareholders
Shareholders	Holders of debt	
Other receipts (e.g. sales of assets)	Shareholders	
	Government	

* To some extent, of course, creditors and the government can place restraints on a firm's use of its funds, and in this sense they, too, can be said to exercise a type of control over the funds of the firm.

As a firm grows, shareholders become increasingly less important in I, may be treated routinely and conventionally in II, and may drop out entirely in III. When a firm is completely self-financed over a period shareholders as well as long-term lenders drop out entirely in I. In these circumstances the firm receives its funds from customers;[3] it is important not to confuse the suppliers of funds with those having legal claims on funds once they are obtained.

[1] Other minor sources of funds are sales of assets and collected debts, which we shall ignore here, and occasionally government subsidies.

[2] The term 'shareholder' to indicate a source of funds is used here to include all purchasers of new shares whether or not they were already shareholders at the time of purchase.

[3] This was part of the basis of a comment by Enrico Bonomi of ENI on a paper by B. A. C. Sweet-Escott of BP at the Fourth Arab Petroleum Congress cited below: 'I feel it is questionable to contend that under the present system, in which around 90 per cent of new investment is financed through profits and depreciation, private enterprise provides the risk capital. It is not the shareholders of the oil companies who put up this risk capital but in fact the consumer.' Enrico Bonomi, 'Comments . . .', (Fourth Arab Petroleum Congress).

The fact that the retained income of firms is not income for shareholders and does not come from them has considerable significance; what happens to money as it changes hands, and to income as it is spent, depends very much on the desires and habits of the people who receive it at each 'round' of expenditure (or income). If shareholders received the entire net income of firms it would be astonishing if they chose to spend the funds received as these would have been spent in the hands of the firm. Some funds might well be given back for reinvestment, but some would probably be spent in other directions; moreover, even if the total sum available to all firms taken together did remain the same, the distribution of funds among firms would almost certainly be different.

The central argument for the 'necessity' of retaining earnings as a source of finance rests upon the proposition that a firm cannot afford to pay out its profits and then go back to its shareholders for funds because the receivers of dividends would not reinvest enough of their receipts in the firm. It is also alleged that if shareholders were in fact given an effective choice, less funds would be available to industry and probably still less for foreign investment. If this is true, then the existing arrangements cause a greater proportion of corporate profits to be saved and also invested abroad than would otherwise have been the case;[1] in addition, they give established firms with large net incomes an additional financial advantage over new and smaller firms.

It should be noted, however, that when producers try to justify high prices on the ground that funds so raised are 'needed' for investment to maintain future supplies, they are in effect arguing that it is right for present consumers to be forced to produce the funds required to meet the needs of future consumers. If, instead, new equity funds had been raised, present savers would have voluntarily produced the required finance; if money had been borrowed, then it would be paid back from the profits made from sales to the consumers who bought the goods financed by the borrowed money. In a growing industry the maintenance of prices to permit the financing of new investment from internal funds always puts a disproportionately high burden on present consumers.[2]

'*Shareholders' Funds*' *and* '*Taxpayers' Money*'

The retained or undistributed earnings of a firm are part of the shareholders' equity in the firm, and are often referred to as 'shareholders' funds'. These

[1] In an earlier paper I have argued that the risk attached to foreign investment is reduced when the investment takes the form of the reinvestment of retained earnings by subsidiaries of domestic firms operating in foreign countries, and that this, together with the tendency of parents to permit their subsidiaries to retain funds for investment, probably not only increases the total amount of foreign investment taking place but also influences the direction of it. See my 'Foreign Investment and the Growth of the Firm', *Economic Journal*, June 1956.

[2] The suggestion is sometimes made that if the profit margin in the price of a commodity is negligible, for example, the profit made on a gallon of petrol, the 'burden' on consumers is also negligible. This conclusion does not necessarily follow, however, for clearly the total sum paid by an individual consumer may be considerable, even if the absolute margin is small. It is, of course, possible that taxes are much the most important single factor in the price of a product, as they are for petrol in many countries. The consumer taxes collected by five of the largest American international oil companies exceeded the combined net income of the Companies by over $700 million in 1965.

privately owned funds, belonging to the shareholders of a firm, are often contrasted with publicly owned funds belonging to the government and raised from taxpayers, in that shareholders' funds are said to be voluntarily provided while government funds are compulsorily provided by taxpayers. Because of this it is asserted that some types of investment are inappropriate uses of public funds, or 'taxpayers' money'.

On these grounds it is widely argued that governments should not invest in the petroleum industry, particularly in exploration, because such investment is highly speculative and risky, and is not an appropriate use of public funds; risk-bearing should be left to private enterprise. This argument needs careful examination in so far as the reason for the distinction lies in the difference in the degree of compulsion.

Let us first clarify the nature of the relationships on both sides. In both cases there are three groups of people involved: these are the management of firms, the shareholders of firms, and the customers of firms on the one hand, and the management of government enterprises, the general voting public (in a democracy), and the taxpayers on the other. The decisions of management, whether it be that of private or of government enterprises, are similar in nature in that they are taken in the light of general policies, specific plans and operating necessities.[1] Neither the shareholders of large firms, nor the voting public, has any direct part in these decisions, although both may be said 'voluntarily' to acquiesce in the decisions if the shareholders continue to hold their shares and the public does not vote the government out of office. There is, to be sure, a very important distinction to be made: Individual shareholders of firms can always sell their shares (not necessarily without loss) providing there are willing buyers, and in this sense the acquiescence of shareholders in the decisions of management is more easily tested than is the acquiescence of the voting public in the decisions of government, since individual members of the public cannot easily alter their political allegiance. But if an enterprise (public or private) is successful there is surely no presumption that the public would acquiesce less readily than shareholders in the way the funds have been used. If it fails, both stand to lose, but it is not clear that one can reasonably say that shareholders in large diversified firms 'voluntarily' accepted the risk of failure of particular ventures *before the event*.

But, and more important for the present argument, what is the position of those from whose pockets the money is actually obtained by both firms and governments? These are consumers of a firm's products on the one hand, and taxpayers on the other. Among the most important sources of tax revenues for governments are taxes on the goods and services consumers buy—import duties, purchase taxes, taxes on services and financial transactions, etc. In other words, the government obtains much of its funds from roughly the same people—consumers—as do business firms—and in much the same way. Whether prices to a consumer of particular commodities are higher because

[1] Government decisions may of course give more weight to 'non-commercial' considerations than private firms can afford to give. But the distinction is by no means sharp, for large firms often undertake unprofitable or non-commercial activities in the 'public interest'. The chief difference may be that private firms cannot afford to make losses to the same extent that government firms can. But this has little to do with the principles we are discussing here.

of the policies of private firms or the taxes of government does not much affect his ability to avoid paying them. In both cases he can only abstain from purchasing as much of the products in question, and this 'voluntary' abstinence is just as much of a restraint on the amount of money the government can raise as it is on the revenues a firm can raise.[1] If consumers should be willing to buy about as much of a product, e.g. petrol, at higher as at lower prices, then governments have found a productive source of tax revenue, and firms, if they avoid 'excessive' price competition, may be able to raise as much money as they want for further investment.

From very ancient times monopolistic pricing has been recognized as an alternative to taxation as a means of raising money. Today it is widely agreed, at least in some countries, that the funds required for expansion should be one of the factors considered in the regulation of prices of nationalized industries or of public utilities. But it can hardly be denied that the effect is equivalent to an indirect tax, the proceeds of which are earmarked for specific purposes. For governments, indirect taxes are among the easiest ways to raise money, and for firms retained earnings are a more convenient and often cheaper source of money than the voluntary savings of those who purchase new shares. No firm in a monopolistic position could be expected to refrain from attempting to charge prices at least high enough to finance its optimum expansion plans if the market will bear it. And when government enterprises, even if they are monopolies, are profitable and can raise the money needed to finance their own investment, they are more often than not applauded. There seems to be a more general approval of self-financing by both private and government enterprises than of borrowing or calling on the general budget for support.

The purpose of this discussion is not to condemn out of hand this method of raising funds, even by private firms,[2] but rather to make clear its true nature. There are indeed many purposes for which it is inexpedient to use government funds in given circumstances, but in the oil industry the distinction that is usually made rests on the proposition that funds used in risky enterprises should be provided *voluntarily*.[3] It should be clear that the distinc-

[1] A government has, of course, much greater power than firms to restrict the alternatives open to consumers since it can tax substitutes and regulate alternative supplies, especially imports, on a wider scale than can individual firms. Our discussion is not, however, concerned with the *general* power of government to tax in relation to the power of big firms to secure monopoly prices, but only with the similarity from the consumer's point of view of equivalent actions by firms or governments.

[2] It can be argued, for example, that large firms must make very long-term investment plans, especially if a long gestation period is involved for investments, and that they cannot afford to be at the mercy of the vagaries of government monetary policy, which may turn the financial tap off at short notice in the interests of the overall economic stability of the economy. This is a strong argument from the point of view of the firms, but it does imply that large firms should be free to continue investment regardless of the position of the economy as a whole. Government can then influence them only by direct appeals, fiscal inducements of various kinds, or controls. But since the ability continuously to make large investments is a necessary condition of a strong competitive position for a large firm, the importance of retaining earnings is reinforced by alleged 'competitive necessity'.

[3] However, British Petroleum has never been known to argue that it is risking the State's funds in exploring for oil, although the British Government as the majority shareholder presumably 'owns' nearly 50 per cent of the company's retained profits; on the contrary, its

tion between public funds raised by the government through compulsory taxation, and private funds raised by firms, is far from sharp. If private funds are raised by firms charging higher prices for the purpose, the choices facing consumers are exactly the same as they would be if the higher prices were the result of indirect taxation on the same products imposed by government. Any implication that when oil companies explore for oil they risk funds voluntarily provided by shareholders, while governments risk funds compulsorily provided by taxpayers is clearly false.[1] Shareholders may voluntarily acquiesce in a firm's use of 'their' funds, but they are presumably content to do so simply because the firm has been able to secure them in the first instance by selling products so profitably that internal finance is available for investment.

INTERNATIONAL RELATIONSHIPS

In the light of the preceding considerations let us now examine certain aspects of the large firm when it operates on an international scale. An international firm can be defined for present purposes as a firm that has substantial operating assets in a number of countries, through branch offices, subsidiaries or affiliates owned wholly or in part. We shall not be concerned with firms that merely export or import. The firm (sometimes called a 'group') consists of the parent company and the subsidiaries or affiliates that are owned or controlled by it.[2] Since many affiliates may be owned jointly with other firms, the boundaries of any one 'firm' are not clear-cut. The most obvious measurements of the scope of a firm can be found in the consolidated financial and operating statements. These are rarely complete, however, since they will generally not include the full interest of the firm in affiliates owned 50 per cent or less, and for this reason the size of large firms, however defined, is usually understated.[3] Moreover, the financial information provided by international firms is subject to special difficulties of interpretation.

Significance of the 'Nationality' of a Firm

All 'international' firms have nevertheless an overriding nationality—that of the parent company—regardless of the local nationalities of affiliates. The dominant nationality, so to speak, is important from two points of view: in the first place, most of the shareholders of the parent are likely to share its nationality, and secondly, the foreign investment of the firm, or 'group', is treated as an investment of the country in which the parent is incorporated.

---

officials argue that 'to finance the hazardous business of exploration . . . surely cannot be a sound use of taxpayers' money. . . .' B. Sweet-Escott, p. 8. Behind this, of course, is the notion that the money might be lost and if so the taxpayer should not suffer. At the same time, however, consumers might be asked to pay higher prices in order to meet the 'increased costs'.

[1] The above discussion is entirely neutral with respect to the question whether any particular government should or should not invest in the petroleum industry.

[2] As stated in the note on p. 9, the terms 'firm', 'group', and Company (with a capital C) are used interchangeably to refer to the consolidated entity. For a discussion of the problems of identifying the boundaries of a firm, see my *Theory of the Growth of the Firm*, Chapter II.

[3] Such financial understatement exists for every one of the major international petroleum companies.

If there is little local ownership in the firm's foreign affiliates, there will be a flow of funds in the form of dividends from the firm as a whole, largely to the citizens of the parent's country. The particular geographical distribution of dividend outlays is, at least partly, a result of historical and institutional considerations related to the organization of the world's capital markets, and to the relative ease with which different investors from different countries can buy shares of the firm in question. The distribution of the nationality of the shareholders of a firm is not easy to discover, but there may be some tendency for ownership to become more widely dispersed internationally as the operations of international firms reach more deeply into more countries and it becomes possible for their shares to be quoted and bought more widely in foreign centres. The shares of the Royal Dutch/Shell group, in many ways the most 'international' of all of the oil Companies,[1] are held by a very mixed group of nationalities: 39 per cent by United Kingdom citizens, 19 per cent by Americans, 18 per cent by Dutch, 12 per cent by French, 10 per cent by Swiss, and 2 per cent by others. The bias is clear, but citizens of the parents' countries hold only slightly over half the total shares.[2]

If we can with some justification treat dividends as a kind of cost (or Marshallian 'normal profit') for the very large firm, then the fact that they tend to flow to the country of the parent firm may be regarded in much the same way as we regard other payments made by a firm which are influenced by its nationality. Trade follows the flag in more ways than one, and foreign firms often use equipment, personnel, and technology from their own country, not necessarily because they are concerned with nationality as such but because they either get better service or simply are accustomed to certain ways of doing things.

The retained profits of the consolidated firm are a different matter. In national income statistics these usually appear as the saving (investment) of the parent's country simply because of the nationality of the firm. Retained corporate income is saving by definition, but an important part of the income of an international firm may well have arisen because the firm had sufficient market power to impose, through its pricing policy, the equivalent of an indirect tax on consumers in countries other than the parent country; these consumers were thus forced to give up purchasing power to the firm, which retained it as savings. The reduction in consumption (or alternative investment) which made the specific saving possible will be spread over a number of economies other than that of the parent firm.

At the same time, however, the investment of retained earnings is also spread over a number of economies, either because the parent firm permits its foreign subsidiaries to reinvest some of the profits attributed to their own operations, or because the parent itself makes new net investments. Investment in some countries will exceed the 'forced saving' of their consumers; in others it will be less, for an international firm tends to distribute its investment outlays internationally in such a way as to augment its own retained profits.

Foreign investment is often looked at as a transfer of capital from one

[1] This group is in the unusual situation of having two parent companies of different nationality—see Chapter IV. It is reasonable, I think, to treat Royal Dutch/Shell as a firm with two nationalities, both of which may at times be important.

[2] Table VII.

country to another—the use of savings generated in one country for capital formation in other countries. US foreign investment, for example, is often spoken of as a contribution of the rich US *economy* to the rest of the world. But a large part of this investment is undertaken by international firms based in the US and obtaining much of their income from foreign operations, and to this extent it is wrong to attribute the earnings and therefore the 'saving' solely to the US *economy*, although it is correctly attributable to corporate nationals, so to speak, of the US. Imagine, for example, a large American-owned international firm which is entirely self-financed. Assume the firm made enough profit in the past from operations in the US to establish subsidiaries abroad on a small scale, which were then permitted to grow through the reinvestment of the profits made on their operations in foreign markets. To sharpen the picture, assume that competition in the US market has been intense and profits very low for a long time, while foreign operations have been very profitable, accounting for by far the greater part of the firm's total profits. In these circumstances, dividends will be largely paid from profits originating from foreign operations, and most of the firm's retained earnings will be reinvested abroad in foreign countries. All of the saving and investment, however, will appear as US savings and investment simply because of the nationality of the parent company.

From the point of view of national income accounts constructed on a national ownership basis, reinvested earnings are properly attributed to the country of which the investing Company is a national; from a geographical point of view, they are attributable to the country in which they were made. For a receiving country it is correct to say that such earnings represent a return on foreign-owned capital (and perhaps payment for foreign managerial skills) and are thus foreign factor incomes which, if reinvested, represent savings supplied by the owners of the foreign factors and made available to the local economy.

But one should beware of assuming that the 'new' net investment resulting from retained earnings represents a new and additional availability of foreign skills, foreign equipment, or foreign exchange to the receiving country. Suppose, for example, that all the profits in a subsidiary had been paid out as dividends to the foreign parent and then immediately reinvested in the same subsidiary. The net effect on the availability of real external resources would be zero in spite of the new foreign investment. The economy gains only in comparison with the situation that would have prevailed had the reinvestment not been made. This is, indeed, a gain, for although the reinvestment provides no additional real resources as such, the cessation of investment would necessarily have resulted in a flow of resources in the other direction.

Thus the legal ownership, which governs the statistical classifications of national income accounts, obscures one important aspect of economic reality, and may lead the unwary to assume that the foreign investment of the firm is a contribution of goods and services to foreign economies which is made possible by savings generated within the US geographical economy. In fact, the term 'United States foreign investment' refers only to ownership of the investment, and ownership implies that both income and capital may be at any time transferred to the US (unless prohibited by the foreign government). The specifically

American contribution to the foreign economies may be little more than the willingness of the American parent to refrain from transferring to the US profits that its subsidiaries have made in foreign markets. The operations of the subsidiaries may not require American management or technology any more than they require new funds raised from the American economy by their American parent. On the other hand, the managerial, technological, or financial contribution from the parent may be considerable and genuinely make new real resources available to the local economy. The point to note is that the financial size of the foreign investment of a firm is no reliable indication of the extent of the firm's positive contribution to the foreign economies other than its willingness to refrain from repatriating profits.

The Significance of Integration

All international firms as defined here are, by definition, integrated across national frontiers either horizontally or vertically or both, for they are conducting similar operations in several countries or are engaged in different stages of the same industry in different countries, or both. A high degree of integration inevitably introduces an important element of arbitrariness in the allocation of overhead costs to different operations and in the setting of the prices at which goods and services are transferred between the subsidiary entities of the firm. If we assume that firms attempt to minimize taxes in their efforts to maximize retained earnings, we can infer that they will attempt to use the scope thus provided to allocate overhead costs among their foreign branches, subsidiaries and affiliates, and to adjust their transfer prices, in order to reduce their total tax outlays.[1]

International firms have open to them a wider variety of opportunities to evade taxation in a given country than do domestic firms, since they are often in a position to determine in some measure the amount of profit arising in the different countries in which they operate. In the extreme case, a firm may be able to arrange for products to be 'sold' to agents or subsidiaries in specified countries who resell them at a profit, in such a way that much of the total profit on their international operations appears to arise in the countries in which such transactions occur. These countries are known as 'tax havens', and are those in which taxes on profits are low or non-existent. The 'sales' are paper transactions—sales of titles—and the physical movement of the actual goods is not affected by the circuitous routes which the titles to them may take.

This kind of 'tax minimization' is perhaps not so important for many large firms as that which results from the 'normal' pricing and allocations of costs among affiliates. The prices charged to any given affiliate for the use of research results, 'know-how' and patents, for managerial and other services, and for goods supplied by the parent or by other affiliates, can be adjusted over a wide range depending on the particular circumstances. Costs must be allocated and transfer prices must be set, but the choice of appropriate criteria is necessarily arbitrary from an economic point of view. Moreover, just what charges are made is often very easily concealed, even from internal revenue authorities, and especially in countries whose administrative expertise is very limited or

[1] For a discussion of these aspects of transfer pricing in a variety of different types of firm see James Schulman, *Transfer Pricing in Multinational Business*, 1966.

who are in a weak position to stand up to large international firms.[1] When the international operations of a firm are characterized by a high degree of vertical integration, the opportunities for adjusting internal transfer prices may be very great indeed, and may have significant consequences for the distribution of income among countries in which the firm operates through its effect on their balance of payments.

Let us consider for the moment a model of vertically integrated international firms in an industry with characteristics similar to those of the international petroleum industry or the bauxite-aluminium industry. The model is extreme but illustrates the underlying logic of the problem. Imagine firms with great market power, producing their raw material in one group of countries and manufacturing and distributing the finished product in others. Assume complete vertical integration. In these circumstances, the price put by the firms on the raw material as it moves from their raw-material producing affiliates to their manufacturing affiliates is an internal transfer price, and any price above cost will attribute profits to the production of the raw material. But since the raw material is not sold outside the firms, such profits are only accounting profits, and if the firms have a high degree of monopoly in the production of the raw material, they are rather arbitrary profits as well. At the same time, however, the transfer prices are export prices for the countries from which the raw material is exported, and import prices for the importing countries.

If profits were not taxed anywhere and were fully transferable, if all currencies were completely convertible at fixed rates of exchange, and if vertical integration were complete within each of the large international firms, it would make no financial difference to any firm nor any country at which stage of their operations the firms chose to show their profits, and inter-affiliate pricing would be determined with respect to the requirements of efficient administration in the light of the conditions of competition. If, for example, profits were largely attributed to raw-material production but were repatriated by the firms, any credit to the exporting countries on export account would be offset by the debit arising from the transfer of profits. Similarly, if profits appeared at the point of final sale in the importing countries but were repatriated, the lower costs of imports to that country would be offset by the greater factor payments abroad.

If profits were taxed, but the effective rate of company income tax (including taxation of dividends remitted to parent companies by affiliates) were the same

[1] In advanced countries the internal revenue authorities keep a sharp watch on this kind of thing, and the criteria they apply may differ substantially from those the firm finds most appropriate for the conduct of its internal operations. For example, I know of one leading international firm (not in the petroleum industry) which prefers to transfer its products between its affiliates in different countries at cost including a managerial fee. The internal revenue authorities in one country where manufacture takes place insist that some profit be attributed to the manufacturing operations for tax purposes. There is also some outside equity interest in the manufacturing subsidiary in that country. The firm must therefore prepare three sets of accounts: one for internal accounting purposes using cost prices, one for the internal revenue authorities, and one for shareholders. In underdeveloped countries, the governments are not always in a strong position to deal with the pricing practices of international firms. We shall see later some examples of their difficulties in the international petroleum industry.

in each country in which international firms operated,[1] then again it would make no financial difference to the firms where they showed their profits, but it would make a great deal of difference to the countries concerned.

If effective tax rates differed in different countries, then the location of the firm's profits would be important both to the firms and to the countries. The higher the rate of tax in any country the greater will be the local contribution made by high profits, but the greater also will be the incentive of firms so to price those items imported from affiliates in other countries that their profits appear, wherever possible, to arise in countries with lower tax rates. Exactly the same considerations are also relevant for the international allocation by the firms of overhead costs.

Clearly there is a wide scope for substantial conflicts of interest between firms and governments and between importing and exporting countries. Because of these conflicts, governments often must intervene to protect their own interests. If, for example, most profits were attributed to raw-material production in the above illustration, even competition from non-integrated suppliers would not necessarily force reductions in the prices to importing countries. Tax considerations might still make it profitable for the firms to maintain their inter-affiliate import prices in the face of falling selling prices and to take accounting losses in their manufacturing and distributing affiliates in particular markets, in spite of the administrative difficulties that might be entailed. It would be thoroughly unreasonable to ask any government to leave firms in such an industry free to operate in accordance with 'ordinary' commercial (or market) considerations, since minimization of taxes is one of the most important of these considerations and necessarily involves discriminating among countries if tax rates vary.

In principle, if vertically integrated international firms were maximizing consolidated profits and could freely determine their internal pricing structure without fear of political repercussions or of government pressures (governments may be powerless to intervene or may be unaware of the discretionary nature of the firms' pricing), then the inter-affiliate price structure adopted by any firm would be such as to leave zero profit[2] in those countries in which the effective tax rates were highest, after making allowance for the effects of any offsets against their home tax liability that their home government allows.[3]

Such is the underlying logic of the position of large international firms if we abstract from all political considerations as well as from other institutional factors that in practice restrict their freedom of action. But the abstraction is

[1] This implies that no special tax treatment is given to any stage of the industry. For example, depletion allowances could not be permitted in the petroleum industry or there would be a tax incentive for firms to price in order to show profits at the level of crude-oil production, which would automatically favour the crude-oil producing countries.

[2] Or, at best, it would leave only the amount of profit required for administrative convenience.

[3] In a number of countries, including the US and the UK, income earned abroad is subject to domestic taxation but relief is given for income taxes paid to foreign governments. This relief is generous under US tax laws, and US companies can entirely escape domestic taxation of their foreign income so long as their foreign tax credits equal or exceed their domestic tax liability. It is when foreign credits begin to exceed their domestic tax liability that companies need to make special efforts to ensure that profits arise in those countries where tax rates are lowest.

severe: the historical position at any time, government surveillance, political necessities, administrative arrangements, incomplete integration (which increases the significance for internal transfer prices of competition at different stages of the industry), and oligopolistic interrelationships between large firms in the same industry, all severely restrict the ability of the individual firms freely to manipulate their inter-affiliate pricing in order to minimize their tax burden.

Vertical Integration in the International Petroleum Industry

Technically, the international petroleum industry is an oligopoly[1] of a type where the few large sellers that dominate the industry are surrounded by a fringe of 'independents' whose activities often have considerable significance, and who have even forced substantial changes in the industry. The seven 'major' oil Companies which are among the largest industrial Companies in the world, are composed of a network of affiliates and subsidiaries, many of which have a high degree of autonomy and conduct their affairs as quasi-independent entities, but firmly within the administrative framework and broad lines of policy laid down for each group as a whole.[2]

Although the major Companies are not organized in any kind of cartel, they are closely associated through joint ownership of affiliates in the production of crude oil, in refining and sometimes in marketing. They are all vertically integrated, operating at all stages[3] of the industry beginning with the exploration for and production of crude oil and ending with the distribution of the finished product to the final consumer; they have also moved into petrochemicals, fertilizers, and other industries which use petroleum derivatives as raw materials. The integration of the Companies extends across national frontiers, and crude-oil production outside the US is concentrated in areas in which both consumption and refining are relatively small. It is widely held that extensive integration is a necessary condition for the efficient operation of the industry; particularly integration of crude-oil production, refining, and distribution.

The economic advantages alleged to flow from extensive integration are (1) assured outlets for crude, leading to a steadier and more efficient planning of output over time; (2) more efficient operation of refineries as a result of an assured and managed flow of crude oil; (3) a more flexible and efficient adjustment to short-run changes in the demand for different products in different

[1] The word 'oligopoly', perhaps because it is an inherently ugly word, seems to evoke an emotional response even from those who have very little idea of what it means. It merely describes a situation in which sellers are so few that each is aware of, and therefore takes account of, the repercussions of its own actions on other similar sellers, and adjusts its own strategy in the light of anticipated or actual actions of its rivals. It does not imply unusually high profits.

[2] See p. 9 for the names of the companies.

[3] A 'stage' of production is an ambiguous concept, and if narrowly defined, vertical integration will be found to exist in almost any firm. For our purposes, however, it will suffice to accept the traditional five stages: exploration and development, production (of crude oil), transportation, refining and the marketing of products. See the discussion of the conceptual problem in J. G. McLean and R. W. Haigh, *The Growth of Integrated Oil Companies*, Chapter I; and M. G. de Chazeau and A. E. Kahn, *Integration and Competition in the Petroleum Industry*, pp. 19–20, 38 ff.

areas, which can then be quickly reflected in the inflows of crude oil; and (4) a consequent avoidance of disruptive fluctuations in prices which would raise costs to both producers and consumers.

Oil is a dirty liquid, and in some of its forms it is volatile and dangerous. It is best managed if allowed to flow continuously from the well, through pipelines, into tankers if necessary, and through refineries, unseen and untouched. Storage is expensive and dangerous.[1] There is a wide variety of crude oils with many different properties, which can be combined to form a large variety of products serving different purposes. In large markets the pattern of output of a large refinery or group of refineries must be continuously adjusted to changing demand, and this requires such complicated calculations for major refinery complexes that the use of computers has been spreading very rapidly in the industry. The calculations of the computer can reach down from the refineries to regulate the quantities and qualities of the crude oils flowing in, on the seas, or flowing from the wells. Moreover, the fact that costs fall rapidly as refinery throughput rises up to near capacity output makes any interruption or irregularity in supply extremely expensive.

There can be no disputing that refiners would be wise to acquire their own sources of crude oil if there is some likelihood that they would otherwise be deprived of a regular flow of supplies or charged monopolistic prices; also crude-oil producers and refiners would sensibly seek to acquire their own outlets if they would otherwise be faced with a monopolistic combination among buyers. And indeed, in the historical development of vertical integration in the industry these were originally the dominating considerations, since competition often took the form of some firms attempting to gain advantages over others by obtaining some sort of monopolistic position in a key sector of the industry. Thus whether or not vertical integration was a particularly efficient way of organizing the industry, it became a competitive necessity for at least the leading firms, its existence providing its own justification.[2]

The fact that the 'urge to integrate' has evidently been strong at all times in the history of the industry is sometimes presented as evidence that integration is peculiarly appropriate to it. I do not doubt that this can be taken as evidence that integration has seemed singularly profitable to the firms in the industry, but clearly, given the imperfections of the market, the degree of monopoly, and the legal arrangements surrounding the acquisition of concessions and the production of crude oil (including the tax systems), it would be a rash economist indeed who would hold that the most profitable organization was also necessarily the most efficient from the point of view of the allocation of resources in the economy as a whole.[3]

[1] See P. H. Frankel, *The Essentials of Petroleum*, Chapter I, for a discussion of these questions. Frankel presents an ingenious discussion of the 'economics of a liquid' in which he examines the technological factors which he considers to be the chief reasons for the prevalence of vertical integration.

[2] De Chazeau and Kahn, p. 43.

[3] More recently it has been argued that the superior efficiency of vertical integration has been demonstrated by the fact that the smaller companies that have discovered more crude oil than they can use in their own operations have had to cut prices severely to sell it and are proceeding to integrate forward as fast as possible. Here there are two factors involved: the companies with established markets and several sources of crude have less need, at least

The question of the economic efficiency of vertical integration cannot be answered by examining the historical record of integrated and non-integrated companies. Most of the integrated companies are not completely 'balanced', for some have to buy much of their crude oil or products from outside while others sell surpluses. The arguments for integration based upon the necessity of planning the flow of supplies between the two phases of the industry would presumably imply that incomplete balance leads to inefficiency somewhere, but there is no way of demonstrating that existing inefficiencies stem from lack of 'balance'. Moreover, integration may also be responsible for inefficient operations.[1] By contrast, there are a few non-integrated companies which have thrived in spite of their specialization. Even this provides no evidence about the economic desirability of integration taken by itself, since such companies have been operating in highly imperfect markets and under the umbrella of the great integrated Companies. The historical record is thoroughly inconclusive, although it is quite clear that some non-integrated refiners and distributors have in fact done very well.

The only other type of evidence so far presented consists of a qualitative list of the advantages to be obtained in principle from vertical integration because of the close co-ordination of operations made possible by managerial control.[2] This does not help in appraising the relative advantages of vertical integration over other types of market arrangements, but assumes away the problem. Almost all such evidence consists of general statements imputing advantages to vertical integration as contrasted to a different type of organization which does not in fact exist but which, it is specified, would not have the advantages listed if it did exist.[3]

---

early in the game, to cut prices to sell newly-discovered oil, since they can make room for new supplies by reducing their offtake from established areas. Clearly this is partly a question of vertical integration, but partly also of market control: if markets are controlled by vertically integrated concerns, then other producers may well have to integrate as well—the point made in the text. See Chapter V for further discussion.

[1] After 1958, when the major companies began to face increasing competition in product markets, they were able to maintain profits in spite of falling prices largely through improvements in efficiency and reductions in costs. See also the references to the types of mistakes and inefficiencies that can arise *because* of integration, in De Chazeau and Kahn, pp. 268–9, footnote 11.

[2] The oil Companies confine themselves to general assertions; the studies of McLean and Haigh show the strategic advantages of vertical integration to Companies but not the superior social efficiency of integration; De Chazeau and Kahn, in the only other detailed study of the problem, point out that in the US the controls that the majors 'secure over supplies of crude, pipeline transport to refinery or market and market outlets are real advantages to them, but they may have been secured at the expense of others: vertical integration itself augments the risk which it then serves to neutralize', p. 268.

[3] On the other hand, a strong case can be made to support the argument that the development of the industry would not have been as rapid as it was, especially in the inter-war period, if crude-oil producers had not themselves planned and expanded refinery capacity and markets along with crude-oil production. In the absence of integration, prospective refiners might well have felt more uncertain about the availability of supplies, and crude-oil producers about the availability of refinery capacity on which their markets depended. In such circumstances, the rate of investment at both stages of the industry might have been reduced by the greater subjective risk, which would tend to be accentuated by any miscalculation at either stage of the rate of development of the other. In other words, coordinated

On the other hand, it is reasonable to point out that there are combinations of alternatives which could conceivably permit the same assured flow of supplies and flexibility of operations as is achieved by vertical integration.[1] That long-term contracts do not apparently interfere with the operational efficiency of those companies relying on them for all or part of their oil supplies provides some evidence that such contracts could be adequate substitutes for control through ownership or common management. We may agree that supplies from a number of different sources must be assured over considerable periods and that considerable flexibility is required for day-to-day operating adjustments, but it is difficult to see why supply contracts between refining and producing companies could not do the job, as they have in fact done over considerable areas of the market for some time. It is not even inconceivable that a skilled group of specialist oil brokers could perform this function with equal efficiency.

Moreover, as we shall see in the discussion of the offtake arrangements actually existing in the Middle East,[2] the type of vertical integration that involves the joint ownership of crude-oil producing companies by parents who are themselves in competition in product markets is inherently inefficient from an economic point of view since it involves a peculiarly unfortunate type of price discrimination and distorts the allocation of investment. It may also have serious drawbacks from the point of view both of some of the parents and some of the crude-oil producing countries.

The fact is that we have no way of determining in the abstract whether or not vertical integration is of superior operational efficiency. Nor have we reason to believe, in the concrete circumstances of the international industry, that the further erosion of the dominance of vertically integrated operations would, in itself, result in greater economic inefficiency than at present exists. There is some evidence that it might reduce it. On the other hand, the advent of the computer as a tool of management may so improve the efficiency of 'logistic' programming, that within the limits of the computer the case for some types of integration may become stronger, especially where great flexibility is required to meet frequent short-run changes. The computer has made possible the combined programming of the operations of a group of highly complex refineries serving different types of market or different areas. Under such programmes tankers full of crude oil can be quickly diverted while in transit to alternative refinery destinations, and the operations of the several refineries can be so dovetailed that the use of capacity and the matching of crude flows to product requirements can be brought to a fine art.

---

planning might well have permitted a more rapid rate of expansion of the industry than the market would have produced. This is a similar argument to that put forward in support of 'balanced growth' by some writers on the economic development of poor countries.

[1] '. . . one cannot neglect the fact that many functions undertaken by the vertically integrated firm would be performed in its absence by market specialists. The elaborate paraphernalia of market analysing and logistical co-ordinating machinery of integrated companies could be a burden rather than an advantage, a cumbersome substitute for the specialized services and information that a more freely functioning market provides all participants.' De Chazeau and Kahn, p. 268.

[2] See Chapter V.

Problems Raised by Integration

The various attacks on vertical integration from both exporting and importing countries have been related not to operational inefficiency but to the methods of pricing adopted by the major Companies in their transfers of crude oil across national frontiers to their refining affiliates, or of products to their marketing affiliates. The system adopted has been such as to create a high degree of discrimination between crude-oil production on the one hand and refining and marketing on the other. The consequences of this have had a great significance for the international distribution of the income arising from the operations of the Companies and for the attitudes of governments.

In the major crude-oil exporting countries almost the entire domestic industry is in the hands of affiliates of foreign firms who explore and produce under comprehensive concession agreements. The chief benefits to the countries arise from the net investment of the Companies and the fees and taxes paid by them. The governments of these countries negotiate tax arrangements with the Companies, and among the most important questions for negotiation are the prices to be attributed to crude oil for tax purposes since most of it is not sold in arm's-length transactions, the expenditures to be allowed as costs for tax purposes, and the rate and form of taxation.

Importing countries (who may of course also be minor producers) are faced with a somewhat different type of problem which is also directly related to the vertical integration of the Companies. Countries importing crude oil need refineries. These are expensive to build, but the Companies have been eager to build them in the developing countries, sometimes very cheaply—but on condition that they have the right to supply the crude oil. If all of a country's refineries are owned by integrated international Companies, then the import price paid for crude oil is the internal transfer price as determined by these Companies. This has been very high in the past, since for a variety of reasons that are discussed later,[1] the Companies have preferred to show their profits at the level of crude-oil production. The absence of independent refineries would make no difference to an importing country so long as little competition existed in the market for crude oil; once competition appears, however, the country must face up to the difficulties created by its inability to secure lower prices for crude oil than the foreign owners of the refineries are prepared to grant. Countries that import refined products have a similar problem if their distribution network is owned by internationally integrated Companies.

In other words, the acceptance of foreign investment in this industry can be unnecessarily expensive for an underdeveloped country, not because of profit remittances abroad, but because of the pricing policies of the Companies. Moreover, because the Companies are largely self-financed the underdeveloped consuming countries, along with the Western industrialized countries, have provided the money which has made it possible for the great international firms to maintain their position in this very rapidly growing industry without significant recourse to outside capital or to their shareholders.[2]

[1] See Chapter VI.

[2] If one takes account of the debt associated with the activities of tanker transport as well as pipeline subsidiaries, the debt-equity ratio is much higher than can be determined from the published data. This point is made by Thomas Stauffer. See footnote 1, p. 145, below.

THE POWER OF THE LARGE FIRM

Large size unavoidably brings economic power in its train. By definition, a large firm controls a large amount of resources the use of which is decided by the management; it commands a large market over which it can exercise various kinds of influence; it raises and spends very large sums of money; it is a large buyer and a large employer. This implies the possession of economic power, for economic power in the widest sense consists of the ability significantly to influence the use of resources, the distribution of goods, the prices of products, the tastes of consumers, the development of new technology, and the distribution of income; and the larger the firm, the greater *ceteris paribus* is its power.[1]

Today the problems created by the power of the very large firms do not relate so much to the cruder forms of coercion or deception, the ruthless elimination of competitors, or even collusive restrictive practices (although all of these, and perhaps particularly the last, are at times important);[2] they relate more to the way in which firms, in their normal investment and operating policies use resources and influence the distribution of income. We are not really faced with 'greedy capitalists' exploiting consumers and suppliers and oppressing competitors in order to line their own personal pockets; rather the problem involves an appraisal of institutions whose activities are directed towards making profits from selling goods to consumers all over the world, primarily for the purpose of reinvesting the profits also all over the world, in order to promote their own profitable growth. Especially favourable access to investible funds is one of the most effective instruments of competition at the disposal of large firms. In a growing industry rapid expansion of investment is an essential condition of maintaining or increasing a firm's share of the market, and is thus one of the chief forms of competition.

Partly because large firms do possess great economic power of this kind, governments have increasingly thought it desirable, and perhaps especially in the developing countries, to negotiate directly with foreign firms over the terms on which they are allowed to operate. For crude-oil exploration and production, oil Companies have usually worked under concession agreements; for refining, special refinery agreements are sometimes made. As we shall see,

[1] Some may not like this concept of 'power', but it is fruitless to dispute definitions. In a different sense, it might be said that a small firm with the ability to have its own way in a small town has 'more' power than a large firm that is constrained by all sorts of considerations, even though the actions of the latter affect far greater numbers of people. A very large firm may have little market power in the traditional sense with respect to any of its products, but its administrative decisions governing the allocation of the resources under its control may have widespread importance. For my purposes, this is the relevant consideration, as will be clearer later, and here power is inevitably associated with size. The difficulty with the word 'power', perhaps, lies in the fact that it has too many emotive overtones.

[2] This does not imply that ruthless disregard of ordinary decencies, cynical support of vicious political policies, and thoroughly irresponsible pursuit of profit-making have been uncommon in the history even of the large firms in industry. Indeed, highly questionable behaviour on the part of the oil companies can be found up to very recent times in the absence of informed and active government policies. See, for example, Lieuwen's discussion of the industry in Venezuela up to as late as the 1940s. Edwin Lieuwen, *Petroleum in Venezuela*, esp. pp. 49 ff.

however, it has not always been possible for a government to protect itself adequately by such agreements when the conditions under which they were originally negotiated change substantially. This, together with certain of their commercial practices, have brought the Companies into conflict with the laws of some countries, or invited investigation by governmental agencies. Some of the international firms are partly owned by governments, while others, on occasion at least, work closely in association with the governments of the countries in which their parent Companies are incorporated.

Needless to say, firms deplore government 'intervention' except when it operates to give them tax privileges, tariff protection, political support *vis-à-vis* other governments with which they may be negotiating, and similar advantages. With such exceptions, the firms urge governments to leave matters to 'free enterprise' and the 'market mechanism'. Plainly this raises many issues, for it is by no means clear that the existing system in fact displays the characteristics of the free-enterprise model to the virtues of which appeal is implicitly made. The great fear of the oil Companies is of increased government attempts to control their activities and, in particular, of 'government-to-government' trading; much of their emphasis on the superior economic efficiency of the large integrated group is part of their attempt to convince both governments and the public that they are so large, so widespread in their interests, so 'international' and impartial in outlook, that they can be trusted to take care of everybody's interests.

International firms also raise special problems partly because, against their own interest, their international activities may be perverted to serve the ends of particular governments, and partly because some governments may connive with the firms to advance their mutual interests as against the interest of the governments or economies of other countries in which the firms operate.

On the other hand, international firms can make greater contributions to the economic growth and development of all countries than they are now making provided that some of the sources can be eliminated of international discrimination and of conflict between their own interests and those of their parent countries on the one hand, and the interests of other countries in which they operate on the other. There is a danger that these conflicts will so reduce the international acceptability of international firms that their potential contribution will not be realized. From the detailed analysis in the following chapters of the operations of the international firms in a particular industry, some of the difficulties, and some of the possible solutions to them, should become clearer.

Chapter III

THE DEVELOPMENT OF THE INTERNATIONAL PETROLEUM INDUSTRY

Petroleum has long been known in many parts of Asia: the walls of Babylon were cemented with bitumen, crude oil was produced in Japan in the seventh century, and from hand-dug wells in Burma by the thirteenth century. But the modern industry dates only from the nineteenth century, when a series of technological changes following from the use of the drill to dig oil wells made possible the very rapid development of the oil resources of both the USSR and the US.[1]

In the US there were very rich discoveries, and the wide market encouraged a wild scramble to find and exploit the country's oil reserves, using and developing the new technology. As oil production expanded, refineries were built and marketing networks were established. Among the new refiners in the 1860s was John D. Rockefeller who, together with his business associates, very quickly built or acquired control of numerous refineries and pipelines and organized an efficient supply system closely linked with the railroads. Between 1873 and 1882 Rockefeller and his group formed alliances with other oil men and the great Standard Oil Trust was created which obtained extensive monopolistic control over transportation and refining in the US. Not only did Standard Oil dominate the domestic industry for a long time but it reached out to the markets of Europe and of the Orient, quickly becoming the world's most important supplier of oil products.

But there was oil and enterprise in Europe, too, and Standard did not have things entirely its own way by any means. By the 1880s production at the Russian oilfields at Baku was in full swing, and by the end of the following decade even surpassed for a brief period that of the US. The early history of the international industry was largely shaped by the rivalry between Standard Oil and its powerful European competitors and by the policies adopted to mitigate this rivalry in their common interest.[2]

[1] Drilling had long been used in the salt industry (the Chinese drilled for salt in the twelfth century, and their methods influenced European technology). In the US E. L. Drake was credited with the 'invention' of drilling for oil, having drilled a successful well in 1859. He was the first to apply this technique in the US, but it had in fact been used earlier in oil production in Europe. See R. J. Forbes, *Studies in Early Petroleum History*, pp. 175 ff, and R. J. Forbes and D. R. O'Beirne, *The Technical Development of the Royal Dutch/Shell*, Chapter I.

[2] *Pioneering in Big Business* 1882–1911 by R. W. and M. E. Hidy, Volume I of the *History of Standard Oil Company (New Jersey)*, gives a vivid picture of this rivalry from Jersey Standard's point of view.

The purpose of this chapter is very briefly to recapitulate the main outlines of the development of the international industry,[1] primarily since the First World War, selecting and emphasizing those aspects that are more important for an analysis of the issues relevant to the subject of this study—the international firms in relation to the developing countries. To some extent the discussion will be one-sided: in particular the development of technology in the industry will be largely ignored and the impressive technical and physical achievements of the oil Companies in exploration, production, refining and distribution will be taken for granted. On the other hand, I shall not dwell on the extensive waste accompanying the activities of the Companies, especially in production in earlier days (notably in the US, Mexico and Venezuela), and as a result of the type of competition they engaged in; nor will the competitive tactics, political deals, and sometimes shady financial and commercial negotiations and agreements, which helped some of the great international Companies to achieve their dominating position in the industry, receive undue emphasis. This type of thing may be looked on as a necessary, but by no means sufficient, condition of their success, and should not be allowed to obscure the fact that the major Companies, in varying degrees, did undertake the solid work involved in the phenomenal development of the international industry that is unexcitingly portrayed in the statistical record.

Most of the pamphlets and 'house' histories put out by the Companies describe, sometimes in dramatic or romantic language, the hardships, risks, and often heroism, associated with oil exploration—now as well as in the earlier days; they detail the enormous investment, placing emphasis (where appropriate) on the low rate of return relative to the risk incurred, or on the long wait for any return at all; and they outline, with superb pictorial illustrations, the technological achievements in drilling, pipeline construction and refining, and the architectural beauties as well as the convenience of their retail service stations. Although such materials are designed to serve the purposes of advertising and public relations, they are essentially accurate as far as they go, and the story they tell deserves, with some exceptions, to be admired and appreciated. A complete history, or even a balanced summary of it, should give appropriate weight to these developments.

The evolution of the individual Companies will be touched on as lightly as possible in the following pages, for this is the subject of the next chapter; but the rivalries, as well as the co-operative arrangements among them, set against the background of a rapid expansion of output and an even more rapid expansion of reserves, are necessarily one of the main themes.

EARLY WESTERN RIVALRIES: THE FAR EAST

Kerosene was the most important of the industry's products until the second decade of the twentieth century. In the 1860s more than half of the kerosene refined in the US was exported; by the 1880s petroleum and its products, most of which went to Europe, ranked fourth in the value of exports from the US.

[1] For a survey of developments in the several countries of Latin America see Chapter XI by Dr P. R. Odell.

At that time Standard Oil had an overwhelmingly dominant position in the export trade, achieving in the beginning a near-monopoly in Asian markets, although in Europe the Nobel brothers and a group of Russian independents and the Rothschild interests of Paris were of great importance. Rivalry was intense, and European producers and marketers with access to the prolific Russian fields and having advantages over Standard Oil in transport costs throughout Asia, Eastern Europe, and the Mediterranean, gave the Company stiff competition.

In the Far East a small Indonesian company was chartered in 1890 to exploit the oil of the Dutch East Indies, and in 1900 a Dutchman with the commercial aggressiveness and brilliance to match that of Rockefeller himself became its manager—Henri Deterding. Deterding soon brought the three largest foreign rivals of Standard—Royal Dutch, Shell and Rothschilds—into a single marketing organization, Asiatic Petroleum, to supply Asian markets; in 1907 he made an outright alliance with one of them—the British-owned Shell Transport and Trading Company—to form the Royal Dutch/Shell. Until shortly before the outbreak of the First World War Dutch interests, especially the Royal Dutch under Deterding, and the Dutch Government, blocked Standard's efforts to obtain concessions to explore for oil in the Dutch East Indies. And in Burma and India, British interests, notably the Burmah Oil Company, successfully opposed similar efforts. Burmah Oil was an active producer, refiner, and distributor, particularly in the Indian and Burmese markets. After the British annexation of Upper Burma in 1885, Burma became administratively a part of India and the tariff of 1888 gave Burmese products, and thus Burmah Oil, a preferential position in the Indian market. By 1904–5 Burma was the source of over one-third of the kerosene trade in India.[1]

Just before the First World War, Standard finally succeeded in obtaining a toehold in the Dutch East Indies through a Dutch subsidiary that it owned indirectly through its Dutch marketing affiliate, but it was not until 1922 that it discovered oil in commercial quantities in south Sumatra, and not until 1928 that the Dutch Government finally gave way to the persistent attempts of the Company, assisted by the American State Department, to obtain additional concessions in the territory under its control. In the following decades other American Companies were then permitted to work in South-east Asia through Dutch subsidiaries.

The struggles for Far Eastern markets among Standard, Royal Dutch/Shell, and Burmah Oil were intense and were marked by a number of 'price wars'[2] followed by market-sharing agreements of one kind or another and the re-

[1] See B. Dasgupta, 'Oil Prices and the Indian Market: 1886–1964', Chapters II and III, for a full discussion of the Indian market in this period.

[2] There is a distinction to be made between price competition through which each seller attempts to sell as much as he can so long as his incremental receipts exceed his incremental costs, and 'price wars' which, like all wars, are one of a number of weapons used to achieve a particular result. Price wars may be waged by the strong to drive weaker competitors to the wall by selling at any price necessary to achieve the object; or they may be waged merely to inflict heavy enough losses on competitors to force them to make agreements about market shares, etc. Once the object of a price war is achieved, prices may again be raised and the market brought back to 'normal'.

storation of prices to levels considered 'normal'.[1] After a series of price skirmishes in the Indian market between 1897 and 1905, Asiatic Petroleum came to an agreement with Burmah Oil to determine their respective marketing territories in the Far East. With this agreement began the alliance between Burmah Oil and Shell which led in 1928 to the formation of Burmah-Shell, a joint marketing company through which both Companies operate to the present day in India.

There were other price wars, notably 1910–11 and 1927–8, both of which saw the rivals of Standard Oil combined against it. The price war of 1927–8, although it began in India, quickly spread to other parts of the world as Standard of New York and Royal Dutch/Shell fought it out. It formed the background oı the 'Achnacarry' or 'As Is' agreement, the first large-scale attempt at a cartel agreement among the major companies to regulate world markets.[2]

In the second decade of the century, oil from Persia, where expansion had been interrupted by the war, began to lay claim to a share in the Indian market. Anglo-Persian Oil Company (now British Petroleum) was closely associated in its origins with Burmah Oil[3] and the Indian market was clearly part of its 'natural' market territory.[4] Until 1928 the greater part of Anglo-Persian's exports to India, which had risen rapidly after the war, had been of fuel oil—kerosene, the more valuable product, was not exported to India since APOC had no share in the kerosene pool through which the rivals of Standard operated their market-sharing and price-maintenance arrangements. In order to ensure Burmah Oil's support for its struggle against Standard Oil, Asiatic made way for APOC in the pool, and Burmah-Shell became, on its formation, the marketing agency for APOC, an arrangement which still exists in the Indian market.[5]

EARLY WESTERN RIVALRIES : THE MIDDLE EAST

In the Middle East the early rivalry among the great Western firms and, to a lesser extent, among their governments, was over producing territories rather than markets, for Middle Eastern markets were unimportant. In south-west Persia in 1908 a great oil-field had been discovered by W. K. D'Arcy and

[1] The rivalry among the oil companies in the Indian market in this period is discussed in F. C. Gerretson, *History of Royal Dutch*, Volume II, pp. 340 ff, and Volume III, pp. 211 ff, and by the Federal Trade Commission, *The International Petroleum Cartel*, pp. 197 ff. A useful summary and evaluation can be found in B. Dasgupta, pp. 13–74.

[2] See Chapter VI, pp. 179–81 below.

[3] See the discussion of the history of British Petroleum, pp. 109–16 below.

[4] In 1928 the Chairman of APOC, in reporting to stockholders, stated: 'It will have been noticed that India is now included in my reference to the general extension of our markets. This has been made possible by a friendly understanding with other companies who were there before us, but who recognise that the contiguity of our producing-fields to that great market gives us, notwithstanding our late arrival, an indisputable claim to consideration.' *Annual Report*, 1928, p. 5.

[5] At this time, too, Burmah Oil purchased shares in Shell Transport and Trading thus reinforcing the association of the two companies. The formation of Burmah-Shell had repercussions on the position and policies of other marketing companies, notably the American company, Socony (now Mobil Oil). See p. 56 below.

Burmah Oil. The Anglo-Persian Oil Company, partly financed by Burmah Oil, was formed to develop it and soon became a leading contender in international oil markets. In 1913 the British navy converted from coal to oil and in the following year the British Government made heavy investments in the Company, purchasing a controlling stock interest.

In the Ottoman Empire, German, British and Dutch interests were deep in Turkish politics in their efforts to obtain concessions in the lands containing the ancient oil seepages of Mesopotamia. Out of the negotiations eventually emerged in 1911 the Turkish Petroleum Company the shares of which were held by Anglo-Persian, Royal Dutch/Shell and a German financial and industrial group. After the 1914–18 war, the German interest was handed over to the French by the San Remo agreement of 1920 whereby the British and French not only obtained mandates over Mesopotamia and Syria but attempted to ensure that oil rights throughout the area would be in the hands of British, British/Dutch and French Companies.

Throughout Asia, American oil interests were in a weak position in the struggle for control of crude-oil reserves outside the US, since much of the promising territory lay within the political dominion of the British or Dutch Governments, who endeavoured to reserve concession rights for their own nationals. It must be remembered, of course, that it was easy, not only in the US but also abroad, to raise fears—not by any means all unjustified—of the 'Standard Oil Monopoly'. The war had made clearer than ever before the vital importance of oil to the great powers (including France). But at the same time fears were growing in the US that its own domestic reserves would soon be depleted, and the US Government was beginning to look apprehensively at the extension of British-Dutch control over the oil-bearing lands of Asia. The US Government itself was drawn more and more into the controversies over oil, and both the British and French Governments became more directly involved through treaties and through their financial interest in the oil Companies themselves. The Germans lost their rights during the war and after it had no chance of returning to the area.

The inter-war diplomatic and commercial jockeying for position in the Middle East is a story that has been re-told in numerous volumes from many points of view and of varying degrees of reliability. The British, having political hegemony in Iraq and in the Persian Gulf, would have preferred to reserve the exploitation of oil in those areas to British Companies. American Companies wanted very much to get in, and in the event succeeded, with the help of their Government, first in the Iraq Petroleum Company (the post-war successor to the old Turkish Petroleum Company) and later in the Persian Gulf and in the Arabian Peninsula itself. Although by far the largest Companies were Jersey Standard, Royal Dutch/Shell and Anglo-Persian, two other Companies obtained important concessions and made great discoveries—Standard Oil of California in Bahrein and Saudi Arabia, and Gulf Oil in Kuwait; the former slipped in partly because of the restrictive agreements preventing the international Companies associated in the Iraq Petroleum Company from taking individual concessions in the territories of the old Ottoman empire (the famous Red Line Agreement), and partly because of the conservative policies pursued by the Iraq Petroleum Company when it came to bidding for concessions in Saudi Arabia.[1] In Kuwait both Gulf and British Petroleum had been negotiating for a concession, but eventually joined forces to establish the jointly-owned Kuwait Oil Company.

VENEZUELAN OIL

In the meantime the crude-oil production of the other major world producer outside the US had also come under the control of the same oil Companies (except British Petroleum[2]) whose rivalry had been shaping events in Europe and Asia. The first commercial oil well was drilled in Venezuela by Royal Dutch/Shell in 1914, and in the following decade or so competition among oil Companies for concessions in that country was intense, the form it took apparently matching the corrupt nature of the ruling political regimes.[3] There was considerable rivalry between British and American Companies, and the Governments of both countries actively intervened, particularly that of the US, where the double fear was widespread that its domestic oil reserves were dangerously low while the British and Dutch cornered those of the rest of the world. Royal Dutch/Shell, Standard Oil of Indiana (which had purchased the properties and rights of Pan American), and Gulf Oil were the most successful in discovery and production, and after the First World War these three Companies accounted for nearly all of Venezuela's output. Jersey Standard, on the other hand, did not succeed in the search for commercial oil in spite of spending some $20 million in the effort up to 1929.[4]

Nevertheless, within the next six years Jersey had got control of half of Venezuela's production. It had obtained Creole already in 1928; in 1932 it bought out Indiana Standard's interests, including the Aruba refinery; in 1937 by a series of agreements it, together with Shell, obtained control of the greater part of Gulf's production and concessions and established itself as the leading producer in the country with 52 per cent of total output, leaving Gulf with only 7 per cent.[5] Shell produced 40 per cent of the total. The US Federal Trade Commission concluded that one of the primary purposes of these transactions was to eliminate Gulf and to bring about the 'control and regulation of petroleum production of all Venezuela, eastern and western alike, so that Venezuelan output would, at all times, accord with the current world market situation as seen by the producing companies. A production quota system was set up to achieve this goal.'[6]

This interpretation seems to be unnecessarily biased. Undoubtedly the fact that the bulk of Venezuelan production was in the hands of two Companies, both of which had extensive marketing networks, enhanced the possibilities of effective planning of the rate of supply in relation to other sources and to

[1] See S. H. Longrigg, *Oil in the Middle East*, p. 107.

[2] The Venezuelan Government did not permit companies owned by foreign governments to obtain oil concessions. Hence British Petroleum was precluded from operating in Venezuela.

[3] See Edwin Lieuwen, *Petroleum in Venezuela; A History*, Chapter III.

[4] *Ibid*, p. 43.

[5] *Ibid*, p. 85.

[6] Federal Trade Commission, p. 164.

markets. But it is not clear that this could not have been accomplished while Gulf was a significant producer.[1] Moreover, Jersey had been spending large sums searching for oil and presumably welcomed the chance to buy its way in as a producer; at the same time Gulf apparently had insufficient markets for the oil at its disposal and was also somewhat short of capital. In other words, the arrangements made in Venezuela can be explained as a sensible business deal from the point of view of the parties concerned without straining to interpret it as a deliberate and conscious policy of attempting to obtain a monopoly; at the same time one must recognize that the degree of monopoly obtained was presumably by no means unwelcome to either Shell or Jersey Standard.

CO-OPERATION AND CONTROL BY THE MAJOR COMPANIES

In 1938, just before the outbreak of the Second World War, Venezuela was the largest producer outside the US and USSR with 515,000 barrels per day, or one-third of world production outside these countries; Iran and Indonesia were the largest producers in Asia (excluding the USSR), producing 210,000 and 150,000 barrels a day respectively, or nearly a quarter of the world's production outside the US and USSR.[2] Venezuelan production, as we have seen, was largely in the hands of Standard and Shell. Iranian production was entirely in the hands of British Petroleum. Almost all of the Indonesian production was controlled by Royal Dutch/Shell, although two US companies, Stanvac (jointly owned by Jersey Standard and Socony-Vacuum) and Caltex (jointly owned by Standard of California and Texas Oil) had by that time made important discoveries. Iraq was producing some 90,000 b/d which was divided among the owners of the IPC group—Anglo-Iranian, Royal Dutch/Shell, Compagnie Française des Pètroles, and the Near East Development Corporation (composed of Jersey Standard and Socony Mobil), with a 7½ per cent overriding royalty for Anglo-Iranian. Only Bahrein and Saudi Arabia were exclusively in the hands of American Companies,[3] but Saudi Arabian production was just beginning, and output from Bahrein had reached only about 20,000 b/d by 1938. Production from Kuwait had not yet started.

During the war, little further development of crude-oil production could take place in Asia and the Middle East; Burma and the Dutch East Indies

[1] Lieuwen, however, notes that Gulf's subsidiary, Mene Grande, had discovered and begun to develop a lighter type of crude oil than was characteristic of Venezuelan output and that this was creating concern: 'Until this time, Venezuela had produced only heavy crudes suitable chiefly for the production of fuel oil. The appearance of undetermined quantities of light crudes from which a high percentage of the more valuable light products could be refined, was of great concern to the world's two largest oil corporations, Standard Oil of New Jersey and Royal Dutch/Shell. What would happen if Gulf, which hitherto had sent nearly all its Venezuelan production to the US, should suddenly be able to offer large quantities of light products in an already saturated world oil market?', p. 84. Unfortunately, Lieuwen gives no evidence to support the suspicion here voiced.

[2] World production in 1938 amounted to 5·6 million barrels a day, of which 3·5 million were produced in the US and 570,000 in the USSR. British Petroleum, *Statistical Review of the World Oil Industry*, 1966.

[3] Bahrein Petroleum Company (BAPCO) is incorporated in Canada and is therefore legally a Canadian company, but is owned by American Companies.

were occupied by the Japanese, and shortages of shipping and materials stopped all activity in the Middle East not connected with the war effort, although after 1944 the output of refined products rose. It was clear, however, that after the war vast new supplies of crude would be available, especially in Kuwait and Saudi Arabia and that much of it would be produced by Companies whose existing markets were far too small to absorb it. This was obviously a danger to the stability of oil markets and to the balance among the Companies that had been worked out in the inter-war period; large-scale entry into the market by new producers of low-cost crude oil, but without correspondingly large established markets of their own, might have had very serious effects on the price structure of the industry and on the position of the established Companies.

The fortunate newcomers, Gulf in Kuwait and Standard of California in Saudi Arabia, needed to find ways of disposing of very large quantities of crude oil, but they had neither, presumably, the desire nor, perhaps, the ability, to take on the existing 'big three'—Jersey, Shell and BP—in an all-out competitive effort to do so. Moreover, the agreement between Gulf and BP, under which the Kuwait Oil Company was formed, included provisions preventing either company from upsetting or injuring the other's trade or marketing position.

The type of solution to be adopted was becoming clear even before the war. Standard of California, which had few marketing outlets, made two agreements in 1936 with Texas Oil Co., which already had extensive Far Eastern and European markets. Texaco obtained a 50 per cent interest in the Bahrein Petroleum Company and in the California Arabian Standard Oil Company (now Aramco) in return for a half interest in Texaco's manufacturing facilities east of Suez. A jointly owned affiliate (Caltex) was incorporated to conduct the joint operations. After the war (1947) Jersey Standard and SoconyVacuum also obtained shares in Aramco. Thus were both markets and capital assured for the development of Saudi Arabian oil. At about the same time after the war, Gulf made long-term contracts to supply large quantities of Kuwait oil to Shell, and BP made long-term contracts with Jersey Standard and Socony Mobil. Dependable outlets were thereby assured.[1] The fact that almost all reserves in the Middle East were controlled by the same Companies, even though the groupings were different, made necessary a high degree of co-operation among them in the planning of investments and in the rate of development.

As new crude-oil reserves were found and production increased, the oil Companies laid pipelines, enlarged and improved their tanker fleets, built new refining facilities and extended their marketing and distribution networks, for the speed with which it was profitable (or possible) to expand these 'downstream' operations controlled the output of crude oil, which is of little use in an unprocessed form. Where oilfields were far from markets and without easy access to the sea, as in northern Iraq for example, the technical and economic considerations governing the feasible size of pipelines, together with the indivisibility of the large capital investment required, set upper limits to the amount of oil that could be lifted per unit of time. Moreover, apart

[1] These arrangements are discussed in more detail in the next chapter.

from the long-term crude contracts between the majors and for government supplies, the integrated Companies for the most part were primarily interested in producing and refining petroleum for their own marketing networks. Thus each Company's demand for crude was closely related to the amount its own distribution system could handle, together with the supplies required for any long-term contracts. The Companies had little desire to engage in price competition in order to sell to independent producers who might compete in product markets. In any event, the seven majors dominated refining and transport as well as crude-oil production. In 1950, they owned over 70 per cent of the world's refining capacity excluding the US and the Communist countries, as well as every important pipeline and about two-thirds of the world's privately owned tanker fleet.[1] Although some of these had been acquired by purchase, the Companies themselves made an overwhelming proportion of the investment required for the rapid increase in world oil consumption which, even excluding the US—by far the largest consumer of petroleum products—had grown some three and a half times between the wars. This powerful position of the majors, together with their superior financial ability to withstand bouts of price-cutting and depressed market conditions (such as those that prevailed in the later 1920s and early 1930s), as well as the high cost of creating a new distribution system, made the successful entry of newcomers expensive and difficult, quite apart from any market-sharing and cartel agreements that may have existed.

As a result of their dominant position, the major Companies were able not only to regulate the rate of development of supply, but also to exercise a strong control over prices, so long as they refrained from price competition among themselves in the 'free' crude-oil markets (which were small)[2] and in product markets. Except for the so-called 'price wars', they seem to have observed a kind of 'basing point system', remnants of which persisted well into the 1950s for crude oil, and for products into the 1960s.[3] There can be little doubt that the ability of the Companies to regulate the rate of supply in spite of the discovery of vast new reserves made it possible for them to maintain a very high cash flow from their operations, and thus obtain from their earnings the funds required for the enormous investments needed to transport, process, and sell a rapidly increasing output. The combination of the vertical integration of these Companies with oligopoly in product markets provides the single most important key to the understanding of the economic policies of the industry, and of the past and present relations between oil

[1] Federal Trade Commission, pp. 25–8. See also Table II, p. 78 below.

[2] It has been estimated that less than 10 per cent of the crude produced in the Middle East moved outside the network of the integrated majors and much less than that at 'arm's length', i.e. between freely bargaining buyers and sellers. It was reported, for example, with respect to the ECA, which financed aid for Europe after the Second World War, that 'More than 94 per cent of ECA-financed bulk oil shipments to Europe, for the year ending April 2, 1949, were made by six of the seven major international oil companies, nearly all of which took the form of shipments to their own affiliates or subsidiaries.' Federal Trade Commission, p. 365. Eleven per cent of the value of all ECA shipments to participating countries up to the middle of 1950 consisted of oil. See Horst Mendershausen, *Dollar Shortage and Oil Surplus in 1949–50.*

[3] The regulation of crude-oil output is discussed in Chapter V, and the determination of prices in Chapter VI.

Companies and the governments of both the countries exporting and those importing crude oil or its products.

THE SECOND WORLD WAR AS A TURNING POINT

In many ways, the Second World War marks a turning point in the history of the industry, for after the war there was a steady decline in the control exercised by the major Companies, which was not apparent in the ten years or so following its end but which became unmistakably evident in the later 1950s. The war itself had few significant long-run direct effects: there was some destruction of assets, particularly of tanker fleets and of installations, for example in South-east Asia; refinery capacity was rapidly expanded in the Middle East to meet the needs of the Allies in the Mediterranean and East of Suez; the development of the newly discovered fields in Kuwait and Saudi Arabia had to be postponed; and the Companies had to give way to some extent with respect to their jealously guarded freedom of pricing. The destruction of assets was soon made good, the development of the new fields was quickly resumed, and American Government pressure for lower prices soon evaporated. As a result of somewhat confused intrigues on the part of California Standard and Texas Oil to combat alleged British attempts to get a foothold in Saudi Arabian oil and to involve the US Government on the side of the Companies, the American Government made an abortive attempt during the war to participate directly in the Middle Eastern oil industry. Later an Anglo-American oil agreement was proposed, but this came to nothing. On the other hand, for some time after the war the fact that much of the oil available to Europe had to be paid for in dollars created problems for some countries, notably Great Britain.[1] She took the lead in imposing discriminatory measures against dollar oil in the face of sharp opposition from the State Department and American oil Companies, but this, too, was short-lived.

Although the normal activities of the industry were retarded by the war, change of another kind was immeasurably speeded up, for the political consequences of the war were far-reaching. In particular, the war profoundly affected the relations between the dominant Western powers and the territories of the Afro-Asian world and hastened the processes that were soon to lead to the formation of new independent countries, altering not only the political map of the world but also the attitudes of both Western and Afro-Asian peoples and governments as well as the political balance of power. Added impetus was given to nationalist and independence movements everywhere, while at the same time the moral and political authority of the West was further weakened. Independence movements gave confidence to popular anti-(western) Colonialism and the evident political and economic strength of the USSR, a non-capitalist power, encouraged anti-(western) Imperialism and anti-capitalist forces, causing some governments to take more aggressive attitudes toward the oil Companies than they had previously taken.

[1] 'In 1949, the British Government estimated that more than half of the sterling area's current dollar deficit for fiscal 1950 would arise through deficits on oil transactions. Transactions with American oil companies alone would account for more than a quarter of the deficit.' H. Mendershausen, pp. 2–3.

The creation of the United Nations, with its atmosphere of 'one State one vote', further encouraged the aspirations towards equality of the Afro-Asian peoples, even in spite of the special privileges accorded the great powers in that Organization, and the bargaining position of the developing countries was enhanced by the conflicts of interest among the great powers themselves.[1] Associated with these changes in attitudes was a growing concern over the poverty of the Afro-Asian world, a growing resentment in that world of the 'gap' between their own standards of living and those of the rich countries, and a growing preoccupation in the West with the ways and means of promoting the economic development of the poor countries.

The international petroleum industry could hardly be immune to these far-reaching psychological, political, and economic changes (sometimes referred to as 'de-colonization') which marked the post-war decade and were, in part at least, the result of the impact of war on *le tiers monde*. Some of the post-war troubles in the industry were the direct result of the inability of the Companies (or their governments) quickly to perceive and adapt to changed circumstances. On the other hand, some governments in the great crude-oil producing countries misjudged their new-found strength and overplayed their hands, taking action shaped more by emotional attitudes toward the symbols of imperialist monopoly and 'exploitation' than by reasoned consideration of their own economic interest. For although post-war developments in some respects strengthened the position of the crude-oil producing countries *vis-à-vis* the oil Companies, they did nothing to improve the ability of these countries to market their own oil, and they hastened certain other types of change in the industry which were to the disadvantage of the producing countries. In particular, the internal political ferment of peoples attempting to modernize their societies and to change the balance of power, not only among different economic and social groups internally but also with the outside world, enhanced the danger to the oil Companies and to the industrial consuming countries of Europe that was inherent in their near-complete reliance on a few sources for the greater part of their supplies of both crude oil and refined products.

Immediately after the end of the war the development of Kuwait and Saudi Arabian oil was accelerated; the output of these two countries rose from 16,000 and 160,000 barrels a day respectively in 1945 to 345,000 and 548,000 barrels a day in 1950, both far exceeding the increase in Iraq's production of from 99,000 to 137,000 barrels a day in the same period. Royalty payments were made to the several governments of the producing areas at a fixed rate per ton of exports,[2] but in spite of the increased revenues to the governments consequent on increased production, it was becoming increasingly clear that revisions of the concession terms would have to be made and these revenues increased. It must be remembered that concession agreements are originally made before oil is discovered and before, therefore, anyone knows how profitable the venture will turn out to be. If it turns out well, pressures for

[1] See the discussion of these points in E. F. Penrose, *The Revolution in International Relations*, pp. 5–6, 63, 91–3.

[2] For a discussion of the royalty arrangements in the various countries see C. Issawi and M. Yeganeh, *The Economics of Middle Eastern Oil*, pp. 130–1.

revision of the original agreements will inevitably arise, and in the inter-war period there were in fact a number of changes.

Demands for Increased Revenues

Before the war, the Anglo-Iranian Oil Company had at times encountered difficulties with the Persian Government over the payments made to the Government for the oil produced. Trouble arose especially when, as in 1931 because of a decline in prices, and again in 1948 because of the postwar limitations on dividends, payments to the Iranian Government actually declined. The Company recognized at both times that adjustments to the existing agreements were required. Over the period 1930–9, however, royalty and tax payments to the Iranian Government substantially exceeded income tax payments to the British Government, and amounted in total to nearly two-thirds of the net profit after tax of the Company.[1] As can be seen in Table I, only in the period 1935 to 1937 did the Company's profits rise substantially faster than payments to the Iranian Government. Reserves for depreciation, amortization, etc., on the other hand, were equal to about half of the net profit in the period.

During and after the war, the American Companies, too, were faced with demands for increased payments, and the Iraq Government was putting pressure on the IPC. It was the Americans who were to take the action that set the pattern for all the Companies, when in Saudi Arabia Aramco negotiated an arrangement under which payments to the Government were to equal 50 per cent of the profits attributed to crude-oil production. Venezuela had already been taxing the income of the oil Companies, and by 1948 the Companies there were required to pay taxes in one form or another to the Venezuelan Government which amounted to at least half of the net income derived from oil operations, but the Government's right to impose even higher taxes was not restricted. Although the agreement between Aramco and King Ibn Saud gave the Government half the 'profits' attributed to crude-oil production, it also limited the Government's right to impose higher taxes, thus putting it in a less favourable position than the Venezuelan Government. Moreover, the Government's half *included* the fixed royalty payments per barrel of oil.

It is clear that the Saudi Arabian income tax law, the first formal instrument for taxing the income of oil Companies in the Middle East, was drafted in the light of the provisions of the income tax laws of the US. In proposing the tax, Aramco fully expected to be permitted to deduct the taxes paid to Ibn Saud from its US tax liability.[2] In order to avoid double taxation of American Com-

[1] It will be noted from Table I that these figures apply only to Anglo-Iranian and exclude all subsidiary companies. They must be interpreted, therefore, in the light of the reservations that we have already discussed with regard to profits attributed to one part of an integrated firm. In particular, the profits of Anglo-Iranian shown in the Table will have been, in part at least, determined by the policy of the Company with respect to internal pricing, including transport charges.

[2] A representative of the US Treasury had been asked for advice as early as 1948 by Saudi officials, and Mr F. A. Davies, Chairman of Aramco's Board, made it clear in his testimony to a US Senate committee that Aramco expected to be able to deduct tax payments to Saudi Arabia. See US Senate, *Emergency Oil Lift Program and Related Oil Problems*, testimony of F. A. Davies and George A. Eddy.

panies operating abroad and, perhaps also to encourage foreign investment, the US law permits firms (and individuals) to deduct taxes paid to foreign Governments from their US tax liability on their foreign income. For this reason, it was important to the American Companies that increases in payments to the governments of the oil-producing countries should take the form of an income tax rather than of royalties. Royalty payments can be treated as a cost and deducted from taxable income, but this gives a much smaller tax relief for the same payment.[1] Aramco paid very large income taxes to the US Government, and the Saudis had apparently raised the question whether it was possible to transfer all or part of the payments being made from the US to the Saudi Arabian Government.[2] Aramco in its turn realized the necessity of raising the rate of payment to Saudi Arabia and was, naturally, very much concerned to reduce the cost to itself of so doing.[3]

The acceptance of a Saudi Arabian income tax designed to raise the Government's revenue to 50 per cent of the profits attributed to crude oil had an immediate effect on the agreements between the governments and the oil Companies in other countries of the Middle East, for naturally other governments expected the same treatment. The Iraq Government had for some time been dissatisfied with the existing agreement with the oil Companies, and negotiations between the Iraq Petroleum Company and the Government were taking place when the new arrangements in Saudi Arabia were announced. The IPC had been set up as a non-profit-making company in order to minimize the tax burden on the parent companies;[4] hence new negotiations were required before an acceptable formula could be arrived at to introduce similar

[1] If, for example, the US income tax rate were 50 per cent and payments were made to the foreign government in the form of royalties, the relief on the US income tax would be equal to half of the royalty payment; if the same payment is made as an income tax the relief would be 100 per cent.

[2] See testimony of Mr Davies in the hearings cited above, p. 1430. That the government of a crude-oil producing country has the greater claim to such income taxes can, of course, be very convincingly argued. Senator Dirksen, for example, in attacking the critics of Aramco stated flatly: 'It is accepted doctrine that the nation from which the oil is produced and is having its resources depleted should be entitled to income taxes.' US Senate, *Petroleum, The Antitrust Laws and Government Policies*, p. 128.

[3] Aramco officials argued that if more costly arrangements had been made, the company would have been put at a serious competitive disadvantage in the Middle East, but it was never made clear how this disadvantage would arise. Clearly the 'sales' of the company could not have been affected since it 'sold' all of its output to its four parent companies at a price determined by them. Moreover, higher taxes would not necessarily have reduced the funds made available to the company for investment since the amount of these was also controlled by the parent companies. It is true, of course, that in general, 'double taxation' of income derived abroad, imposed by both the foreign and domestic governments, would reduce the net return on foreign investment and could well reduce the flow abroad, but this has nothing to do with the 'competitive' position of a particular foreign affiliate.

[4] The IPC was incorporated in Great Britain and would have had to pay taxes to the British Government on any profits earned; in addition, the American co-owners would have had to pay taxes to the US Government. To avoid this, the predecessor of the IPC—the old Turkish Petroleum Company—was set up as 'a non-profit-making enterprise, with each shareholder taking its *pro rata* share of the crude oil produced at a price sufficient only to cover the cost of production, transportation, and a nominal profit—the profit to be limited to a reasonable return on investment'. Federal Trade Commission, p. 61. The nominal 'profit' was one shilling a ton.

measures, but an agreement was reached to be effective from January 1951, which gave the Government a 50 per cent share of the profits attributed to the sale of Iraqi crude by the Companies.[1] One of the important difficulties arose from the fact that the British Governments did not at that time permit the offset of foreign taxes against the Companies' British income tax liability, and hence an arrangement such as that reached in Saudi Arabia bore more hardly on the British than on the American companies. In Kuwait a similar agreement taking effect almost a year later was arrived at in 'an atmosphere of amity'.[2] In Iran, however, the relations between the Anglo-Iranian Oil Company and the Iranian Government continued to deteriorate, and were by this time hopelessly entangled in internal Iranian politics.

THE IRANIAN CONSORTIUM

The story of the conflict between the Anglo-Iranian Oil Company and the Iranian Government is well known, and I will not re-tell it here.[3] Early in 1948 the AIOC had proposed discussions with the Iranian Government to seek ways of remedying the evident prejudice to Iran of the British Government's policy of limiting dividend payments, since the Company was obliged to pay the Iranian Government a sum equal to 20 per cent of its dividend payments over a specified sum in addition to the royalty of 4 gold shillings per ton. Although an agreement on new terms was reached in the negotiations it was not ratified by the Iranian Parliament, and the relations between the Company and the Government went from bad to worse, culminating in the nationalization of the Company in 1951. The dispute dragged on for two more years, involving the International Court, the United Nations, and the intervention of the US, with the issue of compensation becoming central. Finally, in 1953 came the fall of the Government of Musaddiq, whose fiery intransigence had made settlement virtually impossible, and negotiations were opened under a new Iranian Government. These ended in the creation of an International Consortium to operate the oil installations of southern Iran but did not in principle reverse the nationalization.

The effect of the complicated consortium agreement was, among other things, to ensure that the Iranian Government received half of the net profits attributed to its crude-oil production.[4] Although the great oilfields of southern

[1] '. . . since the national fiscal systems of its different constituent groups were not identical, alternative means to attain a result acceptable to each group and to Iraq were various and intricate, and long negotiations between groups and their lawyers were required. . . .' Longrigg, p. 191. The original agreement with Iraq had provided for a royalty of 4 gold shillings per ton, and the appropriate rate of exchange between gold and sterling had long been a serious point of conflict, since the price of gold in the bazaars came to exceed the Bank of England price. The '50/50' formula had at least the advantage of removing this bone of contention. Each of the owners of the IPC set up a special trading company in Iraq to which profits could be attributed.

[2] Ibid., p. 222.

[3] For discussion of this conflict from different points of view see B. Shwadran, Chapter VI and works cited; Longrigg, pp. 159 ff.

[4] In the earlier negotiations a '50/50' sharing of profits was also apparently discussed without result, for the Persians wanted the profit sharing to be applicable to the world-wide profits of the Company made on refining and marketing as well as to operations within Persia. Longrigg, p. 160. Similar proposals had also been made in 1933.

Iran were thus no longer to be exploited by a single Company, but by a consortium of Companies as in the other countries of the Middle East, their operation was still largely under the control of the seven international majors, who between them owned 89 per cent of Iranian Oil Participants Ltd., the company set up to operate the properties of the old AIOC, with the 'big three' alone having 61 per cent. Compagnie Française des Pétroles owned 6 per cent and the remaining 5 per cent was shared among eight American companies.[1]

Although in one sense the Anglo-Iranian Oil Company lost its struggle with the Iranian Government, it can hardly be claimed that the Government won. To be sure, the Government did 'nationalize' the local operations, although it is not accurate to say that it nationalized the oil which, like other sub-soil minerals, belonged to the Government in any case,[2] and it did break the complete monopoly of AIOC when other Companies were brought into the Consortium. But the Consortium that was set up after prolonged negotiations, great economic losses, and severe political damage, brought little improvement in the direct economic benefits derived by the Iranian economy from the operations of the foreign Companies that could not have been obtained by less drastic means. On the other hand, the creation of a National Oil Company with a greater sense of participation in, and greater responsibilities for, the development of the country's oil industry was a significant step.

Another important result of the Iranian crisis was the warning it gave to the Governments of other oil-producing countries of the danger of heeding those of its people demanding the nationalization of oil Companies. The Iranians failed to obtain effective control of the industry because without the co-operation of the international oil Companies neither crude oil nor products could be sold in any significant quantities. As we have seen, much of the world's processing and distributing facilities were in the hands of the major Companies who refused to handle the expropriated oil, and others were deterred since AIOC threatened legal action against anyone who did so. Thus, the dependence of the crude-oil producing countries upon the international majors was painfully brought home to their governments, and even in the tense aftermath of the war with Israel in 1967, none of the chief exporting countries attempted outright nationalization of the major crude-oil producing Companies.

SIGNS OF IMPENDING DIFFICULTIES

The Iranian dispute was the only serious conflict between the oil Companies and the governments of the oil-producing countries of the Middle East in the ten years after the war. These have been called the 'ten golden years';[3] the Companies were waxing fat, the producing countries seemed reasonably satisfied with the concession agreements, demand for oil was rising very rapidly

[1] See below Chapter V, p. 160.

[2] For example, the American members of the Consortium are permitted by the US Internal Revenue authorities to take a depletion allowance against the taxable income arising from their operations in the Consortium. This is permitted only when the Company's interest is held to be equivalent to a lease under US law.

[3] P. H. Frankel, *Oil: The Facts of Life*, p. 12.

TABLE I

ANGLO-IRANIAN OIL COMPANY LIMITED[1]

PAYMENTS TO GOVERNMENT AND PROFITS 1930–9, £000

	1930	1931	1932	1933	1934	1935	1936	1937	1938	1939	*Total* 1930–9
Income less expenses (excluding depreciation, royalty payments and taxation)[2]	7,423	3,775	5,692	5,734	6,854	7,120	10,864	14,152	12,573	9,213	83,400
Less:[3]											
Royalty and tax to Iranian Government	1,015	135	2,645	1,785	2,159	2,192	2,580	3,545	3,307	2,771	22,134
Depreciation, amortization, etc.	1,587	1,252	1,071	1,302	1,459	1,409	1,765	2,244	2,603	2,071	16,763
UK Income Tax	1,259	671	(83)	305	512	409	911	1,652	1,157	1,956	8,749
Net profit after tax	3,562	1,717	2,059	2,342	2,724	3,110	5,608	6,711	5,506	2,415	35,754

[1] These figures represent Anglo-Iranian Oil Company only and exclude subsidiary companies.

[2] Income less expenses represents the balance of profit after taking credit of dividends from subsidiary, allied and other companies, interest on British Government and other securities and after deducting all expenses except those shown above.

[3] The charges are the net amount after adjustment of reserves made for other years' liabilities.

Source: British Petroleum.

and supplies were assured. But a thoughtful observer could discern clear signs of stormy weather ahead; some oil was being sold at discounts or on special terms to independent companies; some new companies were seeking and obtaining concessions to explore for oil; some importing countries were beginning to question the price at which oil was being sold.

The precariously balanced equilibrium of the international industry had depended on successful strategy in three critical areas: the relations with the producing countries; the attitude of importing countries toward supply and price; and the actual amount of crude oil coming on the market in relation to demand. What seemed to be generous arrangements had been made with the producing countries; the Companies had begun publishing ('posting') their prices for crude oil which, though necessary for the new '50/50' taxation arrangements, were also expected to reassure importing countries that uniform f.o.b. prices were being charged to all buyers.[1] Moreover, great efforts were made to assure buyers that the published prices were truly 'competitive prices.[2] But above all there was little pressing demand from any of the major importing countries for cheaper crude oil—indeed, lower prices for crude-oil imports would have aroused great opposition from coal interests in Europe, especially toward the end of the decade, and from domestic oil interests in the US. Moreover, it was not at all clear that the UK would gain from cheaper oil at the expense of the profits of British Petroleum or Shell. Some lesser importers, however, were becoming concerned about the price of oil and oil products, and there were signs that some sellers were willing to dispose of it at less than ruling prices.

The ease with which increased supply could be produced was clearly shown in the Iranian crisis: Iranian oil production dropped from 640,000 b/d in 1950 to 20,000 b/d in 1952 and was not to exceed 600,000 b/d again until 1957. Yet total production from the Middle East continued to rise throughout the period, Iraq output going up over four times between 1950 and 1954 while Kuwait output nearly tripled. The large reserves of oil available, the low cost of developing them in relation to prices, and the desire of at least some of the producing countries for an increase in their share of total output, could be expected to tempt the Companies to increase their rates of production. These underlying pressures were masked by the effects of the Korean War and by the Suez crisis, which seriously interfered with deliveries of oil from the Middle East to Europe. After the Suez affair, a number of events coincided to bring far-reaching changes and to usher in a new phase in the industry's history.

Before we go on to discuss the changes in the international industry that came to a head in the later 1950s, however, we should take a look at a largely fortuitous event—fortuitous in the sense that the immediate causes had little

[1] According to Shell International the 'main reasons for the establishment of posted prices outside the US' are the following: 'The offer to sell to buyers generally was intended to provide an assurance to the governments of producing countries that oil being exported at publicly posted prices to affiliated companies was not being under-valued. It was intended that posted prices plus appropriate long-term freights would provide the basis on which oil would be imported into consuming areas.' Shell International Petroleum Company Ltd., *Current International Oil Pricing Problems*, August 1963, p. 6.

[2] The phrase 'competitive prices' is, of course, extremely ambiguous and the Companies have never made much effort to explain precisely what they mean by it.

to do with any changes taking place in the industry—which nevertheless for many years created serious difficulties for the attempts of the oil Companies and the governments of the Middle Eastern oil-producing countries to resolve their disagreements. The conflict between the IPC and the Iraq Government after the coup d'état in 1958, which overthrew the parliamentary monarchy and established a military regime under the leadership of Abdul Karim Qassim, raised many of the basic issues surrounding the operation of the oil Companies in the Middle East, but the particular form it took, and the timing of it, were determined by the vagaries of Iraqi politics.

Conflict with Iraq

For some time the Iraq Government had been pressing for modifications of its agreement with the Iraq Petroleum Company, but had been careful to keep the issue out of the 'street', for 'imperialism', personified by the oil Companies, shared pride of place with 'feudalism', personified by the agricultural landlords and some members of the Royal household, as a means of rallying excited street mobs to potentially explosive demonstrations. After the fall of the monarchy one of the first acts of the new regime was to confront the oil Companies with demands for modifications of the existing concession agreements. As time went on the negotiations were accompanied by increasing publicity and popular agitation in which the cry of 'nationalization' was frequently heard. In this period, too, the Algerian struggle with the French, which commanded great sympathy in Iraq, was at its height, and many otherwise moderate people were seriously recommending that at least the French share of the IPC should be nationalized.

The negotiations, which continued intermittently from the autumn of 1958 to their breakdown in October 1961, revolved around four main issues: (1) the relinquishment by the oil company of some of the territory held under its concession agreement with the government; (2) profit-sharing arrangements; (3) the rights to natural gas; and (4) equity participation by the Government in the IPC. The Government refused to come to an agreement on any individual issue, insisting on a 'package deal' which would include settlement of all issues. It seems clear that part of Iraq's purpose in these negotiations was to break the existing pattern of profit-sharing and concession arrangements that had up till then prevailed in the Middle East—but which were already being undermined, as we shall see later.

The concession held by the IPC,[1] which gave it exclusive right to explore, produce, refine and export crude oil, covered the entire territory of Iraq except for a small area around Khanaqin on the Iranian frontier.[2] The Com-

[1] Technically there were three companies operating in Iraq: The Iraq Petroleum Company, east of the Tigris, the Mosul Petroleum Company, west of the Tigris and to the north, and the Basrah Petroleum Company, west of the Tigris and to the south. All of these companies had the same ownership, directors and constitution and for most practical purposes can be considered as one company, which I shall refer to as the IPC. However, only for the IPC production proper did British Petroleum have an over-riding royalty, which it had obtained when the American group were given some of its shares in the Company.

[2] For a discussion of the concession arrangements, see George Lenczowski, *Oil and State in the Middle East*, Chapter IV, and Zuhayr Mikdashi, *A Financial Analysis of Middle East Oil Concessions: 1901–65*.

panies expressed their willingness at the outset of the negotiations to relinquish slightly over half of this territory provided that they could choose the areas to be relinquished. This proviso was unacceptable to the Government, and although offers were subsequently made which would have led to the eventual relinquishment of some 90 per cent of the area, no agreement was reached, partly because the Government insisted on its 'package deal', but also because of the Government's insistence on participating in the selection of the areas to be given up.

Under the 1950 profit-sharing arrangements the Government was entitled to 50 per cent of the difference between the (agreed) cost of crude-oil production and the posted prices, which, in principle at least, were under the control of the Companies. In the new negotiations, the Government demanded a greater share of this difference but never clearly specified the proportion it wanted. While the talks were dragging on, the oil Companies twice reduced the posted prices of crude oil throughout the Middle East.[1] These reductions, by reducing the calculated profit attributed to crude oil, also reduced the Government receipts per barrel of crude.

As we shall see, at this time the falling prices of refined products in world markets began seriously to worry the Companies, especially in Europe and Japan where competition was increasing. If product prices fall while posted prices for crude oil remain unchanged, the *de facto* share of the crude-producing countries in the *total* profits from the integrated operations of the companies will correspondingly *rise* since the proportion of profits attributed to crude production remains unchanged and that attributed to refining and marketing falls. If the Companies either sell crude to outsiders or transfer it to their refining affiliates at prices lower than the posted prices on which they are taxed, the governments of the crude-producing countries will receive more than 50 per cent of the profits that the Companies attribute to crude-oil production.[2] When the Companies reduced the posted prices of crude oil they attempted to explain to the Iraqi Government that in the circumstances the Government was in fact obtaining more than 50 per cent of crude-oil 'profits', but the Government, understandably enough, was not impressed. The suspicion was widespread, though certainly unfounded as subsequent events confirmed, that the Companies had reduced posted prices for crude, not because of competition and lower prices in product markets, but in order to squeeze independents trying to enter crude-oil production, and that therefore the existing pressures on prices would be only temporary.

The Government also wanted the Companies to renounce all rights in the natural gas that was surplus to their requirements for field operations. This gas was far from markets and was almost all flared—burnt off. Although the Government had previously acquired the right to use all the natural gas it wanted for internal purposes, it now wanted rights to all of it, including the right to export it. The Companies suggested that gas should be available

[1] The first reduction, in February 1959, led to a resolution at the First Arab Oil Congress in April that the governments should be consulted about price changes; the second reduction, in August 1960, led to the formation of the Organization of Petroleum Exporting Countries. See pp. 83–4 below.

[2] For an explanation of this problem and its consequences, see Chapter VI below.

either to the Government or to the Companies, depending on who had customers for it, but were averse to meeting the Government's demand for complete renunciation.

With respect to its demand for equity participation in the IPC group, the Government pointed out that the agreement between Iraq and the Companies had always provided that Iraqis should have the right to take up 20 per cent of any new shares that might be issued by the IPC. This provision dated from the San Remo oil agreement in 1920 between Britain and France at the time when Iraq was put under mandate to Great Britain. No such shares had ever been issued by the IPC—this has always been a sore point with the Iraqis—and so they were now demanding their 'right' to 20 per cent participation in addition to the 50 per cent share of profits, with no change being made in the posted price arrangements for valuing the crude oil. In addition to the principle involved, which the Companies were unwilling to accept, was the problem of placing a value on 20 per cent of the IPC; the Government suggested book value, which, not surprisingly, the Companies considered hopelessly unrealistic. In their turn, however, the Companies did offer to form new companies with the Government as a partner, for the exploration and development of the existing unexploited areas.

In October 1961, the negotiations broke down completely and in December the Government published the famous Law No. 80, which expropriated around 99 per cent of the concession area of the IPC group, thus effectively putting an end to further expansion by the group in Iraq without an outright nationalization of its operating facilities.

The Iraqi negotiations were doomed to failure, not only because the Government was striking at the very heart of the arrangements governing the established operations of the major Companies, but also because Iraq was acting alone in so doing. Nevertheless, within the next few years provision for the automatic relinquishment of concession areas, an increased share of oil receipts for the governments of producing countries and provision for equity participation by the government, were to become almost standard in new concession agreements and, with the exception of the last, even incorporated in modifications of the existing agreements. As has been shown, the Companies had been willing to go a long way to meet the Iraqi Government demands during the negotiations, but with the unilateral expropriation by Law 80, they felt compelled to stand on their legal and moral position, and they refused to 'accept' the law, demanding arbitration, as was their legal right.

By 1965, however, events throughout the Middle East had gone far in changing the older pattern of concession agreements (as well as in changing the governments of Iraq) and it may have seemed to the Companies that more might be gained by forgetting their grievances and accepting unpalatable terms than by further resistance to the expropriation. In any event, early in 1965 an agreement was arrived at between the Companies and the Government, under which the Companies to a large extent accepted the terms of Law 80 but were to get back the north Rumaila field in southern Iraq; with the exception of Jersey Standard, they agreed to take up concessions in the expropriated areas in partnership with the Government. This last provision was a very great departure for the major Companies, and might well have led to far-reaching changes in concession terms throughout the Middle East. Most of the concession agreements contain a 'most-favoured country' clause under which better terms offered by the Companies in one country can be claimed by others.

At the time of the outbreak of war with Israel in 1967, however, the signing and publication of the agreement had not yet taken place because of difficulties more related to Middle Eastern and Iraqi politics than to difficulties with the Companies specifically. After that war the Government of Iraq seemed to abandon all further attempts to come to an agreeent with the IPC and passed a further law reserving all the unexploited areas of Iraq for exploitation by the State-owned Iraq National Oil Company, which was empowered to enter into arrangements with foreign companies for the purpose. In November 1967, an agreement was signed with ERAP, a French government company, for the exploitation of some of these areas, but excluding the potentially rich North Rumaila field with respect to which CFP, the French partner in the IPC, was negotiating separately with the Iraq Government. The financial terms of the ERAP agreement were very much less favourable to Iraq than those that had been offered by the IPC.[1]

ACCELERATION OF CHANGE

The very rapid changes in the international industry that began to appear in the middle 1950s and to accelerate after 1958 were very largely the result of three mutually reinforcing developments: (1) the increased competition for concessions to explore for and produce crude oil, including the entry of newcomers and the rise of new sources of crude, which changed the terms on which concessions could be obtained; (2) increased competition in crude and product markets, which directly and indirectly affected prices and brought changes in the policies of some importing countries; and (3) the increased importance of government policies.[2]

COMPETITION FOR NEW CONCESSIONS

Although non-major Companies had always played a role in the search for oil, sometimes obtaining concessions and exploring with the end in view of selling out to larger Companies if their activities were successful, there were only a handful of independents in the Middle East before the formation of the Iranian Consortium in 1954.[3] Under pressure from the US Government, nine independent American Companies obtained shares in the Consortium through the Iricon Agency, and this development gave a considerable impetus to the role of independents in the Middle East. In the second half of the 1950s, Companies with little or no international business began in increasing

[1] For further discussion see p. 218 below.

[2] Georg Tugendhat has analysed the changes in the oil industry since the war from a different point of view. See 'An Outsider's View of the Oil Industry', February 1967.

[3] The only two of importance were Aminoil (a group of eleven US independents) and Pacific Western Oil (now Getty), which had obtained concessions in the Neutral Zone as early as 1948 and 1949 respectively, offering to the governments greater payments than had yet been seen in the Middle East.

numbers to seek concessions to explore in various parts of the world, including the existing major oil-producing areas.

When Venezuela put up to auction new tracts of land in 1956–7, large numbers of American Companies were attracted and paid handsomely for concessions. During the three years following the passage of Libya's first oil law in 1955, concessions covering 55 per cent of the land area of the country were granted to fifteen oil companies or groups, including six of the seven majors, CFP, and American and German independents.[1] In 1964 Iran opened some offshore areas for competitive bidding, which attracted some thirty companies,[2] most of which combined in groups for the purpose. The five groups that succeeded in obtaining concessions included AGIP, the Indian Oil and Natural Gas Commission, five French companies, thirteen US independents, and one major (Royal Dutch/Shell).

Competition for new concessions thus came from the international majors, eager to continue their exploration programmes in spite of the current 'surplus' of crude and in spite of the fact that some of them had very large reserves already under their control, as well as from independents of numerous nationalities, but the latter brought significant changes in the prevailing pattern of concession terms. In the beginning the innovators in the competition for new concessions were the American independents, Ente Nazionale Idrocarburi (ENI)—the Italian State company under the leadership of Enrico Mattei—and the Japanese. In 1957 the Iranian Government gave the National Iranian Oil Company (NIOC) authority to engage in the exploration for and the exploitation of oil reserves, and to take foreign companies as partners. Almost immediately concession agreements were signed with AGIP (a subsidiary of ENI) and Pan American Oil Company (a subsidiary of Standard Oil of Indiana), which breached the cherished '50-50' division of profits, provided for an automatic reduction in the size of the concession as time went on, placed the exploration risk entirely on the concessionaire, and provided for equity participation by NIOC.

At about the same time the Japanese Company, Arabian Oil, obtained offshore concessions in the Neutral Zone from Kuwait and Saudi Arabia which gave the governments 57 per cent and 56 per cent of the profits respectively, provided for a 10 per cent share to the government if oil was found, and also contained provisions for the relinquishment of part of the concession area.[3] By 1961 even one of the majors (Shell) had made an agreement (with the Kuwait Government) for an offshore concession which gave the Government the right to purchase 20 per cent of the producing company after the discovery of oil in commercial quantities. The bidding for this concession had attracted thirteen oil companies, some of which formed partnerships for the purpose, and included (besides the successful Shell) BP and Gulf from the international majors, a number of US independents, ENI, and a Japanese group.

[1] Abdul Amir Q. Kubbah, *Libya: Its Oil Industry and Economic System*, pp. 101–8. Of the ninety-five concessions granted up to the middle of 1963, eighty-four had been granted between 1955 and 1959.

[2] These included Jersey Standard, Mobil, Shell, CFP, 16 US independents, 5 French and 5 German companies, and ENI. See *Petroleum Press Service*, September 1964, p. 331.

[3] These agreements are summarized by Issawi and Yegeneh, pp. 28 ff.

In spite of the dominance of Algeria in French oil policy, French Companies also participated in the competition for new sources of oil in the Middle East. In 1965 the State-owned Société Auxiliare de la Régie Autonome des Pétroles (AUXIRAP) took a concession in the Red Sea zone of Saudi Arabia, agreeing to form a joint company with the Saudi State organization (PETROMIN) on the discovery of commercial oil, in which PETROMIN would have 40 per cent shareholding but equal voting and executive power. The rental and royalty provisions of this agreement were called 'severe' by the trade press.[1] In the following year Enterprise des Recherches et d'Activités Pétrolières (ERAP) went even further in adding to the variety of agreements in the Middle East by accepting an arrangement with NIOC under which it will search for oil at its own expense; if successful ERAP will turn half the discovered reserves over to NIOC for development and lend it funds. Both the loan and the exploration costs will be reimbursed to ERAP over a period, partly in oil if desired. In return ERAP will have the right to purchase a certain proportion of the oil produced at cost plus 2 per cent plus 50 per cent of profits calculated on realized (not posted) prices. In this arrangement ERAP agrees to market NIOC's oil if desired, and in effect acts more as a contractor to NIOC than as a concessionaire.[2]

In all the new concessions, provision is made for the relinquishment according to a predetermined time-table of unexploited parts of the concession areas in order to free them for new bidding, for the lack of success by one oil company does not necessarily discourage others from trying their luck. The older type of concession covering entire countries has virtually disappeared, and the older agreements have been modified, with the voluntary co-operation of the old-established Companies, to provide for the relinquishment of large areas. By 1963, for example, the Kuwait Oil Company, Aramco, and the IPC (in Qatar) had all voluntarily given up a substantial proportion of their respective concession areas, with provision for more to come. The areas thus released increased the scope for the entry of new competitors.

In Africa, oil was discovered in large quantities in Algeria, Libya and Nigeria, and exports from all three countries began to be important between 1958 and 1961. Algeria was the second largest producer on the continent with 550,000 b/d in 1965. The first commercial discovery there was in 1956 and production for export began on a significant scale in 1960. With the exception of Royal Dutch/Shell, which had been in the Sahara since 1952, foreign Companies began exploration only in 1957–8 and operated under the *Code Pétrolier Saharien* of 1957 and in joint ventures with French companies in many of which the French Government had a substantial interest.

With the Evian agreements of 1962 and the coming of independence for Algeria, the role of the foreign oil Companies, including the French Com-

[1] *Petroleum Press Service*, May 1965, p. 185.

[2] This agreement was carefully analysed by Mr Thomas Stauffer in a paper presented to the Sixth Arab Petroleum Congress in March 1967, 'The Erap Agreement—a Study in Marginal Taxation Pricing'. Using a discounted cash flow analysis, Mr Stauffer concluded that on certain assumptions the economic benefits to Iran were less than those obtainable from the standard or 'OPEC type' agreement which includes the expensing of royalties and taxation on posted prices. Mr Stauffer's paper is also published in *Platts Oilgram*, December 30, 1966.

panies, was changed radically. Arrangements with respect to the terms on which oil resources were to be exploited were made between the governments. Indeed, Hartshorn points out that the later Accord d'Alger in 1965, which superseded all previous arrangements and laid down in some detail the financial and other provisions under which the industry operated, was one of the first 'between governments for the exploitation of the natural resources of one of them ever to be concluded'.[1] However, the importance of French oil policy to Algeria, and the extremely close links between the Algerian industry and the French market, as well as the importance of the political relationships between the two countries, make the question of the terms of the Algerian concessions a 'special case'. The large international firms have played a small part, and since the theme of this study relates to the role of such firms in developing countries, the evolution of the industry in Algeria and the changes in the arrangements governing it will not be further discussed here.[2]

OTHER NEW PRODUCING COUNTRIES

The industry in Libya has been developed entirely by private international firms and Libya is the only major exporting country in which independent producers have a really significant role. Although exports began only in 1961, by 1966 Libya was the greatest producer in Africa with about 1,505,000 b/d or some 13 per cent of total production in the Middle East and North Africa together. The country's first oil law was drawn up in collaboration with the interested Companies, and its terms were extremely attractive from their point of view. As the extent and nature of Libya's oil resources became clearer, the Government imposed increasingly stringent terms which, until the 1965 revisions, were by and large readily accepted by the Companies; competition for concessions remained strong.[3] In 1965, however, the Government, in line with OPEC resolutions, decided to require the Companies to pay taxes on the basis of posted prices. This was in violation of existing concession terms and the Companies were forced to bring their existing agreements into line with the new laws.[4]

Nigeria is the third largest producer in Africa, with an output in 1966 of 415,000 b/d, about 70 per cent of which was produced by Shell and BP. Oil was discovered in 1956 and exports began in 1958 but reached significant proportions only in 1961. Only Shell and BP, operating jointly, were in the area before the war, the first non-British Company coming in being Mobil in 1955. After the country obtained its independence in 1960 a number of other Companies[5] entered, two of which quickly found oil (Gulf and Tennessee); Gulf began exporting from its offshore discoveries in 1965. As in Libya in the early

[1] J. E. Hartshorn, p. 264.

[2] For a useful brief discussion see Hartshorn, pp. 262 ff.

[3] In addition to providing generous allowances of several kinds, the oil law permitted profits to be calculated for taxation purposes on realized prices over which the Government had no control instead of on 'posted prices', as was the case in the Middle East.

[4] For a discussion of the significance of this and of Libya's position see Chapter VII.

[5] Gulf Oil, Tennessee Gas in association with Sunray DX and Sinclair (US independents), Amoseas (Texas Oil and Standard of California), Mobil Oil, SAFRAP (a French group), and AGIP.

years of the industry, the financial arrangements with the Companies were very much less favourable to the Government than those of the Middle East, but exploration in the Nigerian delta and offshore is costly, and attractive terms were undoubtedly necessary to bring the Companies in. The success of exploration brought increased competition for concessions and the question of revising existing agreement was taken up in 1966. In January 1967 a decree bringing tax and royalty arrangements more in line with the OPEC countries was announced and discussions began with existing concessionaires. The civil war that broke out in 1967 was still in progress at the time of writing and oil exports had virtually ceased.

Libya, Algeria, and Nigeria, together with Abu Dhabi, a small Sheikhdom on the Persian Gulf,[1] were the most important newcomers to crude-oil production in the early 1960s. But the search for new sources of supply, and with it the demand for exploration licences and permits touched nearly every country in Africa, Asia, and Latin America. Payments made by the Companies for exploration rights vary widely, as would be expected, since different areas offer different prospects of success and different risks. The established areas in the Middle East, especially around the Persian Gulf, were able to attract the largest number of competitors and to obtain the most favourable terms, which inevitably influenced the attitudes of other countries where petroleum reserves had been discovered and developed. Many of the newcomers operated in groups (or consortia) to ease the financial burden and spread the risks. The largest single group of newcomers are American companies; others, like the Italian and French are State-owned, and still others, such as the Germans and Japanese, are private.[2]

Although different groups had different reasons for joining the intensifying exploration activity, their willingness to undertake the costs and risks of exploration was itself *prima facie* evidence that the expected costs of finding and developing new reserves of crude oil were below the expected cost of buying their requirements from the market.[3] Only a few of these Companies succeeded in finding oil in exportable quantities, but enough did so to add significantly to the mounting capacity seeking markets. Thus a 'sellers' market' in oil concessions existed side by side with a 'buyers' market' in crude oil and products.

INCREASED COMPETITION IN CRUDE AND PRODUCT MARKETS

The successful search for new reserves of oil brought new supplies on the market, some of which were controlled by the old-established majors, some by State companies producing largely for local markets, and some by companies new to the international market but without refining facilities or marketing

[1] Production onshore in Abu Dhabi is by the IPC group and offshore by Abu Dhabi Marine Areas, in which BP has a two-thirds interest and Compagnie Française des Pétroles one-third. These last two Companies also own Dubai Marine Areas, which in turn owns 50 per cent of a concession offshore Dubai, a Sheikhdom in the Persian Gulf, where oil was discovered in 1966. (The other interests in this company are Continental Oil with 35 per cent, Deutsche Erdoel with 10 per cent, and Sun Oil with 5 per cent.)

[2] See the useful survey in *Petroleum Press Service*, December 1962, pp. 463–7, and January 1964, pp. 9–11.

[3] M. A. Adelman has developed this point in *The World Oil Outlook*, pp. 99 ff.

networks of their own adequate to handle their output. From 1950 to 1966 the share of the seven majors in total crude-oil production outside of Canada, the US and the Communist countries fell from 85 per cent to 76 per cent, while that of the new independent 'internationals' and of State-owned companies rose correspondingly. At the same time, Russian production increased from a mere 730,000 b/d in 1950 to over five million in 1966, and exports of crude and products from Eastern Europe rose from 20,000 b/d in 1950 to over a million b/d in 1966. To top it all, the US adopted mandatory import quotas for crude oil in 1958 which hit especially hard those US companies that had developed foreign supplies since it reduced substantially the extent to which the US market could absorb greater quantities of the increasing supplies outside of the US.

Although consumption continued to rise at an annual average of between

TABLE II
GROSS CRUDE-OIL PRODUCTION AND REFINERY THROUGHPUT
SEVEN MAJORS AND WORLD
(EXCLUDING US, CANADA, USSR, EASTERN EUROPE AND CHINA)
1950, 1960, 1966

Gross Crude Oil and Condensate

	1950		1960		1966		% Incr.
	000 *b/d*	%	000 *b/d*	%	000 *b/d*	%	1950–66
Standard Oil (NJ)	1,020	25	1,920	17	3,150	18	209
Royal Dutch/Shell	770	19	1,600	14	2,390	13	210
British Petroleum	800	20	1,500	13	2,500	14	212
Gulf Oil	300	7	1,170	10	1,780	10	493
Texaco	240	6	790	7	1,440	8	500
Standard Oil of California	180	4	560	5	1,480	8	722
Mobil Oil	140	3	570	5	950	5	578
Total Seven Majors	3,450	85	8,110	72	13,690	76	268
Other Companies	620	15	3,120	28	4,230	24	582
Total	4,070	100	11,230	100	17,920	100	340

Refinery Throughput

	1950		1960		1966		% Incr.
	000 *b/d*	%	000 *b/d*	%	000 *b/d*	%	1950–66
Standard Oil (NJ)	750	19	1,760	15	3,000	18	300
Royal Dutch/Shell	870	22	1,930	16	2,750	16	216
British Petroleum	600	15	900	8	1,600	9	167
Gulf Oil	60	1	370	3	500	3	733
Texaco	390	10	650	5	1,170	7	200
Standard Oil of California	120	3	320	3	510	3	317
Mobil Oil	90	2	430	4	840	5	833
Total Seven Majors	2,880	72	6,360	53	10,370	61	260
Other Companies	1,120	28	5,540	47	6,630	39	492
Total	4,000*	100	11,900	100	17,000	100	325

* Rough Estimate.
Sources: Annual Reports of Companies and British Petroleum, *Statistical Review of the World Oil Industry*. Refinery Throughput Total supplied by British Petroleum Statistical Department except for 1950, which is a rough estimate. The figures for 'other companies' are residuals.

7 and 8 per cent, the eagerness of both old and new producers to find new markets or retain their established ones was such that by 1960 crude was offered at prices substantially lower than those that the international majors were publicly posting. The major Companies gave discounts on crude to non-affiliated buyers on long-term contracts; the Russians also undercut the so-called 'world prices' in order to get into the market and quickly found willing buyers;[1] and certain independents, especially those that had found large supplies in Libya, offered oil at 'cut' prices. Some of the majors early discounted their posted prices even to their own refining affiliates or gave them other concessions. By 1966 very little oil moved at posted prices.[2]

The competition among suppliers of crude oil was reinforced by the growing number of refiners independent of the international majors,[3] while the increased competition in product markets, which put pressure on product prices, reinforced the natural desire of refiners to obtain crude supplies on the most favourable terms. Crude flowed into refineries under almost every conceivable sort of arrangement. Some was sold outright to 'arms-length' buyers.[4] Russian crude was sold partly to State-controlled companies and partly to non-integrated refiners; and some of the non-integrated independent crude-oil producers sold much of their crude either to their partners in production on long-term contracts or to other Companies, including the majors.

For most of the new producers the importance of obtaining assured long-term outlets was the dominant consideration, and perhaps the most common arrangement involved a close relationship between crude-oil producing companies and refiners, either through complete integration, both producers and refiners being under the same ownership, through partial integration whereby owners of crude obtained part ownership in refineries, or through long-term contracts to supply crude to non-integrated refineries. Traditionally the major Companies disposed of practically all of their crude through their own wholly-owned refineries, and this apparently remained the preferred procedure. Some of the newcomers, notably Continental, also built wholly-owned refineries.

[1] Although some countries, for example the UK, excluded Russian oil entirely.

[2] See Chapter VI for a discussion of prices.

[3] The percentage of European refinery capacity, for example, owned by the seven majors fell from 67 per cent in 1960 to 54 per cent in 1965, with a small additional amount of capacity tied up under long-term contracts. Taken from tables prepared by W. L. Newton and presented in a paper given to the International Petroleum Seminar at the School of Oriental and African Studies, University of London, March 1967. The *Petroleum Press Service* estimated that non-major companies controlled around 43 per cent of the refining capacity outside North America and the communist countries in 1965 as compared to 38 per cent in 1961 (January 1967, p. 22). In the same article it was estimated that the share of the majors in product sales outside the US declined from 69 per cent to 64 per cent in the same period. As will be seen in Table II, however, the share of the majors in world refinery throughput of crude outside the US, Canada, and the communist countries rose between 1960 and 1966. Part of the reason for this was the acquisition of a large German refinery by Texaco in 1966.

[4] It was reported in 1963 (*Petroleum Press Service*, April 1963, p. 132) that '. . . entry into Europe . . . is facilitated for the less fully fledged independents by the existence of large brokers who put suppliers in touch with big customers with their own tankage. Examples are the Hamburg firm of Marquard and Bahls, who deal in several million tons a year, and Stephenson Hardy in the UK, but others operate in Italy and the Netherlands as well as Germany and the UK. Increasing quantities of crude, refined by arrangement, may thus move right outside integrated channels.'

But shared ownership between very different types of partners became more common—majors combined with independents, independents with each other, and either (or both) with governments.[1] In the underdeveloped countries particularly, governments often created joint ventures with private Companies. The majors have in general been reluctant to embark on joint ventures with governments, but the pressures of competition have been gradually eroding even their reluctance.

The desire for markets by Companies with crude to sell led them to assist in the construction and financing of refineries. To some extent, especially in Africa and Asia, the financial terms offered in return for the right to supply crude for a given refinery were so favourable that they must be interpreted as an indirect discount on the current price of crude oil. In other words, refinery financing and the arrangements made for its crude supply were sometimes components of a single 'package deal'.

Thus, the competition for owned sources of crude oil, for refining capacity and for product markets created new patterns in the organization of the production, processing, and distribution of oil and products. In these emerging patterns governments were playing an increasingly important role in shaping the course of events.

INCREASING IMPORTANCE OF GOVERNMENTS

Throughout the history of the modern industry, governments have intervened in a variety of ways to protect or advance what they conceived to be their national economic interests. In the late 1950s and in the 1960s, government policies became of increasing significance for the oil industry in almost all countries. In addition to taxation, governments in the developing countries became more concerned about international prices, foreign exchange receipts and payments, currencies of payment, security of supply, foreign investment, and the location, ownership or control of producing and refining facilities.

Most countries can be classed as either primarily consumers of crude oil (and products) or primarily producers of crude oil. The two major exceptions are the US and the USSR, whose domestic policies are excluded from consideration in this study. Both are major producers as well as consumers, but the foreign trade policy of each has had considerable importance for the international industry. Between 1958 and 1960 exports from the USSR doubled, adding to the pressure of supplies on prices in world markets. By 1965, she accounted, together with the rest of Eastern Europe, for 6 per cent of world exports. Russia was a great oil exporting country up to the First World War, supplying some 19 per cent of the world market, and in the 1930s about 15 per cent of Soviet production was being exported.

The US became a net importer of crude and products toward the end of the

[1] P. H. Frankel and W. L. Newton, in describing these developments, have emphasized the virtual disappearance of non-integrated refineries as a consequence. They argued that 'the need for non-integrated refineries has become less definite since the access to refining capacity at the right place is now being secured by mutual availability rather than by ownership or control'. 'Recent Developments in the Economics of Petroleum Refining.' Paper presented to the Sixth World Petroleum Congress, 1963, p. 11.

1940s. All during the 1950s mounting imports, with the consequent greater dependence on foreign supplies, raised fears of a danger to the country's economic security from a military standpoint. This, plus political pressure from domestic producers of coal as well as oil, led to the imposition of voluntary and then mandatory import controls in spite of the fact that US companies had acquired ownership of large reserves of crude oil in many parts of the world.

In countries with little production, and which were therefore primarily importers of crude oil and products, different policies were pursued according to the particular circumstances. The UK sought security of supply through the ownership of crude oil by British companies in territories within the country's 'sphere of influence', and the British Government took a controlling interest in British Petroleum as early as 1914. The overseas operations of the British oil Companies apparently produce a net credit in the British balance of payments, which substantially reduces the overall debit when account is taken of oil imports.[1] The fact that oil is purchased for sterling from British Companies means that the actual foreign exchange cost is considerably below the price paid for imports. Domestic prices have been held up by taxation in order to protect the coal industry when competition between fuel oil and coal pressed hardly on the latter.

The French, the Japanese, and the Italians, all important importers, set out to acquire crude supplies under their own control—that is, in the hands of national (though not necessarily State-owned) companies. The French had acquired a share in the IPC before the war, as we have seen, and in the Iranian Consortium in the 1950s. But the biggest French Government effort was in Algeria, and as Algerian oil became available the French Government required refiners in France to take a specified quota of this 'franc oil'. The Japanese successful search for crude was primarily a private venture in search of cheaper oil but the Government also helped to find outlets for the crude by requiring refiners in Japan to take a stated percentage of their supplies from the Japanese company. One of the chief aims of the Italian State company, ENI, was to find its own sources of crude oil, and this group took the lead in making direct partnership agreements with other governments, both for production and refining. ENI spokesmen openly advocated that government-to-government contracts for crude-oil supplies should be extended in place of the arrangements made by the private international Companies.

In the less developed consuming countries, governments became increasingly less willing to accept without scrutiny the operations and plans of the international Companies.[2] India conducted a full-scale inquiry into the price of oil in 1961 in which the dissatisfaction of the Committee of Enquiry was plainly expressed. Subsequently the Government restricted the freedom of

[1] Figures that would enable one to determine the 'balance of payments on oil account' for the United Kingdom have never been made officially available. The 1964 and 1965 annual reports of both British Petroleum and Shell give some information but its significance is difficult to determine since the method of calculation is not explained. A more detailed statement for 1963 can be found in the Political and Economic Planning Report, *A Fuel Policy for Britain*, p. 209.

[2] The issues involved in the developments very briefly outlined here, and of which very few examples are given, are discussed more fully in the relevant chapters which follow.

action of the refining affiliates of the international majors, forced a discount in the price of crude oil, and began actively to encourage the entry of other refiners willing to act in co-operation with the Government. In Pakistan, too, the Government undertook directly to participate in the industry. Both India and Pakistan promoted domestic exploration for oil by domestic and foreign Companies and by partnerships between them, with some favourable results in India and considerable supplies of natural gas in Pakistan. Ceylon set up a State oil company to market oil and eventually expropriated the facilities of the existing distributors, who were affiliates of international majors, for the use of the State company. Cuba similarly expropriated the international oil Companies. In Africa, the governments went little further than occasionally to obtain shares in refining facilities, although the offers by ENI of State participation were usually accepted.

Political pressures from governments have also been partly responsible for the establishment of refineries in numerous countries the markets of which would ordinarily have been considered too small to support a refinery. Since the Second World War there has been a steady trend toward locating refineries near consuming centres instead of near the sources of crude-oil production. Whereas in 1939 some 70 per cent of refining capacity outside North America and the Communist countries was near the oilfields, this had dropped to about half by 1951 and to about 16 per cent in 1965.[1] There were a number of economic reasons for this, including the growing size of individual markets, especially in Europe, thus making possible the creation of market refineries large enough to take advantage of the economies of scale. But also important, especially in developing countries, were government policies demanding the establishment of local refineries either to save foreign exchange through the import of crude oil instead of products or to promote national industrialization. This led to the development of designs for simple, standard refineries in order to reduce the diseconomies attendant on small size. It also encouraged an increase in the number of refineries which were owned or used jointly by one or more Companies selling in the same area when the market of no single Company was large enough to justify the establishment of even a small integrated refinery of its own.[2]

It was among the crude-oil producing countries where the governments were most insistent on their right to an extensive voice in the conduct of the industry. In Burma the State took over the industry in 1963 (along with other foreign enterprises), but without a head-on collision with the Companies. Indonesia and Argentina permitted foreign Companies to work largely as contractors on behalf of the government; other countries insisted on considerable State participation in producing operations; in still others the State maintains an extensive supervisory and regulatory system, for example, in

[1] Taken from W. L. Newton in a paper given at a seminar on the international petroleum industry at the School of Oriental and African Studies, University of London, 1967.

[2] These arguments are developed in the excellent analysis of the nature and significance of the changes in the location and organization of refineries since the war in the paper by P. H. Frankel and W. L. Newton cited above and in their earlier paper, 'Current Economic Trends in Location and Size of Refineries in Europe' presented to the Fifth World Petroleum Congress in 1959. See also Peter R. Odell, 'The Development of the Middle Eastern and Caribbean Refining Industries 1939–63', p. 208.

Venezuela. In almost all producing countries State-owned national oil companies have been created.

Of great concern to the crude-producing countries, of course, is the revenue the government receives from the industry's operations, and it was a threatened reduction in revenue through a reduction in the prices on which profits were calculated that brought the formation of the Organization of Petroleum Exporting Countries in 1960. There can be little doubt that the establishment of OPEC increased the bargaining power of the crude-oil producing countries, and most observers give OPEC credit for the fact that after its formation posted prices were not reduced as 'market' prices continued to slide, but rather became a subject for negotiation with the oil-producing countries along with other proposals designed to increase their revenues.

ORGANIZATION OF PETROLEUM EXPORTING COUNTRIES

OPEC was established at a conference of the five major petroleum exporting countries (Iran, Iraq, Kuwait, Saudi Arabia and Venezuela) at a conference held in Baghdad in September 1960 on the invitation of the Government of Iraq. Much of the initiative that led to the establishment of OPEC had come from Venezuela who, as the producer with the highest average costs of all the major exporters, had the strongest interest in the prevention of competition which would bear heavily on prices. She had early recognized that stronger control over the rate of supply would be essential if the pressure on prices was to be relieved and had endeavoured to adapt her domestic policies accordingly. Since no other producing country followed suit, Venezuela's policy was doomed to failure. At the ninth OPEC conference in 1965 Venezuela was still attempting to persuade the exporting countries to co-operate to restrict production, and a voluntary programme to regulate supply was agreed upon. By then, however, the problem was even more difficult than it had been in 1960 in view of the clear success of the newer producers, particularly Libya, Nigeria and Abu Dhabi (Nigeria had not yet joined OPEC at the time of writing).

In addition to resolutions establishing the Organization and providing for a common front of the exporting countries *vis-à-vis* the oil Companies, the first OPEC conference agreed to attempt to get posted prices restored to their previous level. In subsequent conferences the arrangements for the constitution and financing of the Organization were completed, provisions for various studies made, new members admitted (Qatar, Indonesia, Libya), and the aims of the member countries were stated together with the demand that negotiations should be entered into by the Companies to achieve these aims. It became clearer as time went on that restoration of the cuts in posted prices was not practicable and the issue was not actively pursued, but a number of other ways of effectively raising the revenues of the producing countries were successfully negotiated. In 1964, agreement was achieved on the expensing of royalties, and the elimination of marketing allowances.[1]

[1] For the OPEC account of the negotiations see 'OPEC and the Principle of Negotiation', a paper presented by OPEC at the Fifth Arab Petroleum Congress, Cairo, March 1965. (Available from OPEC). For further discussion of this agreement see pp. 201–2 below.

CONCLUSION

That the world's crude oil reserves outside the US and the Communist countries ended up after the Second World War almost entirely in the hands of seven large Companies capable as a group of developing them rapidly, yet at the same time refining most of the output and selling it as products within their own marketing organizations, was not entirely due to perseverence in exploration crowned with the luck of discovery; it was brought about quite deliberately by the creation of associations among the major groups in the ownership of crude-oil producing affiliates and marketing companies, and by the negotiation of long-term contracts. In the Middle East, until the formation of the Iranian Consortium, only Anglo-Iranian held in splendid isolation sole ownership of the reserves it had discovered and developed.

It was no coincidence that the Companies, which together obtained control of Middle East oil by the 1950s, also dominated refining and marketing in Europe and Asia. Jersey Standard and Shell had begun their international operations with markets and distribution networks and had sought crude oil to keep them supplied, building refineries as the oil became available; during the 1950s Jersey refined more crude oil in or near Venezuela than in the Eastern Hemisphere, and shipped products all over the world. Socony Mobil, as the successor to the old Standard Oil of New York, which had been charged with foreign marketing in the original Standard Oil Trust before the First World War, had long experience in distribution in Asia and Africa as well as in some countries of Europe. Texaco had been selling in foreign markets since the beginning of the century. Only Gulf and Standard of California, like British Petroleum earlier, owed their entry on the international scene largely to the fact that they had been fortunate in the gamble of exploration and astute in the struggle for control in the Middle East, but they had to make arrangements with others who could use the oil they discovered. Until the 1950s, the Compagnie Française des Pétroles could claim major international status only through its ownership in the Iraq Petroleum Company, but this was a result of the political settlement following the 1914–18 war. The activities of the Company were confined almost entirely to French territories until after the Second World War.

It may well be argued that the result thrown up by history could hardly be expected to have been other than it was, at least in its broad outlines: Western enterprise and Western capital were required if the oil of Venezuela, the Middle East and South-east Asia was to be developed, and the Companies with the greatest interest in the matter were the existing marketers. These could afford to outbid all others in most matters, they could supply capital that others lacked, take the oil that others could not use, and purchase those lesser companies that were weak, or faint-hearted, or merely useful to them. Thus, it was, as has often been noted, 'logical' that the great Companies should prevail—not necessarily all of those that in fact did so, but certainly those that had acquired extensive Eastern Hemisphere markets. It was logical in the sense that the result seemed a thoroughly natural and reasonable outcome of the economic circumstances and needs of the time, and is therefore explicable without the aid of special hypotheses resting on the existence of deep-laid imperialistic plots, conspiracies between governments and Companies, abnormal subversion through bribery or violence, etc.

Nevertheless, the dominance of the industry by a few large firms was also the result of government policies that permitted the growth of monopolistic business organizations. The evolution of the petroleum industry within the US was characterized by the emergence of vertically integrated firms. This was by no means a technical or economic 'necessity', but historically was largely a by-product of the early position and competitive tactics of Standard Oil of New Jersey, which made integration by its rivals a competitive necessity in practice.[1] When US firms reached abroad—and again particularly Jersey Standard—similar competitive relationships were established. If we add to this the terms on which concessions were obtainable in the Middle East, the size of the financial resources of the firms in the international industry, and the policies of governments, we have sufficient conditions for the emergence of vertically integrated oligopoly in the production, refining and marketing of crude oil from the Eastern Hemisphere.

By the 1960s the international petroleum industry had been thoroughly stirred up: rapid changes were taking place in the location of activity, in the organization of the industry, and in the relative position of the governmental and private groups involved in it. The industry was still largely in the hands of the seven international majors, however, for the changes occurring were changes at the margin. Nevertheless, the policies of the majors no longer had a decisive influence on the course of events since their freedom of action was in many ways very much more restricted both by their competitive position and by governments. The Companies were often on the defensive, fighting rear-guard actions, which could only retard changes that seemed inevitable. In the Middle East political tensions were often high, but the most severe crisis began in May 1967 and provoked an attack by Israel on her Arab neighbours. Oil exports to the US, Britain and Western Germany were for a while banned by the major Arab producers in view of the generally pro-Israel position of these powers during the Arab-Israel war, and pressures for nationalization increased in some of the countries.

The Arab ban on exports, which was serious for Britain and British Companies, apparently actually benefited the US balance of payments and some US Companies. The export ban was short-lived, however, and demands for nationalization quickly died down to 'normal' levels. The most important economic consequence of the Israeli war for the international oil industry was the closing of the Suez Canal, which, again, hurt Britain without much affecting the US since the latter does not significantly depend on Middle East oil, and indeed could increase its own exports to the countries affected by the difficulties.

The crisis had not been resolved at the time of writing nor the Canal re-opened. The long-run consequences for the oil Companies were thus not clear. The attitudes of the ruling régime in Iraq toward the IPC hardened further, but the political position of the Companies in other producing countries was unaffected.[2]

[1] See M. G. de Chazeau and A. E. Kahn, pp. 83 ff.

[2] See p. 218 below for further discussion of Iraq's position.

Political disturbances apart, the disturbance of the old order was fundamentally brought about by the increased competition among suppliers of crude oil, which was in part the consequence of the desire of those in control of ever-increasing reserves in the Middle East to secure or enlarge their position in rapidly growing markets, and in part the result of the entry of newcomers into world markets. The spreading effects of increased supplies and sharper competition were also partly responsible for the more active interest of some governments in the operations of the industry, although other factors such as concern for the security of supply, nationalism, 'anti-imperialism', etc., played a significant role. The governments of a number of countries importing crude oil began more clearly to understand the nature of the industry's price structure, and in particular to understand the significance for their economies of the vertical integration of the major firms, and especially of their control of local refinery capacity. These matters are discussed in the following chapters, but first we shall look at the history, structure, and character of the important international Companies.

Chapter IV

THE INTERNATIONAL PETROLEUM COMPANIES

Although there are several dozen oil Companies with significant operations outside their country of origin, and which can therefore properly be classed as international firms, this number was very limited not much over a decade ago. The decisive role of a handful of major international Companies in shaping the history of the international industry is clear from even the brief outline presented in Chapter III. Quite plainly many aspects of the structure of the industry can be traced to the policies of small groups of people in a position to take actions that inevitably had a noticeable impact on the course of events. One cannot explain, or even sensibly discuss, *any* of the important economic characteristics of this industry without reference to the *policies* of the major firms. Conversely, however, one cannot appraise the policies of individual firms without reference both to the technology of the industry (i.e. the character of exploration and discovery, of production, transport, refining, and marketing), to government policies, and to the institutional and market structure in which the firms found themselves and to which they were forced to adapt.

Because of the policies of the leading firms and of the structure of the industry, economists always qualify the word 'competition' when referring to this industry with the adjective 'monopolistic' or 'oligopolistic'. And since the industry is indeed an oligopoly (which merely means 'few sellers'), or more accurately, a 'partial oligopoly' (which means there is a fringe of numerous small independents around the major firms that have oligopolistic market power), any study of it must include a discussion of the individual firms. The 'forces of supply and demand' are not 'impersonal' market forces, but are in fact—particularly on the supply side—very much influenced, and sometimes partly controlled, by the dominant firms. To be sure, the influence has been inadequate from the firms' point of view, and the control far from complete, but this does not alter the principle.

In a critical assessment of the role of the international Companies J. E. Hartshorn asked the question, what are the essential functions of the internationally integrated Companies? He concluded that the Companies had, first, a 'logistic' function—'the disposition of supplies from many sources to meet many different demands'; second, the 'setting of relative prices' (which he also classed, erroneously in my view, as 'logistic'); third, a 'technical and social' function—e.g. the selection and training of managers of all nationalities thus giving them international experience; and fourth, the 'financial' function of specialized investment banks with a long view 'to ensure the steady development of an internationally essential resource'. All these functions he insisted 'need an international viewpoint detached from the special interest of any one nation' so that the Companies act as a 'political as well as an economic middleman between nations'.[1]

The emphasis on the need for an 'international viewpoint' on the part of the Companies presumably reflected the recognition that we accept a great deal if we accept that the international Companies should exercise all these 'functions' free from Government interference (apart from 'reasonable taxation' and 'reasonable' measures to ensure conservation of the natural resources). This is especially true when account is taken of the degree of market power that the Companies still possess in spite of the competitive pressures which have arisen since 1958 (and which the major Companies are vigorously attempting to reduce by further extending their control of refining and by making life difficult for the 'independents'). Since seven Companies alone still produce over three-quarters of the crude oil in the world outside the US and the Communist countries, own nearly 60 per cent of the refining capacity of Europe (the largest consuming centre), and probably do even a greater proportion of the marketing, they have an important voice in the determination of: (*a*) the physical flow of oil and products internationally; (*b*) the level of prices; (*c*) the distribution of the 'income originating' in the industry between exporting and importing countries and among countries within each group (except as modified by taxation); and (*d*) the amount and distribution of investment expenditures (including exploration) both geographically and functionally. To the extent that these Companies are able to maintain prices in order to achieve a level of profits adequate to cover what they feel they need to invest, they also make the decision as to who should pay for the future—present consumers, or present capital markets and future consumers.

All these are matters of considerable international significance given the importance of the oil industry, and it is widely insisted that decisions about them should be made solely on economic and not on political grounds, for they are essentially economic problems. From this flows the argument that the oil Companies should be left free from government interference to make these decisions guided by 'commercial' principles. At the same time it is recognized that 'national interests' cannot be ignored. The resolution of this evident dilemma seems to rest either on the proposition that the Companies' commercial principles coincide with everybody's national interest, or that the Companies are in a better position than anyone else to resolve conflicts of national interests.

The latter is implicit in the notion that the Companies are appropriate 'buffers' between governments—the 'business in between'—which is put forward so persuasively by Hartshorn. It follows, however, as Hartshorn recognized, that the companies must take what are essentially political decisions, and in recognition of this, the argument shifts slightly from 'commercial principles' to expediency, asserting that the private Companies acting in their own interests will in fact make the best political mediators, given existing alternatives, when the national interests of different countries conflict with

[1] J. E. Hartshorn, *Oil Companies and Governments*, pp. 377–8.

each other. This proposition obtains its plausibility from the undoubted fact that the Companies, in forming their policies, must take account of all important considerations, including the expected reactions of any significant political group. Any Company, whether national or international, must do this when it is so large that its actions affect the community—local, national, or international—in which it operates, and since the oil Companies are internationally integrated Companies, they have a powerful incentive to ensure that all the countries in which they work are contented. For this reason there is held (but not by Hartshorn) to be no danger of monopolistic collusion between the Companies and any one group of governments and, because of the competition between the Companies, no danger of monopolistic exploitation of crude or product markets.

The purpose of this chapter is to sketch, so far as it is publicly known, the history and characteristics of the more important Companies in this industry in order to provide a background for a discussion of such issues: Just *how* do the Companies discharge their functions, what *is* the character of this 'international business', and how far *is* it detached from the 'special interest and viewpoint of any one nation'? Indeed, need it be so 'detached'? What are the implications if it is not? We shall begin with the major Companies and then move on to a few of the smaller Companies.

There are broadly two reasons for presenting this description: first, I know of no publication where such information for the important Companies is brought together and conveniently summarized; second, and more important, even the superficial, though I hope not inaccurate, sketch of the development of the international Companies in this chapter will give the reader some idea of the significant differences between them. In subsequent chapters we shall be examining the supply, pricing, and investment policies of the Companies and their relationships with the underdeveloped exporting and importing countries, but by and large we shall have to treat the major Companies as a group, partly for the sake of brevity but largely because of insufficient information about their individual viewpoints and policies. On many of the fundamental issues the major firms have sunk their differences and adopted common policies either for purposes of negotiation or in order not to disturb an uneasy equilibrium in the industry. In consequence, more emphasis must necessarily be placed on the similarities than on the difference among the Companies even though they have in fact very different attitudes and outlooks. These differences have been shaped by the past history of each Company, and some knowledge of the ways in which individual Companies developed, and particularly of the factors influencing the spread of their operations and of the various types of association among them, is essential for an understanding of the evolution of the structure of the international industry.

THE MAJOR COMPANIES

The seven international 'majors' ranked by their crude oil production in 1966, are: Standard Oil Company (New Jersey), Royal Dutch/Shell Group, British Petroleum Company, Gulf Oil Corporation, Texaco, Standard Oil of California and Mobil Oil Corporation (formerly Socony Mobil). Compagnie

Française des Pétroles is much smaller than any of these, as well as smaller than a number of other US Companies, but it very early had a share in Middle East oil and is for this reason often referred to as an eighth 'international major'.

Standard Oil Company (New Jersey)[1]

The Standard Oil Company (New Jersey) is one of the largest industrial corporations in the world, with total assets in 1966 of $13,887 million (£4,960 million) and gross revenue reaching $13,582 million ($4,851 million).[2] It is an American Company with nearly 740,000 shareholders, not one of which, according to the Company, holds as much as 3 per cent of its stock. The largest shareholders are institutions such as insurance and investment companies, pension funds and banks. It is both vertically and horizontally integrated, producing, transporting, refining, and distributing oil and its products and operating in over 100 countries. In 1966 Jersey produced about 14 per cent of the world's crude oil outside of the Communist *bloc* and had another 1·6 per cent on long-term contracts, processed 11 per cent of the crude refined, and accounted for some 16 per cent of the total sales of oil and oil products. Outside Canada, the US and the Communist countries it accounted for 18 per cent of crude-oil production, and refining.[3] If we include the oil purchased under long-term contracts, the Company is self-sufficient in crude oil, producing in 1966 4,109,000 barrels per day, and obtaining another 459,000 under special arrangements, while refining 4,162,000 barrels per day. Although the Company sells some of the crude oil it produces in some areas and purchases the crude oil it needs in others, has numerous 'swap' transactions with other Companies, and must often buy both crude and products to conform with government regulations in some countries, by and large its crude oil is produced for use in its own refining affiliates and not for sale outside the Company.

The first 'Standard' company was the 'Standard Oil Company' incorporated in Ohio by John D. Rockefeller and his associates in 1870. Under the leadership of Rockefeller a widespread alliance of business and oil interests was formed in the following few years whereby the greater part of the refining

[1] The following discussion is based largely on the two-volume *History of Standard Oil Company (New Jersey)*: Volume I, *Pioneering in Big Business, 1882–1911*, by R. W. and M. E. Hidy; Volume II, *The Resurgent Years, 1911–1927*, by G. S. Gibb and E. H. Knowlton. In addition I have drawn on a pamphlet, *Standard Oil Company of New Jersey*, published by the company in 1964, the company's *Annual Reports*, and *The International Petroleum Cartel*, Report of the Federal Trade Commission, 1952.

[2] Total assets as used in this study include fixed assets (less depreciation), investments and advances, and other book assets as reported by the Company, and current assets; it also includes any reported interest of the Company in the assets of non-consolidated affiliates. Since I am using the statistics merely as one indicator of the total size of Companies and not for detailed financial comparisons, I see no reason to deduct minority interests or current liabilities. Gross revenue is defined as revenue from all sources before the deduction of sales, excise or other taxes and includes the Company's share of retained income of non-consolidated affiliates where this is reported. In 1966 Jersey's equity in the net assets of Aramco and the IPC were reported to be about $34 million more than its $123 million of investments in and advances to them.

[3] See Table II, p. 78.

TABLE III

STANDARD OIL COMPANY (NEW JERSEY)

SELECTED FINANCIAL STATISTICS

1950–66

	1	2	3	3/2	4	5	5/4	6	6/4	7	8	8/6
	Total assets[1] *$m.*	*Total receipts*[2] *$m.*	*Total taxes*[2] *$m.*	*Ratio taxes/ receipts*	*Net income before charges*[3] *$m.*	*Cash dividends*[4] *$m.*	*Ratio dividends/ net income*	*Capital expenditure*[5] *$m.*	*Ratio capital expenditure/ net income*	*Net change long-term debt $m.*	*Amounts retained*[6] *$m.*	*Ratio amounts retained/ capital expenditure*
1950	4,448	4,176	608	·15	686	185	·27	352	·51	(28)	487	1·38
1951	5,075	5,119	775	·15	829	292	·35	443	·53	4	524	1·18
1952	5,509	5,749	829	·14	840	303	·36	565	·67	35	523	·93
1953	5,912	6,041	905	·15	911	317	·35	587	·64	29	581	·99
1954	6,615	6,236	1,719	·27	996	322	·32	604	·61	(35)	649	1·07
1955	7,346	6,866	1,948	·28	1,158	381	·33	705	·61	61	746	1·06
1956	8,085	7,789	2,171	·28	1,307	453	·35	927	·71	27	819	·88
1957	8,925	8,509	2,302	·27	1,362	486	·36	1,137	·83	58	841	·74
1958	9,676	8,261	2,365	·29	1,134	503	·44	887	·78	57	584	·66
1959	10,080	8,714	2,526	·29	1,233	507	·41	750	·61	93	676	·90
1960	10,287	8,915	2,617	·29	1,302	512	·39	746	·57	16	743	1·00
1961	10,689	9,356	2,860	·31	1,380	522	·38	844	·61	9	810	·96
1962	11,487	10,576	3,291	·31	1,528	564	·37	1,137	·74	13	894	·79
1963	12,051	11,392	3,694	·32	1,711	623	·36	915	·53	(15)	1,017	1·11
1964	12,523	12,015	3,954	·33	1,764	679	·38	1,141	·61	36	1,010	·88
1965	13,108	12,744	4,415	·35	1,804	712	·39	993	·55	82	1,005	1·01
1966	13,887	13,582	4,720	·35	1,931	747	·38	1,210	·63	104	1,088	·90
Totals and averages	—	—	—	—	21,876	8,108	av. = ·37	13,943	av. = ·63	—	12,997	av. = ·93

Source: *Annual Reports and Financial and Statistical Supplements.*

[1] All assets as reported by the Company, including current assets and from 1965 equity in non-consolidated affiliates.

[2] Including consumer and excise taxes.

[3] Net income after taxes plus depreciation, depletion, retirements, interest and other financial charges and including income due to minority interests. Does not include depreciation of non-consolidated companies.

[4] Including dividends to minority interests.

[5] Additions to property, plant and equipment and net additions to investments and advances.

[6] Net income (Col. 4) less dividends and interest charges.

Number of shareholders 1966: 738,000.

and pipeline capacity of the US came under the control of a very few men, who envisaged the setting up of 'Standard' companies in each State of the Union bound together by a Trust agreement. The present company, The Standard Oil Co. (New Jersey), was accordingly organized in 1882. With the abandonment of the 'trust' method of control in 1892 because of legal difficulties, Jersey Standard became the holding company for the entire group of Standard companies and interests.

In the beginning, the Standard group did not enter into the fiercely competitive business of crude-oil production, but concentrated on refining and pipeline transportation, for Rockefeller saw that control of these two stages of the industry would, in the conditions prevailing in the US at the time, permit the control and 'rationalization' of the entire industry. By 1879 Standard Oil owned 80 per cent of the refining capacity of the US and nearly all the pipelines; by 1904 it refined and marketed over 85 per cent of the crude run through US refineries. The rapid increase of demand to which the expansion of the refining and marketing facilities of Standard was geared, together with increasing competition for crude-oil supplies, had early pointed to the strategic desirability of the company obtaining crude-oil supplies of its own, and by 1911 Standard Oil had become an integrated organization, in some years producing through its affiliates as much as 40 per cent of the crude it refined and nearly 35 per cent of the total crude-oil output of the US. Since the US was becoming the world's most important supplier of oil products, Standard was clearly in a dominant world-wide position.

The Standard Oil Company of New York (Socony) was the chief exporting affiliate in the Standard Oil group at this time, but towards the end of the decade the group began to organize foreign marketing affiliates in Europe, Canada, and Mexico. Some refining was also carried on in Mexico, the West Indies and Europe. Around the turn of the century Standard Oil was rapidly increasing its foreign operations, including production and refining; in marketing it was well established, although its share in foreign markets was probably not more than 60 per cent in 1911. This was less than its share in the 1880s because of the intense competition that had been taking place from Companies obtaining crude oil from outside the US, notably from Russia and the East Indies.

Considerable progress had by this time been made in setting up foreign producing and refining facilities, but both were still of minor importance; in 1912 foreign deliveries took 54 per cent of the total deliveries from Jersey's US refineries. Altogether the Standard Oil Company of New Jersey by 1911 had nineteen affiliates engaged in foreign trade operating through fifty-two subsidiaries and numerous branches in many parts of the world, but largely in Europe.

In 1911 the Supreme Court of the US upheld a lower court's decision that Standard Oil of New Jersey had violated the US Antitrust Acts. The great combine was prohibited from exercising any control over thirty-three of its subsidiaries, which were to become legally separate companies and were enjoined from paying dividends to their former parent. In spite of this, however, the parent Company, today's Standard Oil Co. (New Jersey) remained the second largest industrial Company in the US, and much larger than any of its former subsidiaries. It retained most of its South America and European marketing affiliates, but lost the companies that had been responsible for sales of Standard Oil products in the UK and the Far East. Moreover, it retained very little crude-oil production and few transport facilities; hence the integrated balance of its operations was seriously disturbed.

Although before the end of the First World War Jersey had begun to produce some crude oil in Mexico, Peru and Rumania, and had been exploring in the East Indies, it was not until after the war that the Company seriously set about extending its foreign crude-oil production. The desirability of expanding abroad was reinforced by the fact that the political climate was unfavourable to an aggressive expansion on the part of Jersey Standard in the US, and by growing fears that oil supplies in the US were running out while great foreign Companies, some backed by their respective governments, were competing intensely to secure control of likely oil-producing properties in various parts of the world. Moreover, Jersey's strong marketing position in Europe was being challenged, and the importance to the Company of obtaining a share of the world's oil reserves outside of the US was becoming pressingly clear, for in 1918 Jersey's net crude production was only 16 per cent of the crude oil it used in its refineries. So efforts to obtain concessions, to push exploration and to acquire existing producing companies were intensified.

In Latin America new interests were acquired in Peru, Colombia and in Venezuela, where Jersey bought a major share in Creole Petroleum in 1928. This subsidiary, now 95 per cent owned by Jersey, is one of the world's largest producers of crude oil. In 1932 the Company's production in Venezuela was substantially enlarged when it acquired the foreign properties of Standard Oil of Indiana, which gave it rights over extensive oil lands in Venezuela and Mexico as well as three refineries and a fleet of tankers.[1] In 1937 Jersey bought an undivided half-interest in Mene Grande Oil Company, a subsidiary of Gulf Oil, which is the third largest crude producing company in Venezuela. (In 1938 Jersey's half-interest was shared with Royal Dutch/Shell).

In the Dutch East Indies the long efforts of Jersey to gain concessions and exploration rights in the face of opposition from both the Royal Dutch Petroleum Company and the Government, and the long search to discover oil in the territories acquired, were finally rewarded in 1922 when oil was found in Sumatra. The Company was at last in a better position to meet the competitive challenge of its greatest rival, Royal Dutch/Shell, which was expanding rapidly under the leadership of Sir Henri Deterding.

In the Middle East, Jersey took the lead as early as 1919 in getting the American Companies a foothold. Here, too, the Company was faced with opposition, especially from the Anglo-Persian Oil Company and the Royal Dutch/Shell. In 1920 strong efforts were made to get a concession in Northern Persia with the help of the State Department;[2] in Mesopotamia where Jersey

[1] The Mexican properties were lost in 1938 when the Mexican Government expropriated the properties of all foreign oil Companies engaged in a dispute with the labour union.

[2] The State Department instructed the US Embassy in Tehran to put the American point of view to the Persian foreign office:

'It is assumed that, in accordance with the Department's telegram of August 6, you have discreetly and orally conveyed to the Persian Foreign Office information to the effect that

wanted to take part in oil operations, it was decided that State Department help would be more readily forthcoming if other Companies were associated with the venture, and seven Companies joined in an approach to the Department. After prolonged negotiations an agreement was reached with the Turkish Petroleum Company to enable the US companies to share in the oil of Mesopotamia. The Near East Development Corporation was formed in 1928, in which five of the American Companies participated, and as a group obtained an equal share with Anglo-Iranian, Royal Dutch/Shell, and Compagnie Française des Pétroles in the ownership of the Iraq Petroleum Company, the successor to the Turkish Petroleum Company.[1]

The next major source of petroleum secured by Jersey in the Middle East was through a long-term contract negotiated in 1947 with Anglo-Iranian (now British Petroleum) whereby the Company was to receive either from AIOC's Kuwait or its Iranian production a total of 800,000,000 barrels of crude oil over a twenty-year period beginning in 1952 at cost plus a fixed money profit per ton. (Similar contracts were signed between AIOC and Socony which covered some 500,000,000 barrels over the same period.)[2] These contracts were negotiated at the same time as discussions were taking place between AIOC and Jersey over a proposal that they should jointly build a pipeline from the Persian Gulf to the Mediterranean, and between the IPC and the American Companies over the latter's desire to get out of the 'Red-Line Agreement'.

The Red-Line Agreement was an agreement between the partners of the predecessor of the IPC—the Turkish Petroleum Company—under which each of the participating Companies undertook not to conduct independent operations in a large area comprising most of the territory of the old Ottoman Empire. The purpose of the agreement was to prevent competition among the Companies both in seeking concessions and in the purchase or refining of crude oil produced in the area, and thus to ensure that all members of the

---

the Department believes that American companies will seek concessions in the northern provinces and that the Department hopes that American companies may obtain such concessions. The Persian Government appears to appreciate the undesirability of having an important economic resource monopolized by a single foreign company. The Department has taken the position that the monopolization of the production of an essential raw material, such as petroleum, by means of exclusive concessions or other arrangements, is in effect contrary to the principles of equal treatment of the nationals of all foreign countries. It would seem, accordingly, to be conducive to the best interests of Persia and desirable from the standpoint of international economic relations for the Persian Government to postpone any further grants of its oil resources until opportunity can be given to American companies to enter into negotiations regarding such grants. You may on suitable occasions call the attention of the Persian Foreign Office to these considerations. . . .'

State Department Decimal File, Box 10115. Persia. Oil. Van S. Merle-Smith (for Secretary of State) to Caldwell in Tehran. August 16, 1920. (I am indebted to Mrs Rose Greaves for supplying me with the above extract.)

[1] The original five companies in the Near East Development Corporation included Jersey Standard, Socony Mobil, Atlantic Refining, Gulf Oil, and Pan American Petroleum and Transport Co. Atlantic and Pan American sold their interests to Jersey Standard and Socony (now Mobil Oil) in 1930, and Gulf did so in 1934.

[2] Federal Trade Commission, pp. 148, 152.

IPC would gain equally from activities of any of them in the area. The inclusion of this restriction in the agreement which admitted the American Companies into the IPC was insisted upon by C. S. Gulbenkian, who held a 5 per cent beneficial share in the IPC, and by the French, who wanted to ensure that the other participants in the IPC were not in a position to retard its expansion while turning to develop outside oil to the disadvantage of French interests. The American Companies (Jersey and Socony) became restive under these restrictions, especially when a newcomer to the area, Standard Oil of California (Socal), succeeded in obtaining a concession in Saudi Arabia, where it discovered very large quantities of oil. In 1948 they withdrew from the agreement, thus freeing themselves to take up an interest in Socal's Arabian American Oil Company (Aramco). Jersey purchased 30 per cent of Aramco, becoming assured thereby of 30 per cent of the oil produced from this extremely prolific concession.[1]

In 1954 Jersey took up a 7 per cent interest in the consortium which replaced the AIOC in Iran after the nationalization of oil operations by the Iranian Government and the subsequent breakdown of Iranian oil production. In 1959 a Jersey affiliate discovered oil in Libya and commercial shipments began in 1961. Thus in 1966 Venezuela with 1,380,000 b/d was the primary source of Jersey's crude-oil supplies outside of the US, followed by Saudi Arabia (721,000 b/d) and Libya (536,000 b/d); the Company also produced respectable amounts in Canada, Iraq, Abu Dhabi and Iran. Exploration continued in dozens of areas, sometimes by wholly-owned subsidiaries and sometimes in association with other oil Companies.

As indicated above, Jersey's initial strength in foreign operations derived from its marketing position based on domestic supplies of crude oil. Its search for crude was prompted by the Company's desire to ensure owned supplies for its expanding markets as well as to ensure that the world's oil reserves did not become pre-empted by rivals, particularly the British and Dutch Companies. But Jersey was marketing products—primarily gasoline, kerosene and lubricants—and not crude oil, and consequently the growing supplies of crude required growing refinery capacity. Refining in Latin America was increased from 10,000 to 23,000 b/d between 1920 and 1927, and from 1,000 to 13,500 b/d in the Eastern Hemisphere in the same period; altogether, foreign refinery runs went up from 32,000 to 85,000 b/d, or from 15 per cent to 20 per cent of the total crude oil refined by the company. Nevertheless, in 1927 exports from the US of refined products were still over 60 per cent of the total foreign deliveries by the Company. As foreign sources of crude continued to increase and foreign markets continued to expand, foreign refinery activities were also rapidly expanded, and in 1966 foreign refineries processed around 80 per cent of the oil processed by the Company; more than half of this was handled by refineries in the Western Hemisphere. In 1965 Jersey owned or had a financial interest in sixty-eight refineries in thirty-four countries.

[1] This withdrawal from the 'Red Line' agreement angered the French (as well as Gulbenkian) so much that the matter was taken to court. It is ironical to note that twenty years later the tables were turned and the French took action which similarly angered the Anglo-American-Dutch groups when French companies, including the CFP, finally agreed to negotiate with the Iraq National Oil Company for the exploitation of territories that had been expropriated from the IPC by the Iraqi Government. See p. 218 below.

The only area where the historical efforts of Jersey to discover crude oil were not directly linked to existing marketing needs was in the Dutch East Indies. The light gravity crude of this area had been one of the strongest assets of Jersey's rival, Royal Dutch/Shell, in its competitive drive for European markets and in obtaining pre-eminence in the markets of the Far East. Jersey saw advantages in securing supplies of this high quality crude, in particular, the 'competitive desirability of attacking the Royal Dutch-Shell where it was strongest—in the Orient'.[1] When the oil was found and began to flow, a refinery was built, but no marketing organization existed, and sales were made to Standard of New York (Socony), which had controlled marketing in the Far East for the old Standard Oil group. In 1933 Jersey Standard and Socony-Vaccuum (Standard of New York having merged with Vacuum Oil Company to become Socony-Vacuum) merged their producing, refining, pipeline, and marketing facilities and formed a new company in the Far East —Stanvac—which continued to operate east of Suez on behalf of its two parents until 1960.

The formation of Stanvac was in a sense a re-merging of interests that had been separated by the dissolution decree of 1911 which had upset the balance of operations of the old Standard Oil group, since Socony, as the marketing organization in the Far East, was left with inadequate supplies of crude oil and refining facilities. It also illustrates how the existence of vertically integrated firms in a strong monopoly position forces would-be competitors themselves to amalgamate and integrate—a state of affairs which has been an outstanding characteristic of the petroleum industry. The problem was set out by Socony in a brief presented in a Civil Action brought against the Company by the US Government in 1953:

> Socony in the 'twenties and early 'thirties was in great difficulties through its lack of crude supplies and refining facilities in the Indian Ocean and Far East areas. Its foreign competitors had such sources of supply, both at the centre and on the northern periphery of this area. Without a base of crude supplies and refining facilities of its own, the marketing business of Socony in this area was in danger of further deterioration and even of slowly withering away. In the Fall of 1928, the three major foreign companies operating in these markets, Royal Dutch/Shell, Anglo-Iranian and Burmah, joined in forming a new company to handle this market. Socony was, thereupon, faced with one large joint British-Dutch operation in a marketing area where British and Dutch interests were paramount, and where they had most of the known crude sources and refining facilities. The only other known crude-oil resources and refining facilities in this area were owned by Jersey in Sumatra. These were remote from any marketing facilities of Jersey but adjacent to those of Socony. There was no possibility that Jersey would relinquish these supplies and facilities entirely, but, in 1933, Jersey and Socony each agreed to sell its respective business to a new company, Standard-Vacuum, in which Socony and Jersey each received half of the stock. Standard-Vacuum thus acquired the complementary producing,

[1] Gibb and Knowlton, p. 392.

refining and marketing facilities essential to successful competition in this area.[1]

In 1960 Stanvac was wound up (except in Indonesia), and its assets were formally returned to the parent Companies in 1962 in pursuance of a consent decree accepted by Jersey Standard in a civil anti-trust case in the US courts. Esso Standard Eastern became the subsidiary to conduct Jersey's operations in the Far East.

The Company's rate of expansion has not, of course, been so rapid in recent years as it was in earlier decades of the growth of the industry, although it remains high. Between 1917 and 1927,[2] for example, Jersey's production of crude oil increased about seven times and the amount of crude processed about tripled, as did the total assets of the Company. As we have seen, in this period Jersey was particularly concerned with expanding its own production of crude oil and 'correcting' the imbalances in its integrated structure left by the 1911 dissolution. By the 1940s the Company had largely achieved an overall balance between the demand of its refineries and its own supplies of crude, including the oil received on long-term contracts. Between 1950 and 1966 the various branches of the Company's activities expanded at about the same rate —refinery runs and product sales increasing by nearly 160 per cent, and crude production by 178 per cent (see Table IV).

The most rapid expansion in crude production in both periods took place outside North America. While in North America production between 1917 and 1927 increased nearly sixfold, foreign production increased nearly sixteenfold, largely in Colombia, Peru, and Mexico, with Venezuela and Netherlands East Indies just beginning to become important. Most of its foreign sources of oil, however, came through acquisition and not as a result of exploration and discovery by Jersey. Refinery runs nearly tripled both in North America and outside. Between 1950 and 1966, however, its refining as well as crude production rose most rapidly outside North America. Although the Company still obtains much more of its crude from South America (especially Venezuela) than from any other area, the greatest percentage increases in production between 1950 and 1960 occurred in the Middle and Far East.

After 1960 North African output began to become important, rising from 18,000 b/d in 1961 to 536,000 b/d in 1966. Similarly, the relative contribution of the US and Canada to crude processing fell drastically between 1950 and 1966 from 53 per cent to 28 per cent, while that of Europe rose from 5 per cent to 38 per cent. The Company did not start publishing detailed statistics of the geographical distribution of product sales until 1962, but between 1962 and 1966 the volume of sales in North America averaged about 41 per cent of total sales, sales in Europe exceeding those in the US for the first time in 1963. In 1966 sales outside the US accounted for almost all the increase in the volume of sales.

A similar pattern is discernible in the distribution of the Company's new

[1] This document is published in US House of Representatives, *Current Antitrust Problems*, pp. 881–2.

[2] These and the following statistics for the period 1917–27 are based on Gibb and Knowlton, Appendix 2, pp. 668 ff. For the period 1950–63 data are taken from the Annual Reports of the Company. See Table IV.

TABLE IV

STANDARD OIL COMPANY (NEW JERSEY)

PRODUCTION, REFINING AND SALES

1950–66

	Crude Oil Production[1]						Refinery Processing[2]				Sales[3]			
											Petroleum Products			
	Total 000 b/d	% Latin America	% North Africa	% Middle East	% North America	Supplied under special arrangements[4]	Total 000 b/d	% Latin America	% USA and Canada	% Europe and North Africa	Total 000 b/d	% Eastern Hemisphere	Chemical products $m.	Natural gas 000 m. cu. ft.
1950	1,477	54	—	12	31	—	1,605	37	53	5	1,705[5]	—	—	—
1951	1,748	51	—	14	32	—	1,813	36	52	7	1,933[5]	—	—	—
1952	1,829	50	—	16	32	53	1,916	34	51	11	2,033[5]	—	—	—
1953	1,878	48	—	18	32	83	1,917	33	50	12	2,070[5]	—	—	—
1954	1,933	48	—	19	30	83	2,000	33	47	15	2,154	—	112	394
1955	2,185	50	—	18	29	100	2,274	34	45	14	2,384	—	154	417
1956	2,366	51	—	18	28	91	2,423	34	45	15	2,555	—	178	459
1957	2,432	53	—	17	27	85	2,416	36	43	15	2,571	33	182	483
1958	2,329	52	—	20	24	120	2,464	36	41	17	2,700	34	181	517
1959	2,458	51	—	21	25	137	2,775	34	41	18	2,916	34	256	608
1960	2,516	50	—	23	24	138	2,859	33	38	21	3,068	36	295	628
1961	2,744	48	1	23	25	135	3,104	31	35	28	3,201	38	309	709
1962	3,060	46	4	23	24	130	3,283	31	34	28	3,480	41	340	849
1963	3,412	44	8	22	23	192	3,467	30	32	30	3,650	43	393	954
1964	3,675	40	12	22	23	211	3,622	29	30	32	3,903	44	535	1,041
1965	3,942	38	13	24	22	220	3,934	28	29	34	4,192	46	600	1,266
1966	4,109	29	13	26	23	459	4,162	35	28	38	4,389	48	744	1,480
% Incr.														
1950–66	178	—	—	—	—	—	159	—	—	—	157	—	—	—
1954–66	113	—	—	—	—	—	108	—	—	—	104	—	564	276

Source: *Annual Reports* and *Financial Statistical Supplements.*

[1] Includes 100 per cent of the output of majority-owned affiliates and stock ownership percentages for companies owned 50 per cent or less.

[2] 100 per cent of majority-owned affiliates and quantities processed for Jersey's account by companies owned 50 per cent or less.

[3] 100 per cent of sales of majority-owned affiliates and stock ownership percentage of companies owned 50 per cent or less. Includes supply sales.

[4] 1950–6 supplies under long-term contract; from 1957–66 inclusive figures also include Jersey's offtake in excess of its equity interest in the production of non-consolidated affiliates.

[5] Includes chemical sales.

investment expenditure, half or more of which has been in foreign activities since 1956. As Table V shows, foreign investment was 67 per cent of total investment in refining in 1966 and 56 per cent of investment in transportation; it was even higher in some earlier years. The percentage of investment in crude-oil production abroad fell substantially from a high of 59 per cent in 1957 to 23 per cent in 1962 but rose to 31 per cent in 1966. The geographical distribution of its total assets in 1966 was given by the Company as follows: US 46 per cent; Other Western Hemisphere 22 per cent and Eastern Hemisphere 32 per cent.

TABLE V
STANDARD OIL COMPANY (NEW JERSEY)
CAPITAL AND EXPLORATION OUTLAYS
1966

($ *million*)

Additions to property, plant and equipment	*US*	*Other Western Hemisphere*	*Eastern Hemisphere*	*Total**
Producing	223	61	37	321
Refining	121	57	190	368
Marketing	141	39	172	351
Transportation	30	19	19	68
Other	45	9	43	97
Total*	560	185	460	1,204
Exploration expense	–	–	–	234
Total*				$1,438

* Totals may not add because of rounding.
Source: *Financial and Statistical Supplement to 1966 Annual Report.*

Jersey does not publish its investment expenditures in the chemical and fertilizer industries, which are presumably included in the category 'other' in Table V. In 1960 the Company stated that one-quarter of its research expenditure went into the development of chemical products, and that the major proportion of new investment in this field was planned for outside the US, particularly in Canada, England, France and Sweden. Jersey has extensive world-wide interests in petrochemicals, synthetic chemicals and fibres (including even the spinning of synthetic fibres in Argentine), fertilizers (fourteen plants in twelve countries in 1965), plastics, synthetic rubber, paint, petroleum resins, etc. Its investment in the fertilizer industry in developing countries exceeded $100 million in 1965. Its sales of chemical products have more than doubled since 1960, reaching nearly $750 million in 1966, or 5 per cent of total sales and operating revenues. Other activities of the Company include the manufacture of industrial gases and associated equipment, equipment for service stations, oil burners and pumps, and the operation of chains of motor hotels. Its petroleum operations include an extensive production of natural gas, of which sales reached 4,056 million cubic feet a day in 1966. It has found gas in Europe, including the North Sea, and in Australia.

Some of Jersey's capital expenditure goes into the acquisition of all or part of the shares (or assets) of existing companies, although there are no published figures on which to estimate the proportion. Almost every Annual Report of the Company announces significant acquisitions. In 1964, for example, a large American company manufacturing industrial gases and equipment was acquired, and arrangements were in hand for the purchase of the Potash Company of America as well as the manufacturing and marketing facilities of Tidewater Oil in western US. The latter met opposition from the US Department of Justice and was abandoned. The Courts disapproved of the purchase of the Potash Company, which was to have been bought with Jersey stock. In 1965 Jersey bought the distribution network of AGIP, the marketing subsidiary of Ente Nazionale Idrocarburi (the Italian State company) in the UK, thereby acquiring seventy-nine operating service stations and thirty-one others in the process of construction; it also bought Brazil's largest paint manufacturing firm.

In most years expansion has been financed through retained earnings.[1] In the early 1950s around a third of Jersey's net income (after taxes, but before depreciation) was paid out in dividends. This proportion rose to a peak of nearly 44 per cent in 1958 but averaged 37 per cent in the period 1960–6 (Table III). In most years since 1950 retained funds were almost enough to finance all investment, and exploration expenditure was charged to current income. However, the published figures for capital and exploration expenditure do not include expenditures made by affiliates owned 50 per cent or less. In other words, the Company's share of the annual investments of Aramco, the IPC, Stanvac and similar affiliates is not included. In Annual Reports for the three years 1957–9, however, this figure was given, and in those years Jersey's share in the capital and exploration expenditure of such affiliates ran between 8 per cent and 10 per cent of the capital and exploration expenditures of the Company and its consolidated subsidiaries. Since the division of Stanvac in 1960, however, the proportion of investment in non-consolidated affiliates must have been appreciably smaller.

Not only was total investment much greater than cash dividends in the period 1950–66, but foreign investment alone exceeded dividends in every year of the 1950s up to 1958; since then foreign investment has been, on the average, only slightly below dividend payments. Moreover, foreign investment in the period 1953–65 was slightly over half of total investment. How much of it went to the different countries in which the Company operates is not published.

Thus, it is clear that Jersey Standard is indeed a remarkably international Company, with operations outside its home area steadily becoming more important. It has evidently put a high emphasis on self-sufficiency, preferring to finance itself from its own resources and to produce, transport, manufacture, and distribute its own products rather than to rely on outsiders for a substantial portion of its supplies or services at any stage of its operations.

[1] The measures of the degree of self-financing of the oil Companies that are presented here and for other Companies are taken from the figures published in their Annual Reports. They understate the amount of outside finance if full account is taken of the financing mplicit in leasing arrangements, etc., and particularly with reference to transport financing. See also foot note 1, p. 145 below.

Having become too big too early in the US and suffered the consequences of public fear and suspicion, it took the world as its most important area of operations and continued in this larger sphere to grow, checked only here and there by fears and suspicions similar to those directed towards its 'Standard' predecessor. It seems to have preferred to operate through wholly-owned subsidiaries abroad, until recently an almost universal characteristic of US corporations in general. But in some areas this is no longer acceptable to the local governments, and the Company has bowed to the inevitable and accepts local government and private participation in some of its refining subsidiaries but very rarely in producing affiliates.

In spite of the international character of its operation, Jersey is very American in its outlook, with all that this implies for its foreign policy, for its attitude towards government social and economic measures, and for its views of the relation between big business and government (not all of which are as enthusiastically received abroad as they are in business quarters in the US). Thus, the Company considers that the spreading of appropriate ideology is one of its tasks: 'Our affiliates as corporate citizens communicate their ideas on sound business policy to the people and governments of the countries in which they operate. The public statements made by our management, our written communications, and our advertising seek to emphasize the benefits of free competitive enterprise and private international investment.'[1] At the same time, the Company emphasizes that its operations are conducted to contribute as far as possible to the US balance of payments, pointing out that in every year since the war the flow of new investment abroad has been more than offset by the inflow of income from earlier investment. In 1964 the Company's net contribution to the US balance of payments was reported to exceed a third of a billion dollars (*Annual Report*, p. 5) with still greater 'improvement' reported in 1965.

Jersey has contributed extensively to a wide variety of social welfare activities in the areas in which it operates abroad, subsidizing schools, hospitals, highways, and other amenities, promoting technical training, granting scholarships, etc. In Venezuela it has set up an investment corporation to make investments in private enterprise not related to the oil industry, with the stated objective of contributing to the expansion and diversification of the country's economy and promoting the 'principle of private enterprise'.[2]

Royal Dutch/Shell[3]

Royal Dutch/Shell, the second largest oil 'Company' in the world, is not, strictly speaking, a Company at all, but represents literally a 'group' with two

[1] *Annual Report* (1962), p. 18. In its 1963 Annual Report Jersey expressed concern about increasing government activities in the oil industry. In discussing the Company's 'international problems' the Report mentioned '. . . disturbing shifts in political, economic and social philosophies' of some of the countries—even those with stable governments—in which it operates. 'Some of these changes have led to increased controls and restrictions on private enterprise. Occasionally they have resulted in actual government participation in certain phases of the oil business.' (P. 3.)

[2] See Annual Reports of Creole in Venezuela.

[3] The following discussion is based primarily on Annual Reports and on short accounts produced by the Company; it also draws on F. C. Gerretson, *History of Royal/Dutch Shell*.

TABLE VI

ROYAL DUTCH/SHELL

SELECTED FINANCIAL STATISTICS

1954–66*

	1	2	3	3/2	4	5	5/4	6	6/4	7	8	8/6
	Total assets[1] *£m.*	*Total receipts*[2] *£m.*	*Total taxes*[3] *£m.*	*Ratio taxes/ receipts*	*Net income before charges*[4] *£m.*	*Cash dividends*[5] *£m.*	*Ratio dividends/ net income*	*Capital expenditure*[6] *£m.*	*Ratio capital expenditure/ net income*	*Net change long-term debt £m.*	*Amounts retained*[7] *£m.*	*Ratio amounts retained/ capital expenditure*
1954	1,818	1,715	527	·31	289	34	·12	232	·80	(14)	249	1·07
1955	2,006	1,911	584	·31	329	41	·12	259	·79	23	282	1·09
1956	2,254	2,152	639	·30	374	49	·13	327	·87	(1)	319	·98
1957	2,498	2,417	718	·30	420	53	·13	302	·72	10	360	1·19
1958	2,722	2,379	743	·31	347	55	·16	291	·84	(13)	285	·98
1959	3,046	2,603	861	·33	382	67	·18	334	·87	(15)	308	·92
1960	3,178	2,705	894	·33	392	74	·19	311	·79	24	310	1·00
1961	3,354	2,819	972	·34	398	82	·21	280	·70	27	308	1·10
1962	3,563	3,029	1,090	·36	425	97	·23	293	·69	3	319	1·09
1963	3,804	3,326	1,199	·36	456	109	·24	290	·64	46	337	1·16
1964	4,017	3,519	1,324	·38	457	119	·26	413	·90	(14)	327	·79
1965	4,324	3,795	1,433	·38	487	116	·24	510	1·05	107	356	·70
1966	4,783	4,070	1,539	·38	521	127	·24	511	·98	84	375	·73
Totals and averages	—	—	—	—	5,277	1,023	av. = ·19	4,353	av. = ·82	—	4,135	av. = ·95

Source: *Annual Reports* and *Financial and Operational Information.*

[1] Fixed plus current book assets.

[2] Including sales, taxes, excise duties and similar levies.

[3] Sales, etc., taxes, income taxes and 'other' taxes.

[4] Net income after taxes plus depreciation, depletion and amortization, interest charges, and including income applicable to minority interests. But does not include group proportion of associated companies depreciation.

[5] Including dividends to minority interests.

[6] Including investments in and advances to associated companies less provisions.

[7] Net income (Col. 4) less interest charges and dividends, i.e. net cash income less dividends to parent companies and minority interests.

* Data for years before 1954 not comparable.

Number of shareholders 1966: Shell Transport and Trading, 368,000
Royal Dutch, estimated over half a million.

parent Companies, the Shell Transport and Trading Company Ltd., which is British, and the Royal Dutch Petroleum Company, which is Dutch. Royal Dutch and Shell Transport together appoint the Directors, hold the shares and receive the income, in the ratio of 60 per cent and 40 per cent respectively, of two other companies, the Shell Petroleum Company Ltd. (British) and Bataafse Petroleum Maatschappij N.V. (Dutch). These two companies in turn own the shares directly or indirectly of the other 500 or so companies comprising the Group. The activities of the Group companies are co-ordinated by four service companies: Shell International Petroleum Company Ltd. and Shell International Chemical Co. Ltd. in London, and Bataafse Internationale Petroleum Maatschappij N.V. and Bataafse Internationale Chemie Maatschappij N.V. in the Hague. These four companies not only co-ordinate the operations of, but also give advice and render services to, the operating companies of the Group.

The total assets of the Royal Dutch/Shell group (which I shall refer to simply as 'Shell') equalled £4,783 million ($13,394 million) in 1966 and gross revenues reached £4,070 million ($11,396 million). Shell Transport shares are largely British-owned, but those of Royal Dutch, despite the fact that it is a Dutch Company, are primarily owned by nationals of other countries. The estimated distribution of shareholdings by nationality of shareholder was as follows in 1966:

TABLE VII
ROYAL DUTCH/SHELL
DISTRIBUTION OF SHAREHOLDING BY NATIONALITY
1966

	Shell Transport *per cent*	*Royal Dutch* *per cent*	*Combined* *per cent*
United Kingdom	92	4	39
United States	2	32	19
Netherlands	—	30	18
France	5	16	12
Switzerland	—	16	10
Others	1	2	2

Source: Company communication.

Shell, like Jersey Standard, is a vertically integrated group conducting operations at all stages of the industry and in almost all parts of the world outside the Communist countries. In 1966 it produced about 13 per cent of the crude oil of the non-Communist world outside the US and Canada[1] (and had another 5 per cent on long-term contracts) and refined 16 per cent. It produces less crude oil than it refines, gross production in 1966 totalling 2,981,000 b/d while refinery processing equalled 3,650,000 b/d. Most of the balance of the Group's requirements was obtained under special supply contracts, of which the most important is with Gulf for Kuwait oil,[2] under

[1] See Table II, p. 78.

[2] For a discussion of the terms of this contract see the Report of the US Federal Trade Commission, pp. 161–5.

which Shell received on a profit-sharing basis the greater part of the 861,000 b/d that it bought under special contracts in 1966.

The merger of the interests of two independent Companies, Shell Transport and Trading and the Royal Dutch, and the beginnings of the corporate structure outlined above, took place in 1907. It was a direct outcome of the competitive struggle between the two Companies and of each with the old Standard Oil group which, as we saw in the preceding section, had been and still was at the time, the dominant marketer of petroleum products throughout the world.

Shell Transport had its beginnings in the export-import business in England of Marcus Samuel dating from the early part of the last century. Kerosene in cases became an export sideline of the business and by the 1880s this trade was developing into its major activity. In the 1890s Samuel began shipping kerosene refined from Russian oil (then the chief source of supply outside the US) to the Far East by tanker through the Suez Canal. But Standard had a near-monopoly of Asian markets, while the European markets were tied up by arrangements between Standard, a group of Russian independents, and the Rothschild interests of Paris. Samuel hoped to gain a sufficient advantage from the lower costs of bulk transport and local packaging to enable him to compete successfully with Standard in Asia. His success was rapid and during the first year of bulk shipments, imports of Russian kerosene into India increased by one-third.[1]

In the meantime, however, oil had been discovered by a Dutch group in Sumatra, which had begun drilling in 1884; in 1890 the company that was to be the predecessor of Royal Dutch was formed and set about producing crude, building a refinery and pipelines, and creating its own marketing organization, which competed with Samuel in the Far East. To meet this competition Samuel sought and obtained a concession in North Borneo in 1896, where he also found oil and built a refinery. In 1897 Shell Transport and Trading Co. Ltd. was established and Samuel handed over his tankers and tank installations in the Far East to the new company. Shell was primarily a marketer of oil products, but commercial rivalry and political uncertainty made it seem undesirable for the Company to rely on the purchase of Russian kerosene, and efforts to obtain secure sources of crude oil were intensified. Moreover, the Company's Borneo crude had little gasoline content and could not be used as a means of breaking into the growing gasoline market. Contracts were made for supplies on both sides of the world—in Sumatra and also in Texas, where oil had been discovered in 1901.

Both Shell and Royal Dutch expanded rapidly; the competitive struggles with Standard, as well as with some of the lesser European groups, were sharp

[1] The advent of the ocean-going tanker and the concomitant development of bulk storage and other installations for the bulk handling of kerosene sharply reduced the costs of transport and handling and gave the large concerns a substantially increased competitive advantage over the local importers and traders. (See B. Dasgupta, *World Oil Prices and the Indian Market, 1884–1964*, pp. 18 ff, for further discussion of the significance of this development.) Kesseler, managing director of Royal Dutch, remarked in 1897, 'I see a big change coming in the petroleum trade in the near future. As a result of the erection of tanks, this article will soon be handled in the East only by a few powerful interests.' (Quoted in F. C. Gerretson, Volume 1, pp. 222.)

and painful, broken occasionally by uneasy truces. Shell was closely tied up with the Rothschild interests from which it bought its Russian oil, and in 1903 an arrangement was made under which the Asiatic Petroleum Company, a marketing company in which Shell, Royal Dutch, and the Rothschilds all had an interest, would undertake the transport and marketing of the groups' products in the Far East and thus eliminate their rivalry in this market. This was a prelude to the later amalgamation of the two Companies and paved the way for the transfer of Rothschild's Russian interests to Shell in 1912. In 1907 Shell's Rumanian affiliate had begun production and by 1910, enlarged by merger, the affiliate was the largest producer of crude oil in that country.

Standard Oil made several attempts to obtain control of both Shell and Royal Dutch without success, but a number of co-operative moves occurred between the latter two which led in 1907 to their merger and the creation of Royal Dutch/Shell with the present 60-40 division of interests. Both Companies were fully integrated when they merged, although Royal Dutch had begun as a producer and refiner and moved forward to marketing and transport, while Shell had begun in marketing (Shell owned sixteen tankers compared to the three of Royal Dutch when Asiatic took over transport and marketing facilities of both) and moved backward first into crude production and then refining. Nevertheless, their operations were to a considerable extent complementary, and through merger the strength of each could offset the weakness of the other. Royal Dutch/Shell continued the search for crude oil, and new supplies were found in South East Asia, Mexico, Venezuela, and even in the US, for Henri Deterding, the early guiding genius of Royal Dutch/Shell, realized the competitive importance of entering the US market. By 1920 the Company was producing some 10 per cent of the world's supplies of crude oil and its refinery capacity had been correspondingly extended.

Before the First World War Shell had been, through its wholly-owned subsidiary, Anglo-Saxon Petroleum Company, one of the original shareholders in the old Turkish Petroleum Company. Consequently, when oil was found in Iraq in 1927 by the TPC, Shell obtained a 23·75 per cent interest in it because of its shareholding in the company, which subsequently became the Iraq Petroleum Company. In 1938 Shell lost its Mexican properties, which were reported to account for about 17·5 per cent of its total production outside the US and 62 per cent of Mexico's total production, when the Mexican Government expropriated the foreign oil companies.[1] In the same period it expanded its small operations in Argentina and acquired new properties and production in Venezuela, particularly through the purchase of half of Jersey Standard's 50 per cent interest in Mene Grande, the third largest crude-oil producing company in Venezuela. At the outbreak of the Second World War most of Shell's crude came from Venezuela and South-east Asia; the Company's Middle East production was not to equal that of South-east Asia until the 1950s.

Shell's marketing was often done in association with other Companies, notably with Anglo-Persian (British Petroleum) in the United Kingdom, the Near East and in Africa, and with Burmah Oil in India.

[1] Federal Trade Commission, p. 181.

After the Second World War production was rapidly increased, particularly in Venezuela and the Middle East, where Shell not only had its interest in the Iraq Petroleum Company, but also received large quantities under a special contract made in 1956 to run until 2026 with Gulf Oil Company for supplies of Kuwait oil on a kind of profit-sharing arrangement;[1] in addition Shell obtained a 14 per cent interest in the consortium in Iran. In the 1930s exploration was begun in Nigeria but much later in other parts of Africa. By the middle of the decade commercial production was under way in Nigeria through a company jointly-owned with British Petroleum, and in the Sahara. In 1964 oil was discovered in Abu Dhabi on the Persian Gulf by a company, the ownership of which is the same as that of the IPC and in which, therefore, Shell has 23·75 per cent interest, and by another Company in Oman in which Shell has an 85 per cent interest. In 1966, however, nearly 75 per cent of the Group's production outside North America came from Venezuela, Iran, Iraq and Nigeria.

Total crude produced by the Company rose 108 per cent between 1954 and 1966 while the amount of crude processed rose 98 per cent. Of Shell's crude supply, taking into consideration long-term contracts, North America contributed about 15 per cent in 1966 and South America 28 per cent. Thus Shell drew some 57 per cent of its oil from the Eastern Hemisphere in 1966. After the Second World War the Middle East rose significantly as a source of supply for the Company, from less than 16 per cent of the total in 1950 to over 40 per cent in 1966, if the oil obtained under the Gulf/Shell contract is included. By 1966, however, Algeria, Libya and Nigeria together were contributing over 450,000 b/d. Refining is also largely carried on in the Eastern Hemisphere, 41 per cent of Shell's crude processing taking place in the Middle East and Europe in 1966. The rate of increase in refinery runs has also been very much greater in the Eastern Hemisphere than in the Western Hemisphere.

Overall and in each Hemisphere Shell's owned crude oil and refining runs are just barely in balance if we treat the oil received from Gulf as equivalent to owned crude; in the Western Hemisphere crude production and refinery runs are roughly equal, and in the Eastern Hemisphere the Company processes only slightly more than it produces. However, the Company evidently markets rather more oil and products than it produces, as well as a substantial amount of chemical products.

Except for expenditures in production (the acquisition of oil rights is very expensive in the US) by far the greater part of Shell's capital investment has taken place in the Eastern Hemisphere, primarily in Europe (including the UK and Eire), as can be seen in Table IX. Between 1955 and 1965 there was a marked shift of investment from the Western to the Eastern Hemisphere, but some of this reflected difficulties with Venezuela in the period. Chemical investment accounted for 7½ per cent of total investment in the period 1954–66; in 1966 chemical sales accounted for 12 per cent of total sales net of sales taxes, including a wide variety of petro-chemicals, synthetic rubbers, agricultural chemicals, fertilizers, etc. This seems to be the chief direction of Shell's diversification.

[1] Shell apparently takes over half of Gulf's Kuwait output under this contract.

TABLE VIII
ROYAL DUTCH/SHELL PRODUCTION, REFINING AND SALES 1954–66*

Crude Oil Production[1]

	Total 000 *b/d (excluding supply contracts)*	% *Latin America*	% *Africa*	% *Middle East*	% *North America*	*Supplied on long-term contracts* 000 *b/d (Middle East)*	% *on long-term contracts*[4]
1954	1,430	48		13	23	275	16
1955	1,555	47		15	22	315	17
1956	1,725	49		15	22	294	15
1957	1,895	54		12	21	356	16
1958	1,782	47		18	22	434	20
1959	2,012	49		18	20	452	18
1960	2,021	46	1	21	21	548	21
1961	2,055	44	4	21	21	539	21
1962	2,244	44	5	20	21	640	22
1963	2,305	42	5	22	21	702	23
1964	2,537	43	6	24	20	770	23
1965	2,722	41	7	25	20	869	24
1966	2,981	36	12	27	20	861	22
% incr. 1954–66	108	—	—	—	—	—	—

Refinery Processing[2]

	Total 000 *b/d*	% *Latin America*	% *US & Canada*	% *Europe*	% *Far East*
1954	1,840	30	29	28	12
1955	1,986	28	29	28	13
1956	2,108	27	29	28	14
1957	2,187	27	28	28	14
1958	2,194	29	26	30	13
1959	2,376	28	25	32	13
1960	2,549	26	24	34	13
1961	2,665	25	24	36	13
1962	2,835	24	23	36	14
1963	3,023	23	25	35	14
1964	3,364	22	25	38	12
1965	3,560	21	24	37	14
1966	3,650	22	25	39	10
% incr. 1954–66	98	—	—	—	—

Sales[3]

	Products *Total* 000 *b/d*	% *East Hemis.*	*Crude Oil* 000 *b/d*	*Chemical products* 000 *metric tons*	*Natural gas m.cu.ft. daily*
1954	1,741	49	70	2,100	834
1955	1,862	51	116	2,850	864
1956	1,988	50	124	2,800	917
1957	1,993	49	206	2,900	950
1958	2,097	50	221	2,850	957
1959	2,327	51	168	3,300	1,052
1960	2,459	54	208	3,500	1,240
1961	2,605	55	230	3,500	1,344
1962	2,829	56	341	3,950	1,605
1963	3,083	55	318	4,300	1,786
1964	3,387	57	356	4,900	2,010
1965	3,644	57	368	5,500	2,149
1966	3.848	58	387	6,450	2,462
% incr. 1954–66	121	—	453	207	195

Sources: *Annual Reports. Financial and Operational information.* Information from Company

[1] Includes 100 per cent of production of consolidated companies and ownership share of associated companies.

[2] Including processing for others but excluding processing by others.

[3] Excluding deliveries to other oil companies under reciprocal purchase and sale arrangements which are in the nature of exchanges.

[4] Percentage of total *including* supply contracts.

* Data for years before 1954 not comparable.

TABLE IX
ROYAL DUTCH/SHELL
CAPITAL AND EXPLORATION OUTLAYS
1966

(*£ million*)

	USA and Canada	*Rest of Western Hemisphere*	*Eastern Hemisphere*	*Total*
Oil rights and production	81	13	48	142
Manufacturing—oil	42	6	93[1]	141
Manufacturing—chemicals	27	—	23[2]	50
Marketing	45	5	78[3]	128
Transportation	—	—	—	66
Miscellaneous	—	—	—	6
Total	200	31	244[4]	533
Exploration expenses	—	—	—	73
Total				606

[1] Europe £81 million
[2] Europe £23 million
[3] Europe £60 million
[4] Europe £180 million
Source: Royal Dutch/Shell, *Financial and Operational Information, 1956–66.*

Shell, like Jersey, has financed its expansion largely from retained earnings, and in recent years it also has paid out in dividends a much higher proportion of its net income than in earlier years, the ratio of dividends to net income after taxes but before depreciation rising from 12 per cent in 1954 to 24 per cent in 1966. The Company retains around three-quarters of its income, compared to around three-fifths retained by Jersey. Moreover, depreciation plus retained earnings (net income less cash dividends) exceeded investment expenditures, often substantially, in every year from 1954 to 1963.

Like Jersey, also, Royal Dutch/Shell has been a net contributor to the balance of payments of the parent countries in spite of its large annual investments abroad. It reported that its 'net credit'[1] to the UK balance of payments averaged over £90 million in the six years 1961–6;[2] and to the Dutch balance of payments apparently some £36 million in the decade 1956–66.[3] Thus the group has been able not only to pay for all foreign investment out of retentions from foreign receipts but to repatriate substantial amounts of income as well. Nevertheless, because of the restrictive policies pursued by both the US and UK Governments after 1964 in the attempt to strengthen their respective currencies, the Company began to resort to capital markets for funds (some £150 million in 1965 and over £100 million in 1966).

This survey indicates that Shell also is a remarkably broadly based international Company with a broadly based regional concentration of interests.

[1] This 'net credit' takes no account of imports and presumably includes a 'saving' on oil imports due to the nationality of the importer.

[2] Chairman of Shell Transport at the Annual General Meeting, May 9, 1967.

[3] Statement of Mr L. Schepers, former managing director, reported in *The Times*, (November 28, 1966).

Moreover, with Shell, as with Jersey, the international emphasis is not a new development but has characterized the Company from its very beginnings. Although Venezuela is the greatest single source of crude oil and Europe the greatest processing centre, Venezuela neither produces nor does Europe process as much as half of the Company's oil. As the Far Eastern countries develop their industrial activities and become increasingly more important consumers of oil products, we can probably expect further geographical diversification in the future. Shell is participating with governments and private groups in building new refinery capacity in many areas. At least three new refineries in Africa and Asia in which Shell has an interest went on stream in 1964 in addition to those in Europe. In 1966 the Company estimated the geographical distribution of its net assets as follows: Eastern Hemisphere 65 per cent; US 14 per cent; other Western Hemisphere 21 per cent.

With the high degree of balance between the various branches of Shell's operations both world-wide and regionally, most of its products ex-refinery, as well as most of its crude-oil production, are 'sold' within the group by one affiliate to another. In other words, activities at every stage of production are basically geared to the anticipated needs of the Company's markets for products. The significance of this vertical integration has already been touched upon, but Shell has gone further than any other major Company in attempting to explain to the public the nature and importance of vertical integration in the industry, particularly with respect to the price of crude oil.[1] And in many ways its officials appear to be willing to engage in serious public discussion of the industry's problems as distinct from defensive and uncritical propaganda. The international composition of its top personnel, as well as of its operating executives spread around the world, give it perhaps a more truly international outlook than that of any other oil Company. Shell also has an intensive programme of projects to assist the development of the countries in which it operates, including support of research in agriculture and student training.

British Petroleum Company Ltd.[2]

British Petroleum is considerably smaller than either Jersey Standard or Royal Dutch/Shell, with total assets in 1966 of £1,581 million ($4,426 million) and gross revenue of £1,453 million ($4,068 million). The Company was established as the Anglo-Persian Oil Company in 1909 after the discovery of oil in 1908 on a concession in southern Persia which had been obtained in 1901 by William Knox D'Arcy, who used a fortune his father had made in gold mining in Australia to finance oil exploration in Persia. When D'Arcy ran into financial difficulties in the years between 1901 and 1908 before oil in commercial quantities was discovered, the Burmah Oil Company came to his rescue by taking shares in the Company. In 1914 the British Government,

[1] See, for example, the Company's pamphlet 'Current International Oil Pricing Problems', August 1963.

[2] Unless stated otherwise the following discussion is based on the Annual Reports and other information issued by the Company, and on H. Longhurst's history of British Petroleum, *Adventure in Oil*. Statistical information not from the Annual Reports was largely made available to me by the Company.

TABLE X

BRITISH PETROLEUM COMPANY

SELECTED FINANCIAL STATISTICS

1957–66

	1	2	3	3/2	4	5	5/4	6	6/4	7		8	8/6
	Total assets[4] £m.	*Total receipts*[4] £m.	*Total taxes*[1,4] £m.	*Ratio taxes/ receipts*	*Net income before charges*[1,5,6] £m.	*Cash dividends*[2] £m.	*Ratio*[6] *dividends/ net income*	*Capital expenditure*[6] £m.	*Ratio*[6] *capital expenditure/ net income*	*New long-term debt inc. Ass. Cos.* £m.	*Net change l-term debt BP group* £m.	*Amounts retained*[3,6] £m.	*Ratio*[6] *amounts retained/ capital expenditure*
1957	595	796	273	·34	108	16	·15	114	1·06	23	10	101	·88
1958	685	835	312	·37	123	18	·15	138	1·12	52	31	114	·83
1959	775	864	322	·37	123	23	·19	141	1·15	38	3	112	·79
1960	872	921	361	·39	127	26	·20	109	·86	20	4	112	1·03
1961	948	985	402	·41	131	26	·20	124	·95	13	7	118	·95
1962	1,129	1,037	424	·41	139	31	·22	125	·90	10	1	122	·98
1963	1,123	1,127	474	·42	163	36	·22	129	·79	5	2	142	1·10
1964	1,241	1,243	551	·44	163	37	·23	169	1·04	16	14	141	·83
1965	1,377	1,335	601	·45	173	37	·21	207	1·20	74	64	144	·69
1966	1,581	1,453	662	·46	180	42	·23	206	1·14	42	30	158	·77
Total and averages	—	—	—	—	1,430	292	av. = ·20	1,462	av. = 1·02	—	—	1,264	av. = ·86

Source: The figures used are those shown in the published accounts for the applicable year or, where the accounting treatment has changed, as provided by British Petroleum.

Number of shareholders 1966: 91,000.

[1] 1965/66 UK transitional relief and previous year's tax recovery, and benefit of investment allowances for all years have been excluded.

[2] Cash dividends are the net dividends including preference dividends excluding tax withheld, but including payments made out of capital reserves. Dividends paid to minority shareholders are excluded.

[3] Amounts retained consist of group companies retained income, increase in capital reserves and depreciation, group proportion of associates depreciation, etc.

[4] 1957/59 income and expenses (including depreciation, interest and taxes) are based on the Group Income Statement. Receipts include customs duties and sales taxes.

[5] Net income comprises net income after tax but before minority shareholders' interest; plus depreciation (including group proportion of associated companies depreciation) and interest.

[6] Columns 4 and 6 (and consequently 8 and the relevant ratios) are not comparable with those in the Tables for other companies because they include the BP Group's proportion of depreciation and capital expenditure of associated companies, and similar figures are not available for the other Companies. Comparable figures for 4 and 6 are given below.

	1	2	2/1	
	Net income before charges £m.	*Capital expenditure* £m.	*Ratio capital expenditure/net income*	*Ratio amounts retained*/capital expenditure*[3]
1957	98	108	1·10	·84
1958	112	115	1·03	·90
1959	110	117	1·06	·85
1960	110	99	·90	·96
1961	108	110	1·02	·86
1962	116	92	·79	1·08
1963	136	98	·72	1·17
1964	135	153	1·13	·74
1965	145	176	1·21	·66
1966	153	171	1·12	·77
Total and averages	1,223	1,239	av.=1·01	·88

* Amounts retained here is column 8 above *less* Group proportion of depreciation of associated companies.

having decided to convert the navy from coal to oil, made a long-term contract for fuel oil with the Company and took a controlling interest in it for £2·2 million. Thus, 51·65 per cent of the shares of British Petroleum were owned by the British Government and 24·53 per cent by Burmah Oil (now 48·9 per cent and 23·23 per cent respectively); in 1966 the remaining shares were held by 91,000 shareholders, making it the most closely held of all the international major oil Companies.[1]

Like the other oil Companies, British Petroleum is vertically integrated but it produces much more oil than it refines. In 1966 its crude production reached 2,500,000 barrels per day which is about 13 per cent of the world total outside the US, Canada and the Communist countries,[2] while its refinery throughput (including oil refined under processing agreements) was only some 1,600,000 b/d. It is therefore a net seller of crude oil, much of which is sold on long-term supply contracts to other major companies. Among these contracts are those made with Jersey in 1947 and with Mobil Oil in 1948 mentioned above.

After the discovery of oil in 1908 the next essential step was to build pipelines from the oilfields to the coast and to construct a refinery. By 1913 a small refinery at Abadan on the Shatt-al-Arab was completed and connected to the oilfields miles away by pipeline. Although the newly discovered oil made an important contribution to the fuelling of the British navy during the First World War, the war interrupted the planned expansion of production, and it was not until 1917 that the Company could take full advantage of its resources. Sales contracts with the British Admiralty and also with the Indian and Iraqi railways gave the Company large-scale outlets for its products, particularly fuel oil.

Also before the First World War the Anglo-Persian Oil Company had been negotiating for oil rights in the old Ottoman Empire and in 1909 D'Arcy's representative in Constantinople obtained the promise of a concession. Just before the war broke out an agreement to establish the TPC in which Anglo-Persian had secured a 50 per cent interest had been made. The Deutsche Bank and a subsidiary of Royal Dutch/Shell held the remaining stock and C. S. Gulbenkian, an Armenian financier who had long been interested in Mesopotamian oil, a 5 per cent beneficiary interest. The TCP was subsequently promised a concession by the Turkish Grand Vizier. Then came the First World War, which changed the political map of the Middle East, and after prolonged disputes and bargaining in which the American, British, and French Governments all took an active part,[3] the IPC was set up as the successor to

[1] In 1966/7 BP acquired the chemical and plastics interests of Distillers Company, partly by exchange of shares, thus reducing the government interest to 48·9016 per cent.

[2] See Table II, p. 78.

[3] The extent to which the British Government intervened specifically to advance the interests of the British oil companies (as contrasted to intervention for military reasons or as part of a more general economic policy relating to areas under its control or tutelage) has been widely disputed. Gibb and Knowlton in their history of Standard Oil (especially Chapter 11) see the negotiations as a struggle between 'British imperialism' through the Foreign Office and the American demand for the 'open door'. Elizabeth Monroe, on the other hand, finds very little support for the argument that the government actively supported British 'concession hunters':

'Apart from the efforts that the British Government made before 1914 in the Ottoman

the old TPC, and Anglo-Persian came through with 23·75 per cent interest, plus an over-riding royalty of 10 per cent on the oil production of the IPC.[1]

In 1928 the Company announced an arrangement for the supply of products to the Indian market through Burmah-Shell (a jointly owned affiliate of Royal Dutch/Shell and Burmah Oil) which had been made possible through the 'good offices' of Burmah Oil. Anglo Persian was thus enabled to enter the Indian market without having to engage in competition with other Companies. (It must be remembered that when APOC was formed in 1908 Jersey Standard, Royal Dutch and Shell were well established internationally). By 1939 the Company was marketing in much of Europe and Africa, the Middle East, India and Australia. It had made the Abadan refinery one of the largest in the world and had built new refineries in Britain and France. Up to the Second World War, apart from insignificant production in the UK and Argentina, Iraq and Iran were the only sources of oil for Anglo-Iranian (as the Company was named after Persia chose to call itself Iran), although it had been exploring fairly extensively in many parts of the world.

In 1934, however, AIOC together with Gulf Oil secured a concession covering the entire territory of the Sheikhdom of Kuwait. Gulf had been attempting to obtain the Kuwait concession and was opposed by AIOC, but after long negotiations during which Gulf called for assistance from the US State Department, Anglo-Persian and Gulf got together to form the Kuwait Oil Company for the purpose of obtaining a concession from the Sheikh of Kuwait and exploiting any oil found. In 1938 one of the greatest oilfields of all time was discovered and these two Companies alone found themselves in control of territories the proved reserves of which were still in 1966 estimated to be around 16 per cent of total world proved reserves outside the Communist countries.

It was not possible to develop the Kuwait fields until after the end of the Second World War and exports did not begin until 1946. Between 1948 and 1951 Kuwait production available to BP more than quadrupled and Iraq production more than doubled, with production from new fields in southern Iraq

---

Empire, and the few words that it spoke to the Persian Grand Vizier in 1901 in support of d'Arcy, it seldom used its influence on behalf of concession hunters. Even in places where it was entitled by pledge or treaty to approve the nationality of the concessionaire (as it was in Kuwait or Bahrain), it preferred to leave British aspirants to fend for themselves. The list of occasions on which it gave no help is very long. It gave none, for instance, to the British negotiator in Jidda in 1933 when he sought from Ibn Saud the prize (now the Arabian American Oil Company fields in Eastern Arabia) that he saw pass to an American rival because the Iraq Petroleum Company had authorized him to "speak only of rupees, where gold was demanded". It gave none when the western Arabian concession, later fruitless, was negotiated in 1936; none when the Turkish (later Iraq) Petroleum Company negotiated and re-negotiated agreements with the Iraq Government in 1924–5 and 1931; none in Syria in 1937–8; while in Bahrain in 1929–30 it was amenable about waiving exclusive British rights, and agreed to operations by American firms which took Canadian nationality for the purpose.

'Developments in Kuwait were less straightforward, but the end-product was the same; the British Government never tipped the scales with the Ruler in the two years (1932–3) during which the American Gulf Oil Company and APOC waited on his decision between them. . . .' Elizabeth Monroe, *Britain's Moment in the Middle East*, pp. 105–6.

For a contrary view regarding Persia, see Mikdashi, Ch. I.

[1] This was reduced in 1931 to 7½ per cent. The royalty does not apply to the Basra or Mosul production of the IPC group.

and in Qatar (where the concessions were also held by the IPC group) adding to the total. Since 1951 further acquisitions of oil properties have been made in Canada, Trinidad and the US, and exploration has been continuing in North and South America, Australia, New Zealand, North, East and West Africa, and Europe. Production continued to rise steadily after the end of the war and until 1951, when it dropped sharply as a result of the dispute with Iran, but recovered quickly as output was expanded in Iraq and Kuwait. As a result of the dispute and the subsequent Iranian nationalization the Company lost, as we have seen, its exclusive position in Iran, and a consortium was set up which included all of the seven majors plus Compagnie Française des Pétroles and a group of smaller American independents. BP retained a 40 per cent interest in the Consortium and was paid compensation from both the Persian Government and the members of the Consortium.[1]

Between 1954 and 1966 BP's output of crude oil rose by 257 per cent. Its refinery throughput increased by 220 per cent.

The crude production of BP recovered much more rapidly from the Iranian crisis than did its refinery runs, for crude production could be increased in other areas in which BP had interests, notably Iraq and Kuwait, but the loss of the refinery capacity it had completely owned in Iran could only be replaced by purchase or construction, and in the meantime greater reliance had to be placed on processing contracts with other Companies. There followed a rapid extension of refining capacity and by 1965 over half of BP's refining was done in Europe as compared with around 7 per cent before the war.

It is clear that BP is 'long' on crude, but how much is sold in world markets at market prices and how much moves under military and naval contracts and private long-term supply contracts at special prices is not known. Probably something like 220,000 b/d goes to Jersey Standard at some price unrelated to market prices and at least half as much to Socony-Mobil at prices far below posted prices, in addition to the supplies to a number of 'independents', notably Petrofina. BP also supplies part of the oil of an Italian refinery (in which it has a 49 per cent interest). Until recently the quantities of crude oil sold amounted to more than half that refined by the company or on its behalf.

British Petroleum retains a large part of its net income; the ratio of dividends to income before depreciation but after taxes averaged ·20 over the period 1957–66. Over the same period the ratio of retained funds to capital expenditure averaged ·86. Until recently BP maintained exceptionally large liquid reserves, which implies not only a conservative dividend policy but a conservative investment policy as well, in view of the profitability of the Company in pre-war years. From 1962 to 1965 the Company made extra payments to stockholders from its capital reserves, which were increased as a result of compensation paid to BP by Companies entering into the Iranian consortium after the 1951 nationalization.

In some ways the loss of its Iranian position was as severe a shock to BP in the 1950s as was the dissolution decree in 1911 to Standard Oil. Both Companies had to make substantial changes in the structure of their operations and perhaps in their outlook. For BP this involved an acceleration of the European and UK refining programme. The pre-war history of BP is marked

[1] See Longrigg, Chapter XVI.

TABLE XI
BRITISH PETROLEUM
PRODUCTION, REFINING AND SALES
1950–66

	Crude oil production[1] 000 *b/d*[2] *Total*	Refinery processing[3] 000 *b/d* *Total*	Sales: *Crude oil* 000 *b/d*	*Products* 000 *b/d*	*Total* 000 *b/d*	*% bulk trade and bunkers*[4]	*% UK and Europe*[4]	*Production chemicals*[5] 000 *tons*[4]
1950	800	600	200	600	800	—	—	—
1951	700	500	200	500	700	—	—	—
1952	500	400	200	400	600	—	—	—
1953	600	400	300	500	800	—	—	—
1954	700	500	300	500	800	—	—	—
1955	900	600	300	600	900	—	—	—
1956	1,000	700	400	600	1,000	—	—	—
1957	1,000	600	400	600	1,000	—	—	—
1958	1,200	800	500	700	1,200	—	—	—
1959	1,300	800	500	700	1,200	—	—	—
1960	1,500	900	600	900	1,500	—	—	—
1961	1,600	1,000	700	900	1,600	—	—	440
1962	1,700	1,200	700	1,000	1,700	—	—	610
1963	1,800	1,300	700	1,100	1,800	—	—	737
1964	2,100	1,400	800	1,300	2,100	—	—	1,006
1965	2,300	1,500	800	1,400	2,200	35	45	1,180
1966	2,500	1,600	900	1,500	2,400	33	46	1,600
% Incr.								
1950–66	212	167	350	150	200	—	—	—
1954–66	257	220	400	200	200	—	—	—

Source: *Annual Reports*.

[1] Includes BP share of production in associated companies.

[2] Barrels per day are arrived at by converting long tons at 1 b/d=50 tons/year.

[3] Includes oil refined on BP's account as well as that refined under processing arrangements elsewhere.

[4] Figures are omitted either because they are not available or the figures published are not on a comparable basis with those given in the above table.

[5] Production chemicals is 100 per cent production of all chemical companies in which the BP Group has an interest. It is not the BP share of production as are the other figures above.

by the long duration of its powerful position in the Middle East, the financial backing of the British Government as a shareholder and customer,[1] and its close co-operation with other Companies in the marketing of products. These factors, together with the narrow base of its shareholding may have been responsible for its apparent conservatism.

[1] The relation between British Petroleum and the British Government has been widely misunderstood. Although the Government has a controlling interest and appoints two of the Company's directors it does not interfere with the normal operations of the Company. The Government directors have a right to veto any proposals and the other directors have a right of appeal to the Treasury and the Admiralty, but in practice neither procedure is used. In view of the importance of the Company in the British economy and of the significance of petroleum for both military and civilian purposes, it is highly probable that the practical relationship between the Government and the Company would have been little different even if the Government had had no equity in it.

Since 1950, however, the Company has been extending the geographical spread of its operations more rapidly, although it is still much less widespread internationally than either Jersey or Shell and much more orientated towards Europe, the Commonwealth and particularly the UK.

After 1964 BP's capital expenditures rose sharply in comparison with earlier years and the Company was forced to borrow large sums abroad, partly because the British Government was restricting foreign investment in the interest of the balance of payments. Long-term debt jumped from around £93 million to £187 million between 1964 and 1966. In spite of the large investments abroad BP reported a net credit to the country's balance of payments if no account is taken of the cost of oil imported. In 1965 net earnings on overseas trade less capital investment were reported as £61 million, and the Company estimated that an additional £10 million could be added (by deducting it from the value of capital investment assumed in the calculations) to take account of the export of British materials by contractors working for the Company. Moreover, the debit in the balance of payments for oil imports was estimated to be less by £40 million because oil was imported by a British Company and transported in British tankers.

Like other international Companies, BP is developing new sources of crude oil in many areas—the most successful of which in recent years have been in Abu Dhabi (onshore with the IPC group, offshore with CFP), Libya (with an American interest) and Nigeria (with Shell). Its refining and marketing outside Europe are still largely concentrated in the countries of the Commonwealth, and the Company operates in a smaller number of countries than do most of the other majors. For example, BP seems to have been very slow in obtaining an interest in Japanese refining although it early made loans to Japanese refineries tied to crude-supply contracts, which, admittedly, may come to much the same thing. Nevertheless, in recent years there has clearly been a marked acceleration in the expansion of BP's geographical interest, much of which has been by acquisition of other companies. The Company does not publish the geographical distribution of its income or investments but it did report in 1964 that £450 million of net assets were outside the UK, or about 56 per cent of the total. BP has also been diversifying into petrochemicals, and by 1965 its proportion of the capital expenditure of its associates on petrochemicals equalled about 6 per cent of the total capital expenditure of the Company. As noted above, it acquired the chemical and plastics interests of the Distillers Company in 1966–7.

Like Jersey Standard and Shell, BP supports a wide variety of educational and social projects, often with particular reference to the areas from which it draws its crude oil.

Gulf Oil Corporation[1]

Measured by crude-oil production, Gulf is about the same size as BP, but its total assets, amounting to $5,892 million (£2,104 million) in 1966, were some-

[1] I have drawn for this discussion largely on the Company's *Annual Reports*, a short history it has produced, and the brief discussion of early history in J. G. McLean and R. W. Haigh, *The Growth of Integrated Oil Companies*, and the Report of the Federal Trade Commission.

what larger, as was its gross revenue of $4,717 million (£1,685 million). In 1966 the Company produced 2,293,000 b/d of crude oil and about 10 per cent of the world total outside the US, Canada and the Communist countries, but only some 1,248,000 b/d, or about 54 per cent of what it produced, was processed by it or on its account.[1] In this respect, Gulf's position is similar to that of BP; it too is a net seller of crude. Among its long-term supply contracts is the one to supply Royal Dutch/Shell with Kuwait crude mentioned above. This contract, in effect, gives Shell a contractual share in Kuwait oil up to some unspecified amount, and in recent years Shell has been buying more oil than the minimum specified in the contract. Little is published about these transactions; in 1958 Shell took 58 per cent of Gulf's share of Kuwait's net crude production.

The Gulf Oil Corporation was formed in 1907 to acquire the stock of two other American companies which had their roots in the early history of the oil discoveries in Texas and Oklahoma. It was financed by the Mellon banking group and still is largely controlled by Mellon interests. Its 167,000 shareholders are less in number than those of any of the other international American majors.

As we have seen, in the early days of the US oil industry the dominant refiner and distributor was Standard Oil, and crude-producing companies were thus heavily dependent on Standard to dispose of their crude oil. W. L. Mellon, who had been closely associated financially with the predecessor companies of Gulf and was its vice-president from 1907 to 1909, president from 1909 to 1930, and Chairman of the Board until 1948, was firmly convinced that the only way for an independent company to compete successfully with Standard Oil was to develop an integrated business, and especially to develop its own crude production. Hence the pursuit of integration shaped the development of the Company in the US and it early proceeded to search for crude oil; refinery capacity was more than sufficient to handle its own production.

Gulf's foreign operations began with exploration in 1913 in Mexico, which was the sole source of foreign crude until 1925. But the Mexican production of the Company rose and then fell off very rapidly; having increased from barely a million barrels (net after royalty) in 1917 to nearly 25 million in 1922, it fell to less than 2 million by 1930. In the meantime, however, Gulf's Venezuelan production, which began commercially in 1925, increased sharply to take its place, growing from just over a million barrels in 1925 to a peak of 26 million in 1929. With the successful development in the inter-war period of the discoveries in Venezuela, Gulf's crude supplies far exceeded her capacity to use them, and in the depression of the 1930s the Company found itself in the awkward position of having much crude, falling domestic sales, and a scarcity of funds. In particular, it had insufficient outlets under its own control to absorb the Venezuelan oil potentially available to it and had not the means of expanding outlets rapidly. To solve the problem, the Company sold an undivided half interest in the operations and assets of Mene Grande, the subsidiary through which its Venezuelan operations were conducted, to the International Petroleum Company of Canada, a subsidiary of Jersey Standard, for $100 million in cash and the obligation of International Petroleum to pay

[1] See Tables II and XIII.

half the production costs. In addition, Jersey obtained a number of other interests in Mene Grande's activities, in particular, in its concession contracts. Subsequently Shell bought a half interest in International Petroleum, i.e. an effective 25 per cent of Mene Grande's operations. Gulf also sold at this time its Italian and most of its French marketing interests.

TABLE XII
GULF OIL COMPANY
PRODUCTION, REFINING AND SALES[1]
1950–66

	Crude Oil Production			Refinery Processing[2]		Sales		
	Total 000 *b/d*	% *US*	% *Eastern Hemisphere*	*Total* 000 *b/d*	% *US*	*Products* 000 *b/d*	*Crude produced outside US* 000 *b/d*	*Chemicals mil. tons*
1950	541	44	35	471	—	479	185	—
1951	696	39	42	501	—	533	318	—
1952	752	35	46	532	—	547	368	—
1953	803	33	48	535	—	558	404	·04
1954	828	30	50	588	88	565	460	·07
1955	960	28	53	647	88	596	568	·2
1956	1,052	32	49	753	83	689	534	·3
1957	1,220	31	51	776	79	739	591	·4
1958	1,361	27	57	802	70	798	757	·4
1959	1,407	28	58	779	66	821	782	·5
1960	1,571	26	61	849	65	842	881	·7
1961	1,595	26	61	845	62	860	—	·7
1962	1,773	24	63	968	59	940	—	1·0
1963	1,889	24	63	1,020	58	1,000	—	1·3
1964	2,012	24	64	1,084	56	1,087	—	3·0
1965	2,169	24	65	1,167	54	1,154	—	3·0
1966	2,293	24	65	1,248	54	1,241	—	3·4
% Incr.								
1950–66	324	—	—	165	—	159	—	—
1954–66	177	—	—	112	—	120	—	—

Source: *Annual Reports*. Information provided by company.
[1] Includes Gulf's share in all operations in which it has an interest.
[2] Including oil processed by others for Gulf's account.

Kuwait oil began to contribute a large part of Gulf's total supplies by 1948, and by 1949 its crude oil from Kuwait exceeded that from Venezuela. In addition to Kuwait and Venezuela, Gulf has a two-thirds interest in British-American, through which it receives a small amount of oil from Canada, and it also has a 7 per cent interest in the Iranian Consortium, which it acquired at the time the American Companies were brought into the Consortium.

With respect to its marketing and refining operations Gulf has always turned its attention primarily to the US market. Well over half of its refining capacity is in the US (including Puerto Rico), but it is part-owner of refineries in Venezuela, Kuwait, Iran, Canada, Taiwan, the Philippines, and Korea; in

TABLE XIII

GULF OIL COMPANY

SELECTED FINANCIAL STATISTICS

1958–66*

	1	2	3	3/2	4	5	5/4	6	6/4	7	8	8/6
	Total assets $m.	*Total receipts*[1] $m.	*Total taxes*[1] $m.	*Ratio taxes/receipts*	*Net income before charges*[2] $m.	*Cash dividends* $m.	*Ratio dividends/net income*	*Capital expenditure*[3] $m.	*Ratio capital expenditure/net income*	*Net change long-term debt* $m.	*Amounts retained*[4] $m.	*Ratio amounts retained/capital expenditure*
1958	3,426	3,222	638	·20	614	87	·14	401	·65	(16)	517	1·29
1959	3,562	3,203	652	·20	576	107	·19	336	·58	(18)	469	1·40
1960	3,822	3,241	708	·22	595	110	·18	346	·58	(8)	476	1·38
1961	4,023	3,286	712	·22	591	123	·21	344	·58	(22)	459	1·33
1962	4,244	3,455	809	·23	616	162	·26	556	·90	25	445	·80
1963	4,549	3,612	848	·23	626	176	·28	621[5]	·99	19	470	·76
1964	4,667	3,844	929	·24	687	186	·27	599	·87	(16)	490	·82
1965	5,211	4,211	1,150	·27	761	201	·26	598	·78	144	546	·91
1966	5,892	4,717	1,312	·28	872	227	·26	715	·82	209	620	·87
Totals and averages 1958–66	—	—	—	—	5,938	1,379	av. = ·23	4,516	av. = ·76	—	4,492	av. = ·99

Source: *Annual Reports*. Information from Company.

[1] Including consumer excise taxes.

[2] Net income after taxes plus depreciation, depletion, amortization and retirements, and interest charges, and including income applicable to minority interests. Does not include depreciation of non-consolidated affiliates.

[3] Includes related business investment.

[4] Net income (Col. 4) less interest charges and dividends.

[5] Includes acquisition of a chemical company for $175 million.

* Years before 1958 not fully consolidated and figures not comparable with those for later years.

Number of shareholders 1966: 167,000.

Europe, it has interests in refineries in Denmark, Holland, France, Spain and the UK, but built its first wholly-owned European refinery only in 1961. Gulf's interest in European markets has quickened in recent years and both refining and marketing activities have been expanding.

In spite of the fact that production in the US fell from nearly half of Gulf's total crude production in 1950 to about a quarter in 1966, with Kuwait alone contributing around 58 per cent of the total, over 71 per cent of Gulf's total earnings came from its US domestic operations in 1966. Expansion of the Company's operations in the US was made possible in the 1950s from profits made 'in Kuwait' according to Mr R. O. Rhoades (the 'discoverer' of the Company's Kuwait oil) who resigned as Chairman of the Board in 1960. In 1955 Mr Rhoades told a US Senate Committee that,

> '. . . Kuwait does give Gulf very remarkable profits. Against those profits which we have been able to retain in the company, we have put quite a number, hundreds of millions of dollars, into exploration here in the US and into the development of our reserves here which would not have been possible for us to do but for that income from out there . . .'[1]

In later years, profits and other retained funds attributed to Eastern Hemisphere operations became less important. Gulf's crude-oil production increased more rapidly between 1950 and 1966 than that of any of the other major companies, having risen over four times. Much of the production, as noted above, is sold to Shell under a long-term contract on special terms; total refinery runs and sales of refined products went up only about two and a half times in the period. Gulf, like the other Companies, has financed itself largely from its retained earnings (including depreciation and depletion allowances); its retained earnings plus depreciation just about covering its capital investment over the period 1958–66. This Company, too, has in recent years substantially increased the proportion of net income it pays out and, because of US restrictions on foreign investments, it also borrowed small amounts in Europe in the middle 1960s.

In the Western Hemisphere Gulf has achieved a 'balance' between crude production and refinery runs, but in the Eastern Hemisphere the Company produced in 1960 about eight times as much crude as it refined in that Hemisphere.[2] It carries on an active exploration programme but does not publish statistics of expenditure on exploration by regions, although it reports that most of the expenditure takes place in the Western Hemisphere, particularly in the US. The Company itself stresses that its operations are 'rooted in the US' and that its foreign operations have persistently brought a net credit to the US balance of payments. Nevertheless, it produces in Nigeria, Bolivia and Colombia, where the Patamayo discovery in 1965 was reported to be of 'major importance', and is exploring in other areas of North Africa and the Middle East, including Libya and Turkey, and in the North Sea.

The Company does not publish details of its investment expenditure by region, although in 1964 it reported that about three-quarters of its investment

[1] *Emergency Oil Lift Program and Related Oil Problems*, p. 1324.

[2] 1960 was the last year in which Gulf published the geographical distribution of its refinery runs.

TEXACO

SELECTED FINANCIAL STATISTICS

1950–66

	1	2	3	3/2	4	5	5/4	6	6/4	7	8	8/6
	Total assets[1] *$m.*	*Total receipts*[2] *$m.*	*Total taxes*[2] *$m.*	*Ratio taxes/ receipts*	*Net income before charges*[3] *$m.*	*Cash dividends*[4] *$m.*	*Ratio dividends/ net income*	*Capital expenditure*[5] *$m.*	*Ratio capital expenditure/ net income*	*Net change long-term debt $m.*	*Amounts retained*[6] *$m.*	*Ratio amounts retained/ capital expenditure*
1950	—	1,507	290	·19	250	76	·30	149	·60	(1)	169	1·13
1951	—	1,708	357	·21	301	84	·28	225	·75	7	211	·94
1952	—	1,861	387	·21	325	82	·25	230	·71	(46)	236	1·03
1953	—	1,916	408	·21	354	93	·26	292	·82	(6)	254	·87
1954	—	1,991	410	·21	398	104	·26	299	·75	(5)	287	·96
1955	—	2,216	476	·21	446	118	·26	283	·63	(6)	321	1·13
1956	3,258	2,555	553	·22	507	131	·26	503	·99	155	367	·73
1957	3,690	2,869	614	·21	566	130	·23	389	·69	(59)	423	1·09
1958	4,061	2,913	605	·21	558	138	·25	426	·76	42	408	·96
1959	4,247	3,316	673	·20	616	155	·25	426	·69	(19)	447	1·05
1960	4,508	3,681	877	·24	665	174	·26	433	·65	(6)	477	1·10
1961	4,821	3,825	923	·24	712	193	·27	430	·60	(3)	504	1·17
1962	5,012	4,044	966	·24	775	236	·30	453	·58	(17)	525	1·16
1963	5,395	4,250	1,021	·24	847	269	·32	580	·68	(6)	566	·98
1965	5,815	4,426	1,097	·25	885	298	·34	561	·63	115	574	1·02
1965	6,252	4,716	1,207	·26	972	332	·34	670	·69	3	622	·93
1966	7,370	5,494	1,413	·26	1,112	340	·31	865[7]	·62	297[8]	743	·86
Totals and averages	—	—	—	—	10,289	2,953	av. = ·29	7,214	av. = ·70	—	7,134	av. = ·99

Source: *Annual Reports.*

[1] From 1956 includes equity in non-consolidated affiliates. Figures before 1956 not comparable.

[2] Includes sales and excise taxes collected from consumers.

[3] Net income after taxes plus depreciation, depletion, and amortization and interest charges but including income applicable to minority interests. Does not include depreciation of non-consolidated affiliates.

[4] Including dividends to minority stockholders.

[5] Including investments in and advances to non-consolidated affiliates.

[6] Net income (Col. 4) less dividends and interest charges.

[7] Including $180 million from debentures issued to acquire capital stock of Deutsche Erdöl A.G.

[8] Including long-term debt of Deutsche Erdöl and subsidiaries.

Number of shareholders 1966: 215,283.

expenditures in the previous ten years had been in the US. Of total expenditure in 1965, nearly half was in production, probably reflecting the extensive exploration and development programme in the US. Marketing absorbed slightly over a quarter, and Gulf has developed an extensive travel aid business, including arrangements for the lodging of motorists. It has diversified extensively into plastics, agricultural chemicals and petrochemicals, chemicals absorbing about 5 per cent of investment expenditure. The Company is also searching for deposits of phosphates, potash, and sulphur with the fertilizer industry in view—this kind of diversification will call for skills in exploration and mining as well as in the marketing of chemical fertilizers and other agricultural chemicals. Much of its expansion has been through acquisitions, including the $125 million purchase of a US chemical company in 1963.

Fundamentally Gulf, like BP, has floated along on the enormous amount of crude oil it has had under its control and, like BP, has had its eyes on its home market until the last few years. Neither Company displays the same international spread as do Jersey and Shell. This can perhaps be explained by their easy crude position and the effective maintenance, until recently, of the price of crude oil, making it unnecessary for these Companies to fight for their living in the international markets in the past to the same extent as have the two other Companies. Both BP and Gulf have a great deal to lose when crude prices come under pressure, but Gulf, having so much of its operations in the US, enjoys the advantages of a protected market insulated to a considerable degree from pressures on world prices.

Texaco and Standard Oil of California[1]

Although Texaco and Standard Oil of California (Socal) operate as independent Companies in the Western Hemisphere, their operations in the Eastern Hemisphere were, until 1967, so closely associated through the jointly-owned Caltex that they can in many respects be treated jointly for the purpose of this study, concerned as it is with the international petroleum industry outside the US and particularly in the countries of Asia and Africa. Texaco is much the larger of the two Companies, its crude production in 1966 being 2,263,000 b/d, or about 30 per cent greater than that of Socal (1,731,000 b/d), but this is due almost entirely to Texaco's greater production in the US and Canada. Similarly with respect to refining, the total crude processed by Texaco in 1966 was 1,972,000 b/d, its processing in the US alone exceeding that of Socal in the entire Western Hemisphere (774,000 b/d), while in the Eastern Hemisphere the refinery runs of the two Companies were 647,000 b/d and 511,000 b/d respectively. Texaco's total assets amounted to $7,370 million (£2,632 million) while those of Socal were $4,800 million (£1,714 million), including in both cases equity in non-consolidated affiliates; their total receipts were $5,494 (£1,962 million) and $3,393 million (£1,212 million) respectively.

The California-Texas Oil Co. (Caltex), which engaged in exploration, pro-

[1] The following discussion draws mainly on the Annual Reports of the Companies; Marquis James, *The Texaco Story*, 1902–52; a *Short History of Texaco*, published by Texaco; *A World Wide Enterprise*, published by Standard Oil of California; and, for earlier years, McLean and Haigh.

TABLE XV

STANDARD OIL OF CALIFORNIA
SELECTED FINANCIAL STATISTICS
1950–66

	1	2	3	3/2	4	5	5/4	6	6/4	7	8	8/6
	Total assets[1]	*Total receipts*[2]	*Total taxes*[2]	*Ratio taxes/receipts*	*Net income before charges*[3]	*Cash dividends*	*Ratio dividends/net income*	*Capital expenditure*	*Ratio capital expenditure/net income*	*Net change long-term debt*	*Amounts retained*[4]	*Ratio amounts retained/capital expenditure*
	$m.	*$m.*	*$m.*		*$m.*	*$m.*		*$m.*		*$m.*	*$m.*	
1950	1,464	972	100	·10	229	72	·31	161	·70	(45)	154	·96
1951	1,645	1,159	126	·11	262	75	·29	143	·55	(1)	185	1·29
1952	1,738	1,221	81	·07	275	86	·31	173	·63	(11)	187	1·08
1953	1,915	1,302	107	·08	299	86	·29	224	·75	(6)	211	·94
1954	2,065	1,387	253	·18	333	87	·26	281	·84	—	244	·87
1955	2,266	1,548	279	·18	371	90	·24	347	·94	—	278	·80
1956	2,459	1,760	412	·23	416	113	·27	302	·73	—	301	1·00
1957	2,697	1,951	413	·21	439	129	·29	373	·85	—	308	·83
1958	2,898	1,903	421	·22	408	136	·33	278	·68	97	269	·97
1959	2,979	1,909	434	·23	404	135	·33	328	·81	—	261	·80
1960	3,237	2,053	483	·24	427	135	·32	316	·74	(2)	284	·90
1961	3,568	2,577	493	·19	469	135	·29	377	·80	(1)	326	·86
1962	3,794	2,712	512	·19	508	141	·28	452	·89	13	359	·79
1963	3,960	2,806	534	·19	524	148	·28	388	·74	(2)	368	·95
1964	4,175	2,929	593	·20	562	155	·28	473	·84	33	398	·84
1965	4,187	3,116	635	·20	628	175	·28	547	·87	—	441	·81
1966	4,800	3,393	688	·20	694	192	·28	441	·64	74	502	1·14
Total and averages	—	—	—	—	7,238	2,089	av. = ·29	5,594	av. = ·77	—	5,076	av. = ·91

Source: *Annual Report. Supplemental Financial and Statistical Data.*

[1] Including excess in equity of Eastern Hemisphere affiliates over book value.

[2] Including consumer sales and excise taxes.

[3] Net income after taxes plus depreciation, depletion and amortization and interest charges and including income applicable to minority interests. Does not include depreciation of non-consolidated affiliates.

[4] Net income (Col. 4) less dividends and interest changes.

Number of shareholders 1966: 228,096.

duction, transportation, refining and marketing throughout the Eastern Hemisphere, is owned in equal shares by the two Companies. In May 1967, Caltex was in effect dissolved in twelve European countries (UK, Ireland, Germany, Italy, Switzerland, Belgium, Luxembourg, the Netherlands, Norway, Sweden, Denmark and Greece—*not* in France or Spain). Its assets and operations in these countries were divided between the parent Companies, thus freeing each of them to pursue its own independent policies. Texaco had been aggressively expanding in Europe for several years while Socal seemed more interested in its North America operations. East of Suez, and in East and South Africa, however, Caltex continues to operate for both Companies.

The main producing facilities of Caltex are in Indonesia and Bahrain, and output in 1966 averaged 367,000 b/d. Texaco and Socal each owns 30 per cent of Aramco and each has a 7 per cent interest in the Iranian Consortium. From these three sources they received some 2,062,000 b/d (1,030,000 b/d each). Of this Caltex sold in the form of crude and products 1,187,000 b/d. Crude runs to refineries for Caltex account totalled 1,178,000 b/d.

Although Texaco and Socal are so closely associated in much of their foreign operations, their origins and development were very different and this accounts in part for their association. Standard Oil of California, as the name implies, was originally part of the Standard Oil group broken up by the anti-trust decree of 1911, and operating primarily in California.[1] In the 1920s and 1930s it expanded its production and marketing operations to other areas of the US and joined other American companies in seeking concessions abroad, particularly in Sumatra, Java, and Netherlands New Guinea, the Persian Gulf and Saudi Arabia. In 1928 Socal took over from Gulf an option on a concession in Bahrain, and in 1930 the Bahrain Petroleum Company was organized. Oil was discovered in 1932 and the Company immediately moved to obtain a concession in Saudi Arabia, which it succeeded in doing in 1933. Production in Bahrain rose rapidly from only 31,000 barrels in 1933 to around 1·2 million barrels in 1935. Socal did not have markets in the Eastern Hemisphere for supplies of the kind that seemed likely to be available from this concession and from Saudi Arabia, where exploration had begun. Negotiations were therefore opened with The Texas Company (now Texaco), which had been interested in foreign marketing as early as 1905, and which had in the 1920s and 1930s built up a substantial marketing position in the Eastern Hemisphere.

Texas Oil was one of the few companies that survived the speculative activity following the 'Spindletop' gusher in East Texas in 1901, having been lucky in finding additional oil when the short-lived Spindletop pool was finished. It was an independent company in competition with the then overwhelmingly dominant Standard Oil trust, but from the very beginning it went in for fully integrated operations, building refineries to process its crude oil while continuing to search for more, and acquiring tankers and marketing facilities to transport and sell its products. It began selling in foreign retail markets as early as 1905 and by 1908 had tankers in operation to European ports. Although it acquired 'prospective oil lands' in Colombia and Venezuela

[1] The Standard Oil Company purchased the California company in 1900. Because of Standard's failure at that time to get production in Indonesia, it wanted production on the West Coast of the US to supply its Oriental markets. See Hidy and Hidy, pp. 341–2, 498.

in the 1920s, the development of refining was given preference over crude production, and by 1933 the Company was producing only about 55 per cent of its crude. With the advent of production regulations in the US in the mid-1930s, it became clear that crude oil would be likely to be available in the US only at increasing cost and the Company intensified its activities abroad, where it already had extensive marketing outlets in Latin America and in the Eastern Hemisphere.

In 1936 an agreement was made between Socal and Texaco whereby Socal sold to Texaco half its interest in the companies operating the Bahrain and Saudi Arabian concessions, as well as half its interest in a company holding concessions in the Netherlands East Indies, in return for a substantial cash payment and a half interest in the marketing facilities of Texaco in the Eastern Hemisphere, thus obtaining an outlet for its crude oil. Caltex was formed in 1936 to conduct the jointly-owned marketing operations. In 1947 Caltex purchased Texaco's marketing outlets in Europe and North Africa, completing the overseas consolidation of the two major Companies. At the same time arrangements were made to bring Jersey Standard and Socony Vacuum (now Mobil Oil) into Aramco, since the supplies of crude by Aramco at that time exceeded the marketing outlets of the parent Companies, who, in addition, needed more capital for investment. Texaco and Socal each retained a 30 per cent interest in Aramco, selling 30 per cent to Jersey and 10 per cent to Socony-Vacuum. In 1954 each Company took a 7 per cent share in the Consortium operations in Iran.

Although both Companies always did some marketing in the Eastern Hemisphere independently of each other (Socal of crude, and Texaco of products, primarily in England through Regent—in which it had a 75 per cent holding—and in West Africa), most marketing was done through Caltex until 1967. On the other hand, all of their crude production and most of their refining East of Suez are still carried out by companies in which both participate—Iranian Oil Participants, Aramco, and Caltex. Caltex also operated a large transport fleet which was transferred to its shareholders in May 1967, and it also conducts exploration on behalf of the parent companies through Amoseas.[1]

Texaco's expenditure on exploration outside the US in the three years in which the figures were published (1957–9 inclusive) were about a quarter of its total expenditure on exploration, but in 1965 more than two-thirds of the wells drilled or the rigs operated by the Company were in the US. These figures do not include the Company's share of the exploration activities of Caltex and other non-consolidated affiliates and thus seriously understate Texaco's total investment in foreign exploration, since the search for crude oil in the Eastern Hemisphere is largely conducted by such affiliates.

The international operations of both Companies, but particularly those of Texaco, are increasing in relation to their domestic operations, although by far the greater part of the refining and distribution of both is still carried on in

[1] Caltex is not a publicly owned company and consequently no balance sheets or income statements have been published except for the period 1958–60 in a special report by Socal. The annual reports of the parent companies give very little financial information about their non-consolidated affiliates, among which are Caltex and Aramco, although further information is on file at the Securities and Exchange Commission.

the Western Hemisphere. In 1966 about 75 per cent of Texaco's and 57 per cent of Socal's net income were attributed to Western Hemisphere operations. Between 1952 and 1966 Texaco's refinery runs in the Eastern Hemisphere increased from 25 per cent of its total refining to 33 per cent, and its sales of products from 25 per cent to 38 per cent of total sales, both including its share of Caltex operations. Similarly Socal's Eastern Hemisphere refining increased from 30 per cent to 40 per cent of total refining in the period, and product sales from 28 to 34 per cent. Nevertheless, only 10 per cent of the gross assets of

TABLE XVI
TEXACO
PRODUCTION, REFINING AND SALES[1]
1950–66

	Crude Oil Production			Refinery Processing			Product Sales	
	Total 000 *b/d*	Eastern hemisphere %	US %	Total 000 *b/d*	Eastern hemisphere %	US %	Total 000 *b/d*	Eastern hemisphere %
1950	550	33	56	541	23	28	450	—
1951	662	37	53	697	24	21	500	—
1952	723	37	50	720	25	24	725	25
1953	781	37	49	763	26	—	766	27
1954	828	39	45	801	27	—	781	28
1955	909	39	46	884	26	64	863	27
1956	945	39	43	987	25	60	984	29
1957	1,077	40	40	1,073	25	54	1,056	30
1958	1,129	41	40	1,135	24	53	1,256	26
1959	1,254	39	43	1,231	24	53	1,329	26
1960	1,352	42	41	1,327	24	51	1,368	28
1961	1,463	42	41	1,399	25	48	1,471	31
1962	1,571	42	40	1,467	26	47	1,557	32
1963	1,698	43	40	1,563	26	46	1,669	33
1964	1,815	43	38	1,674	27	45	1,760	34
1965	2,013	46	35	1,814	31	42	1,897	35
1966	2,263	48	34	1,972	33	41	2,099	38
% Incr. 1950–66	311	—	—	265	—	—	366	—
1954–66	173	—	—	146	—	—	169	—

Source: *Annual Reports*.
[1] Including Texaco's share of non-consolidated affiliates.

Socal in 1965 were in the Eastern Hemisphere. Most of the increase in Eastern Hemisphere activities was undoubtedly in Europe, but since neither Company publishes figures which show their position in the different markets of the Eastern Hemisphere no comparison can be made.

The geographical shift of emphasis that these statistics show indicate that these Companies, like the other international American Companies, were responding to the more rapid growth of petroleum consumption in general in the Eastern Hemisphere compared with that in the US, with a consequent widening of opportunities for the American Companies. Texaco is one of the

TABLE XVII

STANDARD OIL COMPANY OF CALIFORNIA

PRODUCTION, REFINING AND SALES[1]

1950–66

	Crude Oil Production		Refinery Processing		Sales				
					Products		*Crude oil sales (less purchases)*[2]		
	Total 000 *b/d*	*Eastern Hemisphere* %	*Total* 000 *b/d*	*Eastern Hemisphere* %	*Total* 000 *b/d*	*Eastern Hemisphere* %	*Western Hemisphere* 000 *b/d*	*Eastern Hemisphere* 000 *b/d*	*Chemicals*[3] $ *m.*
1950	490	37	428	29	479	24	—	23	43
1951	572	43	544	29	562	27	(26)	30	57
1952	608	44	589	30	600	28	(19)	35	55
1953	629	46	654	31	630	30	(33)	36	60
1954	663	49	657	34	643	20	(45)	44	72
1955	724	52	698	33	716	32	(30)	37	90
1956	809	49	758	33	763	33	(6)	39	103
1957	892	49	785	34	765	33	(21)	52	121
1958	857	53	783	35	785	34	(12)	42	123
1959	887	55	790	36	807	34	1	62	143
1960	988	57	836	38	879	33	22	75	156
1961	1,099	57	908	38	1,079	31	40	76	152
1962	1,189	56	922	40	1,113	32	82	88	162
1963	1,255	56	987	40	1,167	32	63	96	168
1964	1,320	59	1,123	38	1,253	32	(11)	155	184
1965	1,517	60	1,207	38	1,334	35	5	192	201
1966	1,731	62	1,291	40	1,423	34	4	221	226
% Incr.									
1950–66	253	—	202	—	197	—	—	861	426
1954–66	161		96		121			402	214

Source: *Annual Reports. Supplemental Financial and Statistical Data.*

[1] Including Socal's share of non-consolidated affiliates.

[2] Excluding royalty oil purchased.

[3] Including excise taxes.

most rapidly expanding of the American Companies, particularly abroad. In 1966 it bought one of the largest of the German oil companies, Deutsche Erdöl, obtaining important refinery facilities, a considerable marketing network, and some crude production in West Germany.

Both companies have financed their expansion largely through retained funds, paying out an average of 29 per cent of their net income (after taxes but before depreciation), as dividends in the period 1950–66. Their capital expenditures over the period came to little more than could be financed from their retained funds. Socal does not publish statistics of the distribution of its capital investment between foreign countries and the US nor according to the type of operation into which its investment goes, but between 1954 and 1961 Texaco did so. In this period Texaco placed in foreign countries 21 per cent of its investment in production, 23 per cent of its investment in manufacturing, and 36 per cent of its investment in marketing. (These figures do not include the equity of the Company in the capital expenditure of the non-consolidated affiliates owned 50 per cent or less, which came to about 19 per cent of its consolidated capital expenditure. Nor do they include expenditure on exploration). It is possible that Socal's figures, if available, would show about the same investment in production and refining abroad, but less in marketing, since its sales in the Eastern Hemisphere have grown only about half as much as those of Texaco. This is perhaps partly accounted for by Texaco's 75 per cent holding in the Regent Oil Company, which refines and markets in the UK. Both Companies are also important producers of petrochemicals. In 1965 Socal's sales of chemicals and chemical products equalled about 9 per cent of its sales revenues (net of excise taxes), and Texaco reported that its sales of petrochemicals had grown an average of 17 per cent per year over the previous five years, which presumably indicated considerable investment in this field.

Mobil Oil Company[1]

Mobil Oil is the smallest of the US international majors. Its total reported assets were $5,511 million (£1,968 million) in 1966 and its gross receipts $5,887 million (£2,102 million), but this understates the true picture, for the Company reported in 1966 that its equity in the net assets of non-consolidated affiliates was 'substantially greater than its investments (less reserves)' but gave no figure. Moreover, it received in dividends $4 million less than its $82 million equity in the net income of such companies. Gross production of crude oil amounted to 1,318,000 b/d, about one-third of that of Jersey Standard's and it had 5 per cent of production outside the US, Canada and the Communist world. Its refinery runs were 1,540,000 b/d (including runs on its account by others), while its sales of products were 1,579,000 b/d. Overall, Mobil produces less crude oil in relation to its total operations than any of the other major Companies, being particularly out of 'balance' within the US, although with respect to its foreign production it produces slightly more than it refines.

Mobil's history is a somewhat complicated history of mergers, the details of which need not detain us for long. In 1966 it changed its name from 'Socony Mobil' to 'Mobil Oil', thus dropping the last hint of its Standard Oil origins.

[1] This discussion is largely based on a short history issued by the company and on the Annual Reports.

TABLE XVIII
MOBIL OIL COMPANY
SELECTED FINANCIAL STATISTICS
1955–66*

	1	2	3	3/2	4	5	5/4	6	6/4	7	8	8/6
	Total assets[1]	*Total receipts*[2]	*Total taxes*[2]	*Ratio taxes/ receipts*	*Net income before charges*[3]	*Cash dividends*	*Ratio dividends/ net income*	*Capital expenditure*[4]	*Ratio capital expenditure/ Net income*	*Net change long-term debt*	*Amounts retained*[5]	*Ratio amounts retained/ capital expenditure*
	$m.	$m.	$m.		$m.	$m.		$m.		$m.	$m.	
1955	2,630	2,764	628	·23	377	87	·23	240	·63	(24)	279	1·16
1956	2,817	3,097	700	·23	434	101	·23	329	·76	(13)	324	·98
1957	3,102	3,343	718	·21	425	118	·28	359	·84	(29)	297	·83
1958	3,234	3,260	755	·23	350	97	·28	331	·95	(5)	244	·74
1959	3,336	3,517	815	·23	365	97	·27	263	·72	(3)	258	·98
1960	3,455	3,646	906	·25	397	97	·24	258	·65	(5)	288	1·12
1961	3,608	3,802	947	·25	426	109	·26	331	·78	(1)	306	·92
1962	4,136	4,446	1,181	·27	483	114	·24	454	·94	3	355	·78
1963	4,660	4,915	1,296	·26	549	128	·23	353	·64	184	396	1·12
1964	4,879	5,079	1,366	·27	575	141	·25	418	·73	(9)	406	·97
1965	5,214	5,517	1,511	·27	634	155	·24	534	·84	29	448	·84
1966	5,511	5,887	1,644	·28	691	167	·24	602	·87	(27)	491	·82
Total and averages	—	—	—	—	5,706	1,411	av. = ·25	4,472	av. = ·78	—	4,092	av. = ·91

Sources: *Annual Reports. Financial and Operating Statistics.* Mobil Oil Company.

* Earlier published data not comparable with those published for subsequent periods.

1 Does not reflect assets of non-consolidated affiliates.

2 Includes consumer excise and sales taxes.

3 Net income after taxes plus depreciation, depletion, amortization, interest and debt expense. Does not include depreciation of non-consolidated affiliates.

4 Including investments and advances.

5 Net income after taxes, before depreciation, depletion and amortization (Col. 4) less cash dividends.

Number of shareholders 1966: 210,300.

'Socony' was the cable address of the Standard Oil Company of New York, incorporated in 1882 as one of the companies of the original Standard Oil Trust. The development of the export business was one of its chief tasks in the division of labour organized within the Standard group of companies, and Socony was particularly successful in the marketing of kerosene in the Far East, where it built up its own marketing outlets. Among its contributions was the provision of cheap, efficient kerosene lamps for China, which were sometimes given away with the first case of kerosene bought by a customer. With the dissolution of the Standard trust in 1911 Socony was left with large marketing outlets abroad and in the north-eastern US, a few refineries, but no crude-oil production. Crude-oil supplies were acquired in the US through the acquisition of several other companies, and markets abroad were extended partly in association with other companies. Crude-oil supplies abroad were also sought in the 1920s, and Mobil's first foreign source came with its acquisition in 1925 of an interest in the IPC, in which it shares a 23·75 per cent ownership with Jersey Standard through the Near East Development Corporation. Only in 1934, however, did Mobil actually begin to receive supplies of oil from Iraq.

In the meantime, the Vacuum Oil Company, which had been formed in 1866 and had sold out to the Standard Oil group in 1879, was continuing to develop a world-wide trade in lubricants, on which it had specialized from the beginning. After the dissolution of the Standard Trust, Vacuum continued this specialization, but after the First World War, the automobile changed the character of the market for petroleum products and the Company extended its operations to include gasoline, since the marketing of lubricants was becoming more closely associated with that of gasoline. In 1931 Vacuum and Socony merged, but the new company, Socony-Vacuum, still had little crude oil abroad to supply the extensive Far Eastern markets it had inherited from the old Standard of New York and had continued to build up. As we have seen, the Far Eastern operations of Socony-Vacuum and Jersey Standard were merged in 1933 to form the Standard-Vacuum oil company (Stanvac), each parent having a 50 per cent interest, which gave the one supplies of crude in the Far East and the other markets, thus creating in Stanvac an integrated oil company in the Far East which soon was operating in 50 countries from the East Coast of Africa to New Zealand.

Socony-Vacuum continued to expand its operations in the US and in Europe and to explore for crude oil abroad. It began production in Eastern Venezuela and, in association with the Texas Company, in Colombia; in 1948 it bought a 10 per cent interest in Aramco in Saudi Arabia. In 1950 it discovered oil in Canada and in 1954 it obtained a 7 per cent interest in the Iranian consortium. The name of the Company was changed in 1955 to Socony Mobil, the trade name under which its products had been marketed for many years, and in 1966 to Mobil Oil. In 1960 the association with Jersey Standard through Stanvac (except in Indonesia) was terminated, and the assets of that company were formally divided between the two parent Companies in 1962.

Mobil Oil has been shifting the focus of its operations into international operations faster than any of the other smaller majors. It operates in Libya in association with other companies (the newest and largest field is jointly

TABLE XIX
MOBIL OIL COMPANY
PRODUCTION, REFINING AND SALES[1]
1955–66*

	Gross Crude Production				Refinery Processing[2]			Product Sales		
	Total	*United States*	*Africa*	*Middle East*	*Total*	*United States*	*Eastern Hemisphere*	*Total*	*United States*	*Eastern Hemisphere*
	000 *b/d*	%	%	%	000 *b/d*	%	%	000 *b/d*	%	%
1955	616	42		34	802	67	32	893	67	33
1956	675	40		33	817	66	33	951	65	34
1957	690	39		32	878	66	33	952	65	34
1958	723	33		37	895	62	36	986	63	36
1959	750	32		39	969	60	39	1,074	61	38
1960	797	29		42	1,020	58	39	1,129	58	40
1961	831	29		43	1,092	55	40	1,188	56	42
1962	872	29	0·7	44	1,141	54	40	1,246	54	44
1963	935	27	1	45	1,228	53	43	1,328	54	44
1964	1,037	25	4	46	1,345	50	45	1,404	52	45
1965	1,165	23	5	45	1,429	47	48	1,503	51	47
1966	1,318	22	10	45	1,540	45	50	1,579	49	48
% Incr. 1955–66	114	—	—	—	92	—	—	77	—	—

Sources: *Annual Reports. Financial and Operating Statistics.* Mobil Oil Company.

* Published data not comparable with those reported for subsequent periods.

[1] Includes Mobil's interest in non-consolidated companies.

[2] Including oil processed on Mobil's account.

owned with the German company, Gelsenberg Benzin A.G.), and it has an interest in the new discoveries in Abu Dhabi through the IPC. Altogether Mobil reported an interest in crude-oil production in some 15 countries outside North America in 1965, including off-shore Nigeria where it has made promising discoveries.

Not only is Mobil's crude production outside the US and Canada some three-quarters of its total production, but its refinery runs abroad accounted in 1966 for over half of its total refinery runs and its sales of products for nearly half of total sales. Over 40 per cent of its net income was attributed to Eastern Hemisphere operations in 1966. In the Eastern Hemisphere both exploration and investment in production and marketing were undertaken by Stanvac until its dissolution in 1962 and the amounts involved were not recorded in the consolidated financial statements of Socony Mobil. The Company also has significant investments in natural gas exploration and production in Germany, Holland and in the North Sea.

As can be seen from Table XX, the single most important item in the Company's capital investment in 1966 was crude-oil production, primarily in the US. Although refining was the second most important item, in marketing the greater emphasis on overseas areas is plain.

The importance of diversification into the chemical industry, again primarily in the US, is also brought out in the Table. As a percentage of total capital expenditures, investment in chemicals averaged 11 per cent over the period 1960–5, reaching 14 per cent in 1966. Sales receipts from chemicals more than quintupled in the four years 1962–5, reaching $347,000 in 1965. The Company has moved rapidly into the production of agricultural chemicals, chemical coatings, industrial chemicals, petrochemicals and plastics, and a variety of industrial chemicals not based on petroleum. It even explores for potash deposits in Canada.

Like the other international majors, Mobil's expansion has involved the

TABLE XX
MOBIL OIL COMPANY
CAPITAL AND EXPLORATION OUTLAYS
1966

	US and Canada $m.	*Other countries* $m.	*Total* $m.
Capital Expenditures			
Producing	164	29	193
Refining	108	38	146
Chemicals	88	4	92
Marketing	53	77	130
Marine	4	2	6
Other	13	6	19
Total	430	156	586
Exploration expenses			
(Including non-productive wells)	70	26	96
Total	500	182	682

Source: *Annual Report*, 1966, p. 22.

purchase of several smaller companies and has been largely self-financed, the ratio of retained funds to capital expenditures averaging ·91 in the period 1955–66, but it too has gone to the market for funds in recent years, and particularly to European capital markets in view of the pressure on American Companies by their Government to reduce the drain of foreign investment on the US balance of payments.

OTHER COMPANIES

The seven Companies so far discussed have been dubbed 'the seven sisters' (allegedly attributed to the late Signor Mattei of ENI) for they have personified for many a cohesive and somewhat sinister group. Their role in the industry has indeed been, and still is, a major one, and to some extent this entire study will be an appraisal of that role and of the claims made by the Companies, or on their behalf, in justification of the appropriateness of it, and of the responsibilities to the world economy which it carried with it. But the seven are not the only international Companies, and some of the others, though much smaller than any of the majors, are of considerable importance either internationally or in particular markets.

In the international industry non-major Companies are traditionally referred to as 'independents', and it will be convenient to divide these into three groups: non-American independent Companies; American independents; and State Companies. We shall select for discussion in the first group, Burmah Oil Co., Compagnie Française des Pétroles,[1] the Arabian Oil Company (Japan), and Petrofina (Belgium); in the second, Continental Oil, Marathon Oil, Standard Oil of Indiana, and Phillips;[2] and in the third we shall consider only the Italian State Company, Ente Nazionale Idrocarburi.

Burmah Oil Company[3]

Burmah Oil is the oldest British international oil Company, having been founded as a joint stock company in 1886 to conduct oil operations in India and Burma, then a province of India. Its first machine-drilled well was in 1889, in Burma, and in the next decade it built pipelines to supply crude oil to its refinery and acquired ocean-going tankers to supply products, particularly kerosene, to India. Burmah was the parent Company of BP, for it was Burmah that financed the D'Arcy explorations which discovered oil in Persia in 1908 and participated in the formation of the original APOC to exploit its Persian concessions. When the British Government acquired its 51 per cent interest in

[1] The Compagnie Française des Pétroles has often been classed as a 'major', probably because it has had a share in Iraq oil as part-owner of the IPC in association with other major Companies from the beginning. It can, perhaps, be treated as a kind of 'minor major' in view of its history, although it is not by any means in the same size class as the majors, and is smaller than many of the American independents.

[2] Every one of these American 'independents' in the international industry is classed as a 'major' Company in the US. All but Phillips had their origin in companies of the old Standard Oil group.

[3] This discussion is based mainly on pamphlets issued by the Company, a communication to stockholders issued by the Company in 1963 on the occasion of the takeover bid by BP and Shell, and the Annual Reports. Some information has also been obtained directly from the Company.

APOC in 1914 Burmah lost its controlling ownership of the company, but it still today holds 23·2 per cent of the company's shares, or nearly half of the non-government holding. Burmah reported assets in 1966 of £263 million. Of this, property and operating assets accounted for only £50 million. Its profit on trading (including relevant dividends and interests) before depreciation and write-offs amounted to £21 million, with an additional £17 million attributed to other investments. It produced 2·5 million tons of crude oil, processed 3·9 million tons and sold 5·2 million tons of products.

In spite of its age, Burmah is a small company, its operations and interests in the petroleum industry before the 1960s having been confined chiefly to Burma, India and what is now Pakistan. In the early 1960s it acquired interests in oil and gas exploration and production in USA (Gulf of Mexico) and Canada; later this geographical diversification extended to Peru, Ecuador, Australia and the UK Continental shelf with small interests in Holland. Castrol Ltd., a company marketing lubricating and industrial oils in a number of countries, was acquired in 1966 partly to enable the Company to get trading income in the UK because of the new corporation tax. In addition to its nearly one-quarter interest in BP, it holds 3½ per cent of the shares of Shell Transport and Trading; evidently the Company prefers to earn profits from investments rather than to use the funds in its own productive operations, for in 1966 its investment income was much greater than its trading profits. Moreover, the Company's property and operating assets were only 19 per cent of total assets.

Although most of Burmah's operations in India are conducted in association either with the Government (through Oil India Ltd., owned in equal shares by the Government and the Company) or with Shell, it still owns and operates oilfields and a refinery at Digboi, the original site of oil production and refining in India, through the wholly-owned Assam Oil Company, in which it bought a controlling interest in 1921. It discovered the Nahorkatiya oilfield in Assam, which is exploited in partnership with Oil India Ltd. A 720-mile pipeline has been constructed to feed the two government refineries. Refining and marketing elsewhere in India are in association with Shell.

In Pakistan, the Company markets in East Pakistan through Burmah Eastern Ltd., 51 per cent of whose shares are held by Pakistanis. It discovered the Sui gas field, one of the largest in the world, and has constructed an extensive pipeline network to transport the gas. Most of its other operations in Pakistan, which include refining, are conducted in association with other companies, or with the Government. In 1962 and 1963 the Company bought the Lobitos oil group which gave it interests in Ecuador, Peru, Canada and the UK. In the US, Burmah is in partnership with others in oil exploration and production offshore Louisiana. It is also exploring in Australia and the North Sea.

Burmah Oil is not a major Company in the international industry. In India it has lost ground relative to other Companies in marketing, but in other respects its importance has increased. Its Rangoon refinery was destroyed during the war; after the war rehabilitation of the Company's properties was retarded by political disturbances in Burma, and in 1963 the Company's association with Burma, which had lasted over three-quarters of a century, was finally terminated when the Government acquired its remaining interests in spite of the fact that BOC had formed a partnership with the Government as early as 1954.

Before the war the Indian market was, as we saw in Chapter III, more or less cosily divided among the international Companies, although sharp price wars broke out on occasion when one or another group tried to improve its relative position. Since the war the governments have been taking an increasingly active role in the industry in both India and Pakistan, and there has been considerable international competition to build refineries and supply crude oil.

Compagnie Française des Pétroles[1]

France entered the international petroleum arena in 1920 when the French request that they be given the 25 per cent interest of the Deutsche Bank in the old TPC was confirmed by the San Remo conference. The French interest thus gained was held by a company formed for the purpose, the Société Française pour l'Exploitation du Pétrole, in which Royal Dutch/Shell had 41 per cent participation and the Banque de l'Union Parisienne the rest. There was considerable opposition in France, particularly on the part of Raymond Poincaré, to the Shell participation, and in 1924 the Compagnie Française des Pétroles was established with capital raised from private banking and industrial interests, especially from the marketing and refining affiliates of the international oil Companies. CFP concluded an agreement with the French Government in 1924 which placed under its control the French international petroleum interests, including its interests in the TPC. The discovery of large quantities of oil in Iraq in 1927 had focused attention on the desirability of a national policy, and a law to encourage refining in France was passed in 1928. The refiners and marketers who held shares in the CFP were opposed to the Company taking up refining and, after an investigation, the Government decided itself to take a 25 per cent[2] share of CFP and to set up a refinery company, the Compagnie Française de Raffinage, in which the State had 10 per cent of the shares and CFP 56 per cent. CFR is part of the CFP group. CFR was at the same time authorized to refine 25 per cent of the needs of oil distributors in France.

CFP's activities were for a long time confined to France and to crude-oil operations, but with the increase of output in Iraq after 1950 its sources of oil increased rapidly and CFP's production rose from less than a million metric tons in 1945 to over 8 million in 1953. In 1945 the Ministry of Industry and Commerce in France set up a Bureau de Recherches Pétrolières (BRP) to develop exploration programmes in France, the colonies and protectorates. This organization in co-operation with the authorities in Algeria created the Société National de Recherche de Pétrole en Algérie (S. N. REPAL). CFP was disturbed at what it considered *étatist* ventures, but agreements were negotiated between the two Companies for exploration in Algeria. S. N. REPAL had access to far larger government funds (Frs. 56 milliards as compared to CFP's Frs. 1·2 milliards) than did CFP, but eventually the two Companies to-

[1] The following discussion has drawn on Jean Rondot, *La Compagnie Française des Pétroles: du franc-or au pétrole-franc; Histoire des Grandes Entreprises*—7 (Plon, 1962) and Annual Reports.

[2] Now 35 per cent.

gether obtained a large concession in Algeria, the costs and any discovered oil to be divided equally between them, and in 1956 the great Hassi Messaoud oil field was discovered, and the Hassi R'Mel gas field in 1957. In 1954 the CFP took a 6 per cent interest in the Iranian Consortium and was one of the first Companies to obtain a concession in Libya.

In 1966 the Company's production amounted to 40·3 million metric tons (806,000 b/d), from the IPC, the Iran Consortium and Algeria, with a small amount from Abu Dhabi and Canada.

As its supplies of crude oil rose, the Company proceeded to integrate forward. In addition to selling crude oil to Europe and Japan, it concluded processing agreements in France and elsewhere in Europe, and it established distribution systems to sell the products in Europe, Africa and even in Australia under its 'Total' brand. In 1966 around 26 million metric tons of crude were refined by the CFP group or on its account and sales of products reached 27 million tons. Exploration is continuing in several parts of the world, including offshore Abu Dhabi and Dubai in association with British Petroleum, and in Canada and Australia.

Arabian Oil Company[1]

In December 1957 a Japanese company signed an agreement for a concession in the offshore waters of the Neutral Zone with the Government of Saudi Arabia. The following February the Arabian Oil Company was established in Tokyo to take over the concession, and in July an agreement was signed with the Ruler of Kuwait, since Kuwait shares with Saudi Arabia an equal and undivided interest in the Neutral Zone. At the time when the agreements were signed the normal profit-sharing arrangements in the Middle East provided for an equal division of the profits attributed to crude oil between the concessionaires and the governments, but under the Japanese agreements 56[2] and 57 per cent of the profits were to go to Saudi Arabia and Kuwait respectively. This innovation caused much head-shaking and criticisms among the international Companies at the time. Moreover, the governments obtained the right to purchase at par up to 10 per cent of the shares of the Company after the discovery of oil in commercial quantities, which right they have since exercised.

The Japanese move into production in the Middle East was a pioneering venture in several ways: it was not made by an established producing company, but by a new organization set up largely by consumer groups of businessmen; its financial backing was not large by international oil Company standards (reported to be something like $10 million);[3] and it offered better terms to the governments than those of the major Companies, including a share in any downstream operations it might develop. Such bold entrepreneurship was widely deplored in the industry and gloomy predictions of the losses to come were widespread. But by January 1960 the Company had struck oil and six-

[1] Except where noted, the following discussion is based on pamphlets put out by the Company, its Annual Business Report to Shareholders and OPEC discussions of concession terms.

[2] This has since been raised to 57 per cent.

[3] See *Petroleum Press Service*, June 1961, p. 224.

teen months later exports began, in spite of the fact that at the first oil well (Khafji No. 1) there was a very serious and prolonged fire shortly after drilling started in August 1959. But by 1962 the Company began to show a small profit, and in 1963 declared its first small dividend of 2 per cent, which was increased to 12 per cent in 1964. In 1965 it produced and sold over 65 million barrels of oil.

In 1964 the total assets of the Company were $220 million and its gross revenue about $84 million. At the end of 1964 it had nearly 4,000 shareholders, including the Kuwait and Saudi Arabian Governments, as well as industrial, insurance, electricity and gas companies in Japan.

The speed with which this Japanese oil Company rose to an important position in the industry, reportedly supplying some 15 per cent of the Japanese market in 1965, was indeed remarkable. To be sure, it was helped in developing outlets for its supplies of crude by the Government, which required refineries in Japan to take some proportion of its Khafji crude. Most Japanese refineries are tied to foreign Companies for their crude supplies, either through ownership, loan contracts, or special supply arrangements, and the Company might have found considerable difficulty in disposing of its oil in the Japanese market through straightforward competition unless the Government had intervened or unless it had built its own refining and distribution networks. In the early 1960s there were signs that refinery capacity had been over-extended in Japan, and the Government was more inclined to restrict than to encourage further expansion. In 1966 the Company made a contract with the Kuwait Oil Company to supply a refinery expected to come on stream in 1967. Under its concession agreements Arabian Oil is required to construct a refinery, and in 1964 the Khafji refinery was started, designed to manufacture mainly bunker oils, and other products for field operations.

The future of Arabian Oil in the international industry will depend partly on the petroleum policies adopted by the Japanese Government to reduce the foreign exchange costs of oil imports by increasing the proportion of the Japanese market supplied by the Japanese company. The general question of the cost of crude-oil imports when they are supplied by the international Companies and of the economics of government intervention is discussed below.[1]

Petrofina[2]

Petrofina is a Belgian Company established in 1920. Until recently it was engaged almost exclusively in refining and marketing, but it now produces small amounts of crude oil in Angola, the US and Canada. The Company refines and markets widely in Western Europe and to a lesser extent in North America and Africa. Its total sales of finished products increased from 7,200,000 metric tons in 1958 to 16,400,000 in 1966. In 1966 it reported a gross income of 56,090 million Belgian francs (£400 million) and total assets of 57,804 million Belgian francs (£413 million), 46 per cent of which were in Europe and 38 per cent in America. Nearly 60 per cent of its capital expenditure in that year was in Europe, a third in America, and 3 per cent in Africa.

[1] See especially Chapters VI and VIII.

[2] Based on *Annual Reports*.

The American 'Independents'

There are well over two dozen US Companies operating or exploring outside their own country. Here we shall discuss a few of the most important from the point of view of their operations in the Eastern Hemisphere. All of them are producers of crude oil and most of them also refine and market products. Outside the US few operate independently to any significant extent at any stage of the industry; that is to say, their exploration, development, and production of crude oil is more often than not carried on jointly with other Companies or they operate in a kind of consortium such as Iricon or Aminoil;[1] their crude is likely to be processed in jointly-owned refineries or under processing agreements, and even marketing is sometimes conducted in partnership with other Companies. In this way, the smaller Companies can pool resources, especially financial resources, and thus increase their ability to undertake the risks and make the investments involved in expanding their operations abroad. For all of them, however, international operations are very much less important than domestic operations. Here I shall consider four of the more important integrated Companies with international production, giving passing reference to some of the even smaller Companies with which they are associated in some stages of operations.

Continental Oil Company[2]

Conoco is among the most important of the smaller American Companies operating abroad. Its average daily net production of crude oil and condensate in 1966 was 398,000 barrels, of which 204,000 were produced outside the US and Canada, primarily in Libya. It reported a gross revenue of $1,915 million and total assets of $2,070 million.

Continental was originally a domestic marketing company of the old Standard Oil group. After a series of mergers and consolidations it emerged in its modern integrated form in the late 1920s. Although it had concessions in Canada and Mexico in the 1930s, it was not until after the war that it obtained foreign production by acquiring a part interest in the Hudson Bay Oil and Gas Company in Canada. Between 1949 and 1958 the Company spent $156 million on exploration and development abroad (including Canada) in accordance with the management's policy that '. . . the company should have diversified sources of foreign crude as well as adequate domestic reserves'.[3] But until 1959, when it purchased control of San Jacinto Petroleum Corporation with production in Venezuela, and a ·417 per cent interest in the Iranian Consortium (through Iricon), it had no foreign production outside the US and Canada. In 1955 it took a concession in Libya through the Oasis Oil Company owned by Continental, Marathon, and Amerada in equal shares. The first exploratory well was sunk in 1958 and commercial production began in 1962. In the six years 1960 to 1965 Continental spent $109 million in exploration and development in Libya, which has paid off handsomely. In 1966 oil was

[1] Iricon, a group of nine companies, has a 5 per cent share in the Iranian Consortium; Aminoil is a consortium of nine US companies, of which the most important are Phillips, Signal and Ashland.

[2] The Annual Reports of the company provided the primary source for the following discussion, but other general works were also drawn on.

[3] *Annual Report* 1958, p. 8.

discovered in Dubai on a concession in which Conoco has a 35 per cent interest. By 1966 the Company had exploration interests in thirteen countries, as compared with six in 1958, but had abandoned exploration in Egypt after spending $7·2 million on exploration and incurring a net loss of $3·3 million,[1] and in British Somaliland (where it had sunk $2·5 million), and in Guatemala.

Continental did not have refining and marketing facilities for its Eastern Hemisphere crude and became known for its aggressive attempts to break into European markets hitherto served largely by the majors. In 1959 it sold an average of only 207,000 b/d of refined products outside the US but seven years later it was selling over 71,000 b/d. To obtain markets, Continental purchased Jet Petroleum Company in 1961, an independent distributor in the UK, with which it proceeded to expand in the UK markets; it acquired a marketing chain of outlets in Germany and Austria and another in Belgium and made arrangements to supply crude to independent refineries and petrochemical companies in Italy and Germany. Processing agreements were made with refineries in these countries and a 20 per cent interest in a new 40,000 b/d refinery at Karlsruhe was obtained (later reduced to 10 per cent).

Thus, within a short space of six years a relative newcomer to the international field successfully established itself in the Eastern Hemisphere as a producer of crude oil and as a marketer, and had begun to acquire refining capacity, including the building of a refinery in the UK. To be sure, in 1966 its total production outside North America and the Communist countries was about 21 per cent of the smallest of the American majors and 6 per cent of the largest, but its output was growing rapidly. In addition to Libya, Continental's exploration in the Eastern Hemisphere continued in over a dozen countries as far apart as New Guinea and Australia in the Far East and the UK and Germany in Europe, and included Pakistan in Asia, Dubai and Qatar on the Persian Gulf, and Tunisia, Mauritania, and Somali in Africa.

Another area of the petroleum industry in which Continental is making an impression is in the preparation and supply of liquefied methane. The Company began research into the problems of supplying methane to the British market as early as 1958 and soon was supplying that market from Algeria through a 40 per cent affiliate, Conch International Methane (of which Shell also owns 40 per cent). The gas is purified and liquefied by Compagnie Algérienne de Méthane Liquide (CAMEL) (50 per cent Conch owned) at Arzew in Algeria for the British market and for Gaz de France and is transported to Britain by a transport company owned 50 per cent by Conch and 50 per cent by the Gas Council of Great Britain. The Company has also acquired chemical interests in Europe, Japan and Argentina. Plant foods, chemicals, and plastics accounted for 16 per cent of sales revenue in 1966. In spite of the fact that only about 10 per cent of the Company's net income was attributed to operations in foreign countries in 1966, around 24 per cent of its capital expenditures went to foreign countries.

[1] This provides a good example of the way in which joint operations plus the effect of tax arrangements reduces the cost to a company of crude-oil exploration. The total amount spent by Conoco and its associates in the search for oil in Egypt was $26·5 million, of which $7·2 million was Conoco's share. The company's net loss, however, was only $3·3 million because of the tax concessions obtained. See *Annual Report*, 1958.

Marathon Oil Company[1]

Marathon is much older than, but is about half the size of, Continental, with total assets of $871 million in 1966 and gross revenues of $626 million. It produced 126,000 b/d of crude oil and natural gas liquids (net), in the US and Canada, and its tanker liftings from Libya came to 216,000 b/d.

Marathon was originally incorporated in 1887 as the Ohio Oil Company, part of the Standard Oil group. It changed its name in 1962 in recognition of a 'fundamental change' in the character of the Company arising out of the international status it acquired with the commercial production of Libyan crude oil.

During 1966 its sales of Libyan crude averaged over 100,000 b/d to affiliated European refineries. Like Continental, with whom the Company is associated through Oasis Oil in Libya, it had acquired Libyan oil without the 'downstream' facilities to handle it and had to arrange refining in Europe. It quickly began to obtain refining capacity and expand its marketing networks, largely through acquisition of existing marketing chains. It is associated with Spanish interests in a 40,000 b/d refinery, which went on-stream in 1964 and produces for the Spanish market, one of the most rapidly growing in Europe; and it has joined with a German firm to build a refinery in Germany. Marathon began marketing products in Europe only in 1964 and by 1966 it was selling 21,000 b/d of refined products. Although actively exploring abroad, its expenditures are much smaller than those of Conoco since it is a much smaller company, but the ratio of exploration to capital investment is somewhat higher; moreover, over 40 per cent of its exploration expenses are outside the US and Canada.

Associated with Conoco and Marathon in their Libyan activities is Amerada Petroleum Company, a company with total assets in 1965 of $347 million and a gross income of $213 million. It is the largest of the American companies engaged wholly in crude-oil production. It has disposed of much of its Libyan crude oil to Royal Dutch/Shell, giving Shell a virtual ownership interest. The output of Oasis exceeded 650,000 b/d in 1966, or about 5 per cent of the combined production of the majors in that year outside the US, Canada and the Communist countries.

Standard Oil Company (Indiana)[2]

Indiana Standard is one of the largest of the US oil companies, ranking sixth in 1966 with total assets of $3,849 million and gross revenue of $3,351 million. Its net crude production was 491,000 b/d, and refinery input was 829,000 b/d.

Indiana Standard was founded in 1889 as one of the Standard Oil group of companies engaged in refining and marketing in the Mid-west. After the dissolution of 1911 it expanded as an independent Company, and through the acquisition of Pan American in 1925, obtained substantial crude-oil properties abroad, especially in Venezuela and Mexico. In the midst of the great depression, however, Indiana sold the foreign interests of Pan American to Jersey Standard concentrating thereafter on domestic operations. Only after the

[1] The discussion is based on Marathon's Annual Reports.

[2] Source: *Annual Reports* and P. H. Giddins, *Standard Oil Company (Indiana): Oil Pioneer of the Middle West*.

Second World War did it begin to look for crude-oil reserves outside North America, but in a very small way up to 1957, for in that year it was exploring only in Venezuela, Cuba and Jamaica. In 1958 the Company announced an acceleration of its foreign activities; it made an agreement with the State oil company of Argentine (YPF) to produce oil, obtained an offshore concession with the National Iranian Oil Company in the Persian Gulf, acquired oil rights in Libya and Algeria, made an agreement to explore in association with another Company in Mozambique, and acquired concession interests in Italy. By 1966 the Company was producing in Canada, Iran, Egypt, Libya, Colombia and Venezuela, had discovered gas in the Netherlands and the North Sea, and was exploring in Mozambique, Norway, Germany, Trinidad, and Australia. It was refining in Italy and Australia, and was participating in refineries in India and Pakistan, with rights to supply crude oil. In addition it had interests in chemical or fertilizer plants in Europe, Iran and Japan.

With few exceptions Indiana operates abroad through wholly-owned subsidiaries ('Amoco' in Europe and 'Pan American' outside Europe). The company gives little financial information about its international activities, although it did report that 8 per cent of its net investment in properties was in foreign countries (excluding Canada) in 1965, and 14 per cent of its new capital expenditure. Its net foreign production (excluding Canada) in 1966 was 48,000 b/d.

Phillips Petroleum Co.[1]

Phillips Petroleum, although a much smaller Company than Indiana Standard (total assets in 1966, $2,673 million; gross revenue, $1,708 million) has more foreign production and wider foreign interests. Of the 367,000 b/d of crude oil (net) produced by the Company in 1966, 45 per cent came from outside the US, although about 76 per cent of its refining runs and 90 per cent of product sales were in the US.

Phillips Petroleum was established in 1917 mainly as a marketer in the US. It soon began to integrate backwards, acquiring its own sources of crude production and refining facilities, but has always been short of crude oil. It is, however, one of the leading producers of natural gas in the US. It acquired a 45 per cent interest in Pacific Petroleum, the fifth largest oil company of Canada in 1962.

After the war, Phillips joined with seven other companies to form the American Independent Oil Co. (Aminoil), in which it has a 33·54 per cent interest, and in 1948 this Company obtained a concession in the Neutral Zone, acquiring the Kuwait half interest in the concession (Pacific Western held the Saudi Arabian half). Production began in 1954. Aminoil also has, through Iricon, a share in the Iranian Consortium. Altogether Phillips was obtaining 37,200 b/d (net) in the Middle East by 1966. In addition discoveries were made in Libya, Algeria and Tunisia, North African output reaching 23,100 b/d (net) in 1965. To find outlets for growing crude supplies. the Company extended its refining abroad, joining in 1965 with Imperial Chemicals in a refining project in Great Britain to process North African crude and taking part in an Indian refinery to use Middle East crude.

[1] Source: *Annual Reports.*

Ente Nazionale Idrocarburi[1]

Of all the Companies entering the international petroleum arena since the end of the Second World War, ENI has been the most colourful and has aroused the greatest controversy. Born in the heat of the post-war discussion in Italy over the structure of the national economy, it grew up under the lively tutelage of an aggressive and independent-minded entrepreneur, Enrico Mattei. Between 1954 and 1965 the value of the group's sales quintupled, approaching 850 billion lire (£185 million) in 1965. Total assets were over 2,000 billion lire (£1,180 million), and gross revenues nearly 890 billion lire (£500 million). Crude-oil production reached only 8 million tons, 6·1 million of which were in Iran and Egypt.

Although Italy had had a petroleum enterprise whose stock was owned by public agencies as early as the middle 1920s—Azienda Generale Italiana Petroli (AGIP)—which had small interests abroad and produced a little oil and gas in Italy, nothing very significant arose until the discovery of gas in the Po valley in the early 1940s. Jersey and Shell both wanted concessions in this area with secure exploitation rights, and the political controversy in Italy over the relative roles to be allotted in the economy to public and private (including foreign) enterprise became acute. In the petroleum sector the solution was to establish ENI as a public holding Company responsible for a number of operating companies, each of whom was supposed to be autonomous and independent for operating purposes, legally looked on as private companies and competitive with private industry. The Government was responsible to Parliament for ENI, ministerial representatives sat on its board, with final ministerial responsibility resting with the Council of Ministers. But no independent control by the State over the operating companies was provided for.

The objectives of ENI as laid down in the 1953 law setting it up were to promote and undertake new activities in hydrocarbon and natural steam operations, to explore for and to exploit hydrocarbon deposits in the Po valley, including the construction of pipelines (these activities were reserved to ENI exclusively), and generally to engage in production, manufacture and commerce in hydrocarbons. The Government assigned to ENI its companies and interests in the hydrocarbon industry (including AGIP, and also SNAM—a gas transport company) and in a variety of other activities unrelated to petroleum.

ENI had little success in its efforts to discover oil in Italy and went abroad for supplies. It wanted to become independent of the major Companies, which meant that it had either to buy oil from independent sources or to discover its own supplies. It did both, moving in 1955 to obtain interests in crude production in Egypt and in 1957 and 1958 to obtain concessions in Iran and North Africa. In 1958 it began importing from the only large independent supplier available—the USSR.

The concession agreement made with the National Iranian Oil Company by ENI's subsidiary, AGIP, was the first to be concluded by any Company in accordance with the Iranian Petroleum Law of July 1957 and, according to

[1] This discussion draws on Charles R. Deschert, *Ente Nazionale Idrocarburi: Profile of a State Corporation* (Leiden: Brill, 1963), articles in various issues of the *Petroleum Press Service* and *Annual Reports* of the Company. See also P. H. Frankel, *Mattei; Oil and Power Politics*, and Dow Votaw, *The Six-Legged Dog*.

Votaw, was 'vigorously' fought by the Consortium.[1] AGIP agreed to bear the entire exploration risk, but when oil was discovered NIOC would pay its share of the costs incurred and a joint company would be set up in which the two companies would have equal interest. The Joint Company (SIRIP) would pay 50 per cent of its profits to the Government as tax, but NIOC as joint-owner would receive half of the remainder. Thus Iran effectively obtained 75 per cent of the profits attributed to crude-oil production but put up 50 per cent of the capital. The first offshore Iranian discovery was made by this group (in 1960).

ENI continued to offer unusually favourable terms to obtain concessions in other countries, including North Africa and Nigeria, and at the same time rapidly extended its refining and distribution in Europe, Africa, and even Latin America. Refineries were built with the joint participation of local governments, and extensive long-term credits for equipment, plant, and services were granted.

The Company was thus a pioneer in the international industry in actively co-operating with the developing countries. This was partly related to the general philosophy of Enrico Mattei, but was also a competitive necessity if the entrenched position of the majors was to be circumvented directly and immediately. Moreover, Mattei did not like the international majors, whom he accused of monopolistic and discriminatory behaviour, and he attacked them for what he considered to be their arrogant economic imperialism;[2] he was determined effectively to assert the independence of ENI and if possible to undermine the position of the majors.

By 1962 there were rumours of a *détente* between ENI and Jersey, and on the death of Mattei in 1962, ENI's attitude rapidly changed. In 1963 a long-term contract was made with Jersey Standard for crude supplies at a price reportedly competitive with the Russian prices, and with Gulf, although at the same time ENI continued to import from the USSR.[3] In 1965 the Company signed another long-term contract with Jersey for natural gas from Libya, and in the same year Jersey bought out ENI's marketing network in the UK, which had not developed as the Company had anticipated.[4] Arrangements were also

[1] Votaw, p. 77.

[2] According to Deschert and Votaw the majors saw to it that ENI's application to join the Iranian Consortium was rejected, and they also refused to co-operate in a European pipeline venture with it. (Deschert, Chapter IV, and Votaw, p. 19). Deschert concluded that in general 'The major companies in the recent past have demonstrated an extraordinary lack of insight into the motives of many in the new politico-economic elites who make decisions on the basis of values and ideology different from those of business.' P. 90. See also Frankel, *Mattei: Oil and Power Politics*, p. 95.

[3] ENI's purchases of Russian crude were roundly criticized in the industry and by the trade press, both because of the source and because of the cut price. The *Petroleum Press Service*, for example, frankly hoped that the 'logic' of ENI's position would encourage the Company to turn increasingly to the majors and away from the Russians. (The 'logic' being: 'ENI requires crude—the majors have abundant reserves'.) Moreover, the journal hinted that the Italian Government was subject to political pressures ('. . . the Italian Government cannot be entirely impervious to the critical attitude of their partners in NATO and EEC'.) The article concluded: 'It may be hoped—though there can as yet be no certainty—that the Italians will in due course accept the whole logic of this situation and accordingly re-arrange their policy of trading.' May 1963, pp. 178–81.

[4] It has been reported that among AGIP's difficulties in the UK was that of establishing a

made for ENI to join the other companies in European pipeline consortia. By December 1965, it was possible for a trade journal, which seems (to an outsider at least) to be largely a spokesman for the major Companies and which had long criticized ENI's relationship with Russia, its attitudes towards the majors, and its attempts to operate independently, to report of ENI that 'the erstwhile rebel is responding increasingly to the realities of the international oil business'.[1]

ENI's expansion has been more rapid than could be financed from retained earnings and equity capital, and it has had to rely heavily on bond issues, which were taken up by the State. Its large bonded indebtedness, combined with indifferent profits, has been widely criticized. In 1964 the Company announced extensive expansion plans for the period 1965–9 (to cost over £500 million) and for which further public finance was expected to be required. Since much of the expenditure was destined to go into further crude-oil exploration, and since the directors of the Company saw little chance of raising sufficient funds elsewhere or through retained earnings, they called on the Government urgently to consider the best way of publicly financing exploration, for the finance of exploration is traditionally considered to be an undesirable use of public funds in view of the nature of the risk involved.[2] ENI's investment plans include extensive exploration in Italy itself, in Africa, and in the North Sea, the expansion of pipeline networks (especially in Europe), and of refining capacity and distribution networks (especially in Africa). It also hopes to obtain further concessions in the Middle East.

A COMPARATIVE RECAPITULATION

Certain of the characteristics of the history, present structure, and policies of the international oil Companies are of central importance in any analysis of the economics of the international industry and of the problems and attitudes of both exporting and importing countries of the Afro-Asian world. The most important of these are the long and almost complete dominance of the industry by a few Companies and the recent rise of a fringe of independents; the vertical integration of the Companies and the differences among them with respect to the 'balance' of their operations at different stages; their financial and investment policies; their rivalry and their co-operation; the interest of governments in their activities; and their attitudes towards the problems of developing countries.

However measured, there is an enormous disparity in the relative size of the Companies engaged in the international industry; the two largest of the seven majors together possess total assets slightly greater than the next five largest put together, and also produce almost two-thirds as much crude oil as the other five combined. However, BP, though much smaller than Jersey and

sufficiently flexible system of supply because it was not permitted to use any of its Russian oil in the UK market. See *The Financial Times* (London), October 19, 1965, p. 15.

[1] *Petroleum Press Service*, December 1965, p. 458.

[2] See the discussion of this point in Chapter II, pp. 37–40 above, and in Chapter VIII, pp. 243 ff below.

Shell, produces more crude oil in the Eastern Hemisphere than either of them, and because of its weight in the Middle East must be reckoned in their class in the world outside North America. These three were historically the great rivals and the dominating forces in the shaping of the structure of the international industry. They are also the dominant refiners and marketers; outside the US the other majors trail far behind in every respect and the later arrivals on the international scene are even further behind.

All of the important Companies in the international field are vertically integrated, but few of them are completely 'in balance'. Texaco makes a special point in its Annual Report of the 'excellent balance' maintained between its various operations, while Gulf and Mobil are at the extremes, the first having a greater surplus of owned crude oil relative to its refining than any other major Company and the other having the greatest shortage of crude oil. Shell also refines significantly more than it produces, BP and Socal produce much more crude than they refine, while Jersey is roughly in balance. The smaller Companies have come into the international field looking for crude oil, some for their US markets. These have had their plans upset by the US import quotas, and most of the successful ones are now 'long' on crude, at least outside the US, and have offered it for sale as crude oil or have sought to integrate forward by constructing or acquiring refining capacity and marketing outlets. In Japan, all refineries are required to take some of the output of the Arabian Oil Company as a matter of government policy, and refiners in France are required to take Algerian oil.

The single most important observation to be made about the financial policies of the major Companies relates to the extent of their self-financing until recently. In the periods for which comparable data are available the retained funds of every one of the major Companies were just about equal to their capital expenditures.[1] In the early 1950s, however, the gap between the financial requirements of some of the Companies and their internal resources widened, and for all, the financing problem was aggravated by restraints placed on foreign investment by the British and American Governments. This forced a number of Companies to borrow abroad, particularly in Europe. One major Company calculated that during the years 1962–6 inclusive, the proportion of the growth of net assets represented by new funds from outside sources was as high as 30 per cent, a considerably smaller degree of self-financing than had been the case earlier.[2]

The high degree of self-financing is in itself an indication of a comfortable

[1] An official of British Petroleum reported that the whole of the capital investment of that Company between the years 1930 and 1952 'was financed by retained profits and without a recourse to outside borrowing of any importance. And broadly speaking the financial history of the other large integrated companies during the period seems to have been very similar'. B. A. C. Sweet-Escott, 'Financing Problems of Integrated Oil Companies', p. 4. For later years see Table X, pp. 110–11 above. For all companies, however, the amount of borrowing implicit in leasing arrangements, especially for tankers, is not shown. In a private letter to the author Mr Thomas Stauffer has indicated some of the tentative findings of his own work, so far unpublished. He finds that 'In general, the effect of leased facilities and chartered tankers results in at least a doubling of the figure for economic versus purely juridical debt. Further, one must include various contingent liabilities, as well as borrowings by pipeline interests jointly held or held by unconsolidated subsidiaries.'

[2] Reported to the author by Shell.

profitability in an industry where very large investments have been required to meet rapidly growing demand; moreover, the Companies even had funds to invest substantially in petrochemicals. It is not useful to attempt financial comparisons for the non-major Companies: CFP does not publish enough information in usable form; ENI is a State corporation engaged in a large number of activities and the financing of its petroleum operations cannot be disentangled from other things; the Arabian Oil Company is too new for the exercise to be meaningful; operations of the other American international independents are still overwhelmingly concentrated on the US and are not yet of great significance in the developing countries.

In spite of the widespread fears expressed by a number of major Companies that the fall in prices after 1958, plus increased taxation, was seriously jeopardising their ability to finance the investment required to meet the rising demand for petroleum products,[1] the rate of return of most of them was higher in 1966 than it was in 1960.[2] All of the majors have been diversifying their activities and expanding rapidly in fields *outside* the traditional production, refining, and distribution of oil and products. The investment policy of Jersey Standard has for some time emphasized 'the finding and undertaking of profitable new opportunities, including several in fields not directly related to petroleum',[3] and most of the Companies seem to have adopted a similar policy in varying degrees. Like that of other firms, the diversification of the international oil Companies has been in directions which, by and large, have had some connection with their primary lines of business; of especial importance have been investments in plants that provide the Companies with opportunities for 'upgrading' the final value of their crude oil, notably in the petrochemical industry. Basic chemicals for industrial use such as plastics, resins, protective coatings, detergent raw materials, synthetic rubbers and fibres, together with fertilizers, are now produced by all Companies in significant quantities. Shell took an early lead in the development of chemicals. In 1965 Jersey announced plans to make a number of finished chemical products instead of selling to other manufacturers.

Although the primary diversification of the Companies has been into the production of petrochemicals and fertilizers, other possibilities have not been entirely neglected. Jersey has interests in the building industry, in motels, an electrical plant in Hong Kong, and in various industrial complexes both in the US and abroad; Shell is involved in the building industry, and Socony in marine and heavy-duty maintenance paints: Socal is promoting the manufacture of asphalt stabilized soil brick and Gulf is a coal producer. Altogether it would seem likely that investment expenditures (including related research

[1] 'The balance, however, is a delicate one. If developments in the coming years prevent oil companies from obtaining sufficient funds on a commercially sound basis to provide the finance that will be needed, expansion on the scale required will not take place . . . the growth of energy consumption would thus be restricted and this would cause a slowing down of the growth of the world economy as a whole.' *Shell Review*, 1961.

[2] BP is the only Company that seems to have suffered a significant fall in profitability, while Shell showed a slight fall. The increased profitability of the American Companies is partly accounted for by the greater profitability of operations within the US consequent upon the protection afforded the US market by government policy.

[3] *Annual Report*, 1963, p. 6.

expenditures) associated with the diversification of the oil Companies may be absorbing sums equal to between 8 and 13 per cent of their new capital expenditures.

An unknown proportion of the capital expenditures of the petroleum firms has gone, not into construction of new capital assets, but into the acquisition of assets from other companies. Large and financially strong firms are in a favourable position to buy up assets of weaker competitors or to establish themselves in new fields by acquiring existing going concerns. When a firm decides to diversify into a new field it very commonly does so by acquiring a company in that field, and often does the same when it moves into a new geographical area. The Annual Reports of the oil Companies do not record all such moves and rarely give the cost of the acquisitions.

There is no apparent relationship between the size of a Company and its profitability. There are variations in the reported 'return on capital' and the ratio of net income to total assets among the Companies but these seem to bear no relation either to size, dominance in the industry, or degree of integration. The reported figures of assets and earnings for different Companies cannot be closely compared because of the differences in accounting and in the impact of taxes on Companies operating under different tax systems; comparisons with Companies in different industries are even more misleading: some oil Companies include as costs an item for depletion and all expense certain exploration expenditures, which may run as high as 30 per cent of capital expenditure. In addition, it is difficult to take adequate account of non-consolidated affiliates.

A very rapid rate of expansion of a firm will itself tend to keep down reported rates of return on assets since an investment may appear as a capital asset in the accounts considerably before profits are earned on it, and the more rapid the rate of expenditure on investment, the lower will be the reported return on assets at any given time for any given level of costs and prices. Nevertheless, there is no evidence that the international oil Companies have made spectacularly large profits since the war, although cash flows have remained comfortable and profits good.[1] The rate of return on average net assets after taxes and depreciation reported in the accounts of the major Companies between 1960 and 1966 have ranged very widely, from a low of slightly over 7 per cent for one Company in 1960 to a high of over 16 per cent for another in 1966. All the American Companies improved their position over the period.

Although the amount of co-operation, particularly in crude-oil production and marketing, among both the major Companies and the newcomers is very great, it should not blind us to the strong rivalries among them. There was very little price competition until recently in this oligopolistic and rapidly growing industry, and investment in expanding output, plus product competition and

[1] Thomas Stauffer, in a doctoral dissertation being prepared at Harvard University, is for the first time attempting to present a complete picture of the cash flows for the several major Companies on a consolidated basis. His results indicate that the rates of return reported by the Companies in their annual reports give a very inadequate picture of the financial results of their operations from an economic point of view, largely because of the treatment of their non-consolidated affiliates. Moreover, he finds that their comparative ranking so far as profitability is concerned is radically altered when account is taken of all factors.

sales promotion, were the chief manifestations of rivalry. The price structure of the industry put a premium on the possession by each firm of its own crude oil. Those Companies with much more crude oil than they could use sold much of it on long-term contracts, largely to other majors, for the free independent market for crude oil was very limited given the dominant position throughout the industry of vertical integration. Each Company had a strong incentive first to ensure its own supplies of crude oil, and secondly to find outlets under its own control for its crude. Much, though by no means all, of the concession-hunting and exploration was explicable by the desire of firms to secure control of crude-oil reserves before someone else did so, either because they wanted more (and hopefully lower cost) oil themselves or for fear that new discoveries would provide competitors with even cheaper supplies.

The early coalitions of groups formed to exploit oil reserves in Iraq and Kuwait were the result partly of these rivalries, in which various national governments, particularly the British and American, had a strong interest. By contrast, the coalition formed in Saudi Arabia, and the long-term crude-oil contracts there and in other areas, were designed partly to redress the imbalance caused by the extensive oil reserves discovered by Companies without adequate marketing outlets. These arrangements only mitigated the imbalance, and some Companies still had too little, and others too much, oil for the needs of their own integrated operations. There is extensive rivalry for markets among the majors, and after 1958 there developed considerable price competition for large supply contracts.

The history of the industry and the policies of the firms composing it have both been influenced to a marked degree by the policies and attitudes of governments and the national interests of the importing and exporting countries. The Companies competing for crude-oil concessions in South-east Asia and in the Middle East were sometimes actively supported by their governments in a kind of imperialistic rivalry; the policies of governments towards domestic oil or coal industries have influenced both the price and the import policies of the international Companies; the governments of the crude-oil producing countries, in their effort to maximize the benefit to their economies from petroleum operations, have influenced the international distribution of income as well as the investment policies of the Companies within those countries and elsewhere; the governments of importing countries have influenced not only pricing, but the distribution of refining facilities and exploration activity as well; and tax policies of all governments have been of great importance. In other words, in no way can the activities of the private firms be considered in isolation from the policies of their own governments and those of other countries with an interest in their activities.

This raises the question of the significance of the 'international' character of the Companies. A Company operating in more than one country is by definition 'international', and the geographical distribution of a Company's activities is relevant for the effects of its operations on the international distribution of income, including effects on the balance of payments of different countries. The international character of its interest may also determine its policy in a particular country, for it must always balance the effect of its decisions in one country, for example pricing decisions, against the expected repercussions in other countries; in this respect an international firm's behaviour in any country may be very different from the behaviour that might be expected from a domestic firm. All of these questions will be discussed in the following chapters, beginning, in the next chapter, with an analysis of the regulation of supply.

Chapter V

CONTROL OF SUPPLY BY THE COMPANIES

In spite of the substantial changes in the conditions of supply consequent upon the rapid development of large low-cost reserves after the end of the Second World War, the rate of production of crude oil was, until the later 1950s, surprisingly closely adjusted to the rate at which it could be processed and sold as products without 'disorderly' effects on prices. As we shall see in the next chapter, this does not mean that the increasing availability of low-cost oil or the various short-term disturbances had no effect on world prices, but the effects were controlled, and few changes were forced on the firms in the industry by competition until the end of the 1950s. Partly for this reason the industry has frequently been described as a cartel, but its development in the period was in fact shaped by a very unstable combination of individualist competitive enterprise and co-ordinated planning.

RIVALRY AND CO-OPERATION

It is clear from the present position and the historical record that the major international Companies are, and have been, genuine competitors (or rivals) in very many fields, but the nature of the competition and the groups involved in it have been different at different times and in different areas. It is also clear that market-sharing and price agreements were often made and that there were many outright fusions of interests through joint enterprises in marketing, refining and production. The problem of deciding whether the operation of the industry in any period is best described in terms of an 'international petroleum cartel' or of the rivalries among the firms engaged in it, is precisely the problem of deciding how much weight to give to the competitive as contrasted to the co-operative elements, and it must be admitted that the latter often seemed to be predominant. For example, six major Companies had an interest in the Indian market for most of the post-war period, but from a competitive point of view these coalesced into three: Socal and Texaco operated as one through Caltex; Jersey and Mobil through Stanvac; and BP sold through Burmah Shell. Moreover, all three pursued exactly the same price policy, and one might be hard put to demonstrate conclusively the existence of any competition at all.

Similarly, the production of crude oil in the Middle East was in the hands of companies whose parents were some combination of the international majors and who therefore had to make their production plans jointly.[1] Since

[1] See Chart II.

most of the production of Venezuela and Indonesia was also controlled by some of the same Companies, it would seem that the circumstances were ideal for the creation of a true cartel through which the several Companies could plan as one enormous monopolist, setting overall targets for world output and for each Company's share therein, deciding jointly the prices that would prevail, and establishing machinery for dealing with the fringe of outside independents and for disciplining any cartel member who failed to adhere to the decisions taken. To have been effective, such a cartel would have had to extend to the markets for products, since without some agreement among the Companies on their respective shares in product markets, a crude-oil agreement would always have been under strain.

Chart II.

OWNERSHIP LINKS BETWEEN THE MAJOR INTERNATIONAL OIL COMPANIES (including Compagnie Française des Pétroles) AND THE MAJOR CRUDE-OIL PRODUCING COMPANIES IN THE MIDDLE EAST 1966

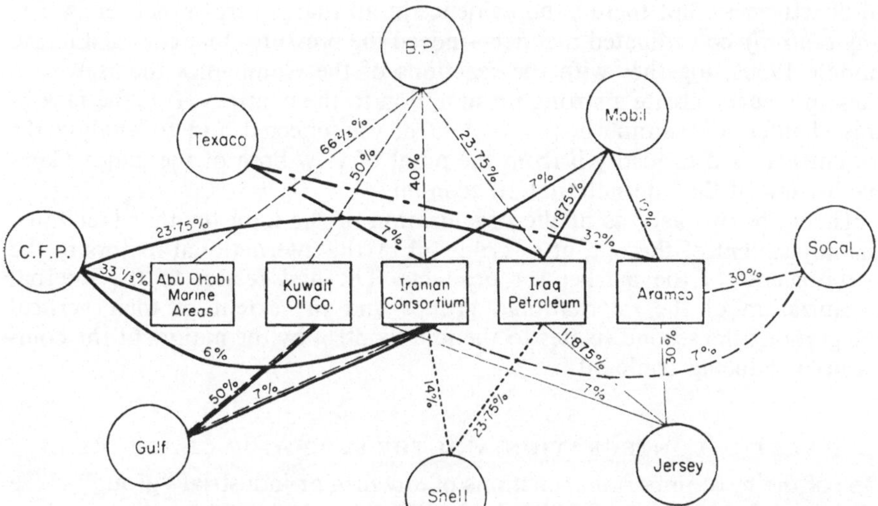

In the nature of the case, any such cartel arrangements would be kept as secret as possible, and much of the evidence for their existence would have to rest on inference from the observed behaviour of the Companies. The Federal Trade Commission in its Report, *The International Petroleum Cartel*, made a comprehensive attempt to collect the available evidence, and, as the title of their report implies, concluded that the designation 'cartel' was appropriate. The Report did show that some of the Companies had attempted to establish marketing cartels in numerous markets before the Second World War, that all were very much aware of the desirability of limiting rivalry among themselves and of respecting each other's markets, and that a world-wide pricing system was effectively in operation for varying periods.

But the Report was unable to show that there was a sufficient community of interest among the major Companies to permit the effective creation of a comprehensive world-wide cartel for the control of crude-oil production or the marketing of products. The situation was much too complicated for such a simple solution. Nevertheless, the fact remains that the pace at which the great new reserves of oil were developed was so well controlled that output could expand rapidly as demand expanded without disrupting the older centres of production and with a surprisingly gentle effect on prices given the extent of the changes in supply. Moreover, it is not at all clear that price movements played any role at all in the supply adjustments consequent on the virtual cessation of production in Iran during the Abadan crisis from 1952–55 and in the subsequent re-absorption of Iranian output after the settlement.

It seems incontestable that the rate of supply was subject to some sort of controlled planning. Each international Company was able to adjust its output of crude oil to its planned sales of crude and products with a high degree of effectiveness, but there is no evidence at all that control was exercised in any *centrally* co-ordinated manner—indeed the pressures that emerged in the middle 1950s, together with the reactions of the Companies themselves to these pressures, create a strong presumption to the contrary. It is the task of this chapter to examine how this control was effected and to analyse the arguments used to justify it from the point of view both of the major Companies and of the international oil economy.

There are two aspects of the organization of the industry that facilitated the adjustment of the output of crude oil by the international majors to the requirements of the market for products. The first relates to the internal organization of the international groups and the extent of their vertical integration; the second relates to the joint control by the majors of the companies producing crude oil.

VERTICAL INTEGRATION AND THE SUPPLY OF CRUDE OIL

One of the most important functions of any firm or industrial "group" is the planning of production and distribution. This involves the co-ordination of a multitude of activities within an administrative framework and with reference to specified objectives. Those in charge of a firm's several activities may possess a very large measure of autonomy, but the existence of some co-ordination and central supervision of a firm's productive activities is the chief feature distinguishing an industrial firm from a financial holding company. The international petroleum Companies are industrial firms in this sense, for their administrative arrangements ensure a high degree of co-ordination with respect both to the production and distribution of oil and products throughout the world and to financial affairs. Each of the international Companies can, therefore, be looked on as a 'planning unit'. The internal organization differs in the different Companies, but in all of them planning encompasses all the stages of the industry from exploration to the delivery of the end products.

Demand for output at intermediate stages of the industry is derived from the anticipated demand for final products, which range from bunker fuel and residual oil to the highly refined products of the most advanced refineries. In a competitive market and in the absence of vertical integration, the prices of products ex-refinery, and the prices of crude oil would be determined by the bids and offers of independent crude-oil sellers and independent refiners, each of whom would be attempting to make profits from his individual operations without reference to the effect of his decisions on the whole industry. But when firms are vertically integrated, the entire integrated group is the relevant unit for the calculation of profitability; any profit or loss that may be attributed to intermediate stages is relevant only for internal accounting purposes or for taxation.[1]

If an integrated firm were completely in 'balance' with respect to its requirements for crude oil (producing all the crude it uses and selling none outside), and with respect to its requirements for products (refining all the products it sells and selling none through other firms), its output of both crude and products would be related only to the demand for its products in markets served by its own distribution network. The adjustment of the firm's supply of crude and products to the demand for products would be a planned adjustment unrelated to market prices for crude oil. The amount of crude oil transferred to refineries or the amount of products 'sold' to the marketing units would be worked out 'logistically', as an 'engineering' rather than as a market problem.

Plainly if all firms in an industry were vertically integrated and all were 'in balance' in the sense used here, there would be no market for crude oil at all, and the rate of total output of crude oil in the industry would be administratively controlled: competition could only take place in the sale of final products or in the purchase of rights to explore and produce (or to engage in refining and marketing where permission was required). It would still be necessary to establish internal transfer prices and to find a 'fair price' for crude and a 'fair' refinery margin for taxation purposes, but this would not alter the fact that with no market there can be no market price.

In fact, of course, such complete and balanced integration does not exist in the international petroleum industry; not all Companies operate at all stages of the industry, and of those that do, few are completely in balance. Nor are the Companies always free to buy and sell as they please. Consequently, there has always been a market of sorts for crude oil. It nevertheless remains true that most crude oil is produced for use in the refineries of the integrated majors as a group; some is sold on long-term contracts between them, and some to governments; but in recent years competition for independent outlets has increased. Output is programmed in advance by the relevant supply departments of each Company in the light of anticipated requirements for refined

[1] Internal accounting prices are important tools of management, and if they are not closely related to the 'market prices' of the same or similar commodities, grave difficulties can be created for a firm. This question is taken up in the next chapter; the point here is that the consolidated profit before taxes is not affected by internal transfer prices. If there is an outside equity interest in one of the subsidiary companies the profit imputed to the company is important for the outsiders.

products and of the amount of crude it has contracted to sell under long-term arrangements to other Companies or to governments.[1]

So long as very little crude oil was in the hands of sellers competing to dispose of it to independent marketers, and so long as the independent demand for crude oil was so small that it could be satisfied without risking substantial independent competition in product markets, the rate of output could be effectively controlled by the vertically integrated majors without there being any evidence whatsoever of deliberate monopolistic control of the output or the market prices of crude oil. Vertical integration alone would not have permitted an effective regulation of the rate of total supply; in addition, it was necessary that the few dominant firms should control the sources of supply of crude oil and be able to use this control to weaken the competitive position of independents. The growth, either of independent supplies of crude on the market, or of a strong independent demand for crude from non-integrated refineries, would necessarily threaten the effectiveness of the control of the dominant firms; the former because of the increase in uncontrolled supply, and the latter because of the strong temptation to those majors with crude oil in excess of their own requirements to increase their profits by selling to the independents.

As we have seen, crude-oil production outside the US and the Communist countries has in fact been largely controlled by a handful of Companies which avoided price competition in product markets as far as possible, and in some cases joined together in refining and marketing. In the Middle East the ownership arrangements that were established for different reasons in the different countries resulted in a system under which the crude-oil producing companies were jointly owned by two or more of the international majors. In spite of joint ownership, however, each of the major Companies was primarily concerned with its own profitability and with its own general competitive position *vis-à-vis* the others; in consequence, the fact that the companies producing crude oil were jointly-owned affiliates of groups which were in competition in product markets meant that some method of controlling the allocation of crude-oil supplies among the joint owners had to be devised.

JOINT CONTROL OF CRUDE-OIL PRODUCTION[2]

The competitive position, and in particular the profitability, of different sellers of refined products is affected by the terms on which crude oil is available to them. If crude-oil producers were independent of refiners and were competing to maximize returns from crude-oil production, then similar crudes from the same sources would tend to be equally available at the same f.o.b. cost (subject perhaps to quantity discounts) to all companies engaged in refining. In fact, when the major groups began in 1950 to publish common

[1] The actual physical movements of crude oil are not as straightforward as might seem to be implied by the above statement. A Company's own crude is not necessarily the most convenient source of supply from the point of view of transportation, nor is it necessarily the most suitable type of crude at a given time and place. There is therefore much swapping of crude among the Companies, which reduces cross-hauling and increases flexibility of supply.

[2] The substance of the discussion in this section was published in my article 'Vertical Integration with Joint Control of Raw-Material Production: Crude Oil in the Middle East'.

prices for which crude oil f.o.b. the Persian Gulf was not only to be sold to independent buyers but also transferred to their own refining affiliates, one of the effects was to provide an illusion of this equality of treatment. But when firms are vertically integrated we need to look behind the illusion.

There are wide variations in the costs of production of crude oil from different sources,[1] and the raw-material component in the cost of oil products will be different for different integrated Companies, depending on the distribution of their ownership rights in the affiliated companies producing crude oil, and on the amount of oil each affiliate is scheduled to produce in any period of time. The production programmes of the affiliates must be drawn up with the approval of the joint owners, and will therefore be influenced by the competitive relationships and conflicts of interest among them.

In the Middle East the pattern of ownership that emerged in the inter-war period left the international groups in different positions with respect to the relation between the amount of crude oil they owned and the requirements of their integrated operations. The regulation of the rate of production of crude oil from the Middle East, therefore, required the formulation of acceptable rules for determining the amount of crude that would be made available to the several owners. All of the arrangements actually made seem to have been derived from two fundamental principles: first, the generally accepted principle that an investor is entitled to a share in the benefits of a joint venture in proportion to his investment in it; and second, that those owners wanting to obtain more crude than they were entitled to by virtue of their share in ownership should be able to purchase at least some additional quantities at a preferential rate.

At the same time, it is clear that because co-owners were competitors in product markets, each stood to gain from arrangements which put any of the others at a disadvantage in obtaining cheap supplies of crude oil. Except in Kuwait, the arrangements made for the allocation of investment to increase the capacity of the producing affiliates, and the arrangements governing the terms on which current output could be 'lifted' by the different Companies, did put at a comparative disadvantage those Companies whose requirements were disproportionately large in relation to their ownership share in the affiliates.

The methods of determining the amount of investment to be made by each of the parent Companies for the expansion of crude-oil output differed among the affiliates, but all of them had to provide answers to the same question: what should be done when a parent Company wants, and is willing to put up the funds for, an amount of crude oil, that, given the demands of other owners, exceeds the amount it is entitled to according to its ownership share?

To illustrate the problem with an imaginary example, suppose we have a producing affiliate owned by parent Companies A, B and C in the proportions 20:35:45, the output of which is being planned for a future period. Of the new increment in production Company A, with little crude oil of its

[1] Although the present discussion is concerned with the cost of crude oil, it should be remembered that oil is not a homogeneous commodity and similar considerations apply with respect to the quality and location of crude oil.

own elsewhere, wants 90 'units' of crude oil annually, B wants 40 units, and C, having plentiful alternative supplies available, wants only 20 units. To satisfy the demands of all parents, the additional output would have to equal 150 units. Since this would require B and C to put up funds for the production of more oil than they themselves would want, a problem clearly arises for which there are a number of solutions.

The output of oil would be the largest if the limit imposed by ownership shares were ignored and A were allowed to put up 60 per cent of the investment funds necessary to produce 150 units, taking 90 units, or 60 per cent of the output. This would create problems with respect to the ownership of the new fixed assets, or would involve a *de facto* lending and borrowing among the parents. The most rigid arrangement would be to require that additional investment from each owner should be accepted in the same proportion as the ownership shares. Thus C's unwillingness to invest for more than its 20 units would limit total output to 44 units, 45 per cent of which would give C its 20 units of crude oil, with A obtaining 20 per cent of the total, or less than 9 units instead of its desired 90.

Within these two extremes are numerous opportunities for arrangements requiring owners to compromise on investment plans and permitting those who want 'too much' to pay more in order to induce their co-owners to make the requisite investments. The particular arrangements that were in fact chosen in the Middle East reflected a number of factors, important among which were the political need to expand crude-oil output of the affiliate concerned at some minimum rate, the actual or potential competitive relations among the parent Companies, and perhaps the amount of oil the affiliate was expected to be able to deliver during the life of the concession.[1]

Kuwait

The most liberal arrangement was adopted by the owners of the Kuwait Oil Company, which has the great Burgan oilfield in the Middle East; the existence of a joint ownership seems in no way to have restricted the amounts of oil that could be taken by its owners, British Petroleum and Gulf Oil. Funds for new investment were apparently put up by the owners in proportion to ownership shares (which are 50 per cent each), but each of the parents was allowed to ask for and to lift as much as it wanted; if one of them lifted more than its 50 per cent share it paid the other the capital cost of the investment required for its production.[2]

Thus, apparently, neither parent exercised a veto over total investment, and therefore over output, in order to restrict the amount taken at cost by the other.[3] In making this liberal arrangement, however, the owners originally

[1] In view of the very large reserves of oil in the Middle East, oil left in the ground has virtually no present value, particularly since supplies are expected to far outlast the period of the concessions. This consideration has therefore probably been of very little significance.

[2] Since the capital cost included depreciation and amortization as well as interest, the 'lending' company recovered its capital in time. The Kuwait agreement is described in US Federal Trade Commission, *The International Petroleum Cartel*, pp. 133–4.

[3] But BP was permitted under the agreement to supply Gulf's requirements from Iran or Iraq if it so wished. Ibid.

agreed not to use the oil so freely available to each to injure each other's position in product markets. This part of the agreement has been abandoned in view of, among other things, the attitude of the US authorities to such agreements. The fact is that at the time of the original agreement between Gulf and BP setting up the arrangements for the exploitation of Kuwait oil, the two Companies did not look on each other as potential competitors.

Iraq

In Iraq the arrangements were more complicated, at least after 1948.[1] The parent Companies with large reserves available elsewhere in relation to their requirements were in a position to set a limit on the amount that could be taken by others; moreover, oil lifted in excess of a Company's ownership share was more costly to the Company lifting it. The investment programme for the Iraq Petroleum Company was set for five-year periods five years in advance. This long planning period made for rigidity in the adjustment of output to changing conditions and probably resulted in a rather slow response to the enormous post-war increases in demand.[2] Investment in the company was made by each parent in proportion to its equity.

The regulations governing the planning of increased output were laid down in a Heads of Agreement document of 1948.[3] For the period up to 1952, when post-war readjustments were presumably taking place and output was primarily determined by the speed with which physical capacity could be made available, there were no stated restrictions;[4] for the period 1952–6 definite quantities were set forth in a schedule to the Agreement.[5] There is no way of knowing how these quantities were determined, but they show Compagnie Française des Pétroles to be a heavy 'overlifter', i.e. obtaining much more oil than its ownership share entitled it to. This excess would have been bought from the other shareholders, chiefly, apparently, from the Anglo-Iranian Oil Company (BP), which was listed as requiring very much less oil than it was entitled to. However, the unexpected events in Iran after 1951 deprived BP of large quantities of its expected supplies, and it made up the deficit partly

[1] I have not seen any discussion of the pre-war arrangements. The ownership of the IPC is British Petroleum, 23·75 per cent; Royal Dutch/Shell, 23·75 per cent; Compagnie Française des Pétroles, 23·75 per cent; and Near East Development Corp. (owned in equal shares by Jersey Standard and Mobil) 23·75 per cent. The C. S. Gulbenkian estate has 5 per cent.

[2] For this reason steps were taken in the early 1960s to shorten the planning period, for no other company in the Middle East had such an inflexible programming policy.

[3] This, and other documents, are published in US House of Representatives, *Current Antitrust Problems*.

[4] For some time after the war available pipeline capacity imposed an effective physical restriction on the output of the IPC from Kirkuk; moreover, after 1948 the southern pipeline running through Palestine (including the line to Haifa) could not be used as a result of the bad relations between Iraq and the new State of Israel. For a discussion of the policies and problems relating to the development of pipelines see Longrigg, *Oil in the Middle East*, pp. 179–82. The difficulties of more rapidly extending the pipelines may account to a large extent for the necessity of restricting the total amount of oil available to each of the major companies, but it can hardly explain the intricacies of the '$\frac{5}{7}$' or the '125 per cent' rules described below.

[5] Schedule B, in *Current Antitrust Problems*, p. 962.

from Iraq. It seems probable, therefore, that the lifting arrangements laid down for 1952–6 in the 1949 Agreement were altered, but the actual liftings have never been published.

For the period 1957–61 (which had to be planned in 1952) there were two options: each of the groups owning the company was limited *either* to a stated increase (26,000 tons for each 1 per cent of ownership) *or* to a quantity calculated in accordance with what was known as the '$\frac{5}{7}$ rule', whichever was the higher. According to the $\frac{5}{7}$ rule the total amount to be programmed would be determined by adding up the amounts wanted by the various groups and then reducing the total demanded if the highest single demand exceeded five-sevenths of the sum of the two lowest demands.[1] The revised total would then be allocated to the groups according to their ownership shares, and arrangements were made to enable a group wanting more than its share to buy from those wanting less.

Thus a limit was set to the total amount that could be obtained by any group, but no group, by setting its demands very low, could thereby keep the amount available to the others below the fixed quantities stated in the first option.[2] In subsequent periods, the first option was replaced by the '125 per cent rule', which in effect enabled the groups that had been allocated the largest quantities in the previous five-year period to obtain at least a 25 per cent increase, again preventing one or two groups from putting their demands very low in order to restrict total output.

The arrangements with respect to lifting from current output, as distinct from planning the investment to provide for future output, permitted each Company to contract to buy oil in excess of its ownership share from others willing to sell, at a price half-way between cost and the 'average market value of the oil'—later the posted price. This allowed the selling group to make a profit on sales to its partners, but by the same token increased the cost of oil to them as 'overlifters'. We do not know how the system actually worked, and the Companies no longer publish the amounts of oil they lift from the IPC, but it seems almost certain that had the Kuwait type of agreement existed in Iraq, the latter's output would have grown very much faster, and perhaps the pressure to explore elsewhere felt by CFP and by Shell and Mobil in

[1] To illustrate the nature of the restrictive effect of the '$\frac{5}{7}$' rule if one or two groups wanted to use it to keep down total output, consider the following example: (I ignore for simplicity the 5 per cent interest of Gulbenkian and assume equal shares for the four major groups). Suppose that the two low groups asked for 150 and 200 'units' of oil; the highest amount any group could get would then be restricted to $\frac{5}{7} \times 350 = 250$. Imagine, then, that the following set of demands existed: group A, 150; B, 200; C, 220 and D, 350—a total of 920. D's demand would be reduced to 250 and total output to 820. Group A, by further reducing its demand to 80 units could restrict all others to 200 units and total output to 680 units. Group A could conceivably consider this a useful way of restricting the competitive power of the others, while at the same time selling its own excess at a higher price to others since, out of a total output of 680, A would be entitled to $\frac{1}{4}$ or 170 units. If A took only 80, having plenty of oil elsewhere, it could dispose of 90 to the others; but even if D took all the excess oil, it would be unable to satisfy its original requirements.

[2] Part of the reason, apparently, for these complicated arrangements was the desire of the Gulbenkian interests, entitled to 5 per cent of the oil, and of the French, who wanted oil that could be paid for in francs, to ensure that limits were set on the extent to which other owners with alternative supplies could restrict total output.

particular would have been reduced,[1] especially after 1952 when a new 30–32 inch pipeline was completed from Kirkuk to the Mediterranean.

The inflexibility of the IPC arrangements, originally due to the very large investments required to expand output, particularly in pipelines, as well as the methods of pricing overlifted oil, became increasingly inappropriate in the late 1950s and early 1960s as conditions in the industry became more competitive. Posted prices no longer reflected the 'average market value' of oil and the half-way price had become unattractive to potential overlifters among the shareholders. In 1960, therefore, discounts began to be recognised in calculating the half-way price. Later other changes were made, but these have not been announced.

Saudi Arabia

The way in which the investment programme of Aramco is determined has never been published, although various writers have described the terms on which the groups are allowed to lift current output.[2] Aramco, unlike the IPC, was set up as a profit-making company,[3] and its profits are distributed to the parents[4] on a complicated system which in effect makes oil lifted by a parent in excess of its ownership share more expensive than its 'owned' oil—how much more depends on the prices realized by Aramco from sales of all oil. Thus, although the parents of Aramco, who control its Board of Directors, have apparently been willing to expand capacity rapidly, and each of the parents has been able to increase its offtake progressively from year to year,[5] they do not make the same profits on 'overlifted oil' which, as in the IPC, is therefore more expensive. So it could come about that Mobil, with only a 10 per cent share in Aramco, has had less owned oil in relation to its requirements than of the other majors.

Although it is not clear how expansion of capacity is determined, and whether the system under which profits are distributed provides sufficient inducement for each parent to invest its proportionate share of the total capital required to meet all demands regardless of the amount of oil it is itself prepared to lift, it is difficult to escape the conclusion that there must have been some restriction on the amount of oil an owner could get at cost,

[1] Some of the parent companies of the IPC, and particularly Anglo-Iranian, were apparently very much concerned to arrange for production to develop slowly (see Federal Trade Commission, p. 105), while others very much wanted more oil. CFP, for example, was largely responsible for pushing up the rate of development of Iraq oil, but nevertheless had to buy large amounts at the 'half-way' price. See M. Laudran, *Le prix du Pétrol brut*, p. 268.

[2] J. E. Hartshorn, pp. 181–2; Wayne Leeman, *The Price of Middle East Oil*, pp. 21–23; Mikdashi, pp. 181 f.

[3] For tax reasons the IPC shows no profit, oil being transferred to shareholders at cost plus a nominal fee.

[4] Standard of California, 30 per cent; Texaco, 30 per cent; Standard Oil (New Jersey), 30 per cent; Mobil Oil, 10 per cent.

[5] According to Hartshorn, whose descriptions of offtake arrangements of the Companies are the most complete that have been published, 'Each partner company normally undertakes to take up to its proportionate requirement of this rising production; there is no penalty for "underlifting", but a group that takes more than its proportionate share in any year is entitled to lift as high a percentage again the next, at the expense of the proportions available to any who chose to underlift.' P. 181.

since at least one of the owners has been acutely short of 'cost' oil. A recent change, however, reflecting the ever-increasing proved reserves in Saudi Arabia, has made it cheaper for partners to take more than their proportionate share of output. The revised arrangement for offtake 'appears to have been particularly useful to Socony Mobil; but it may also have reinforced Texaco's growing aggressiveness in European competition'.[1]

Iran

Finally, another variation on the theme is to be found in the operation of the Iranian Consortium. Again the details have not been published, but according to one of the best informed observers the total programme is determined by the lowest 'nomination' of the shareholders holding some given percentage of the shares in the consortium. Each shareholder proposes a total programme for the output of the consortium for the following year; these are ranked in order, and the programme chosen is the 'lowest total nominated that will cover the estimates put in by shareholders representing a given percentage majority of the shares in the consortium. This total is then redistributed into entitlements proportionate to the holdings of each shareholder in the consortium' and those who exceed their entitlement ('overlifters') have to pay the full posted price.[2] In other words, the proposals ('nominations') by each shareholder for the consortium's *total* output are ranked in order of size and that output chosen which is the *lowest* of the nominations of the groups representing some unknown (to outsiders) percentage of the shareholders.[3]

Whether or not a very few Companies have had the power to determine the total output of the consortium thus depends on what this critical percentage is, for a low nomination by a few groups can be defeated only if those

[1] J. E. Hartshorn, p. 181.

[2] Ibid., p. 180.

[3] To illustrate, suppose the following are the proposals by the shareholders for the total output of the consortium in the following year, and suppose the percentage of shareholding, of which the lowest proposal will be accepted, is 65 per cent. Then the output planned for would be 60.

Shareholder	*% Shareholding*		*Proposal for total output (ranked by size)*
Mobil	7	66% (Mobil, Shell, BP, Iricon)	85
Shell	14		80
BP	40		75
Iricon	5		60
Esso (Jersey Standard)	7		50
Texaco	7		45
CFP	6		40
Socal	7		35
Gulf	7		30

It should be noted that this type of arrangement *need* not result in an output lower than would have been obtained if each company were able to nominate and get just the amount it wanted. It could even be a way of ensuring that total output would be *maintained* at an acceptable level. In practice the programme chosen is apparently based on a figure somewhat above the average nomination.

that want more oil own a percentage of shares in the consortium which equals or exceeds this critical cut-off point, which probably is between 60 per cent and 79 per cent of total shareholding.[1]

The arrangements made for determining the output of the Iranian consortium must undoubtedly have been the result of a compromise between the desires of the Iranian Government (which wanted US Companies to come in), of BP, and of the US Companies themselves, many of which were not at all eager to take on the obligations associated with joining the consortium. There are no provisions for 'over-lifting' except at the 'posted price', however, and consequently there presumably is very little. This in itself would tend to reduce total offtake.

EFFECT ON THE EXPANSION OF CAPACITY

Not only did the joint ownership of crude-oil producing companies make possible the planning of output in the light of the individual requirements of the different owners, but it also opened the way to a special type of restriction on output caused by the conflicting interests of joint-owners that were at the same time competitors in a common market for products. It may therefore be useful to consider the conditions under which the conflict of interest of the parent Companies would in itself restrict the amount of investment made in a particular crude-oil producing affiliate, and thus keep the expansion of its capacity below the rate which would have maximised its net income.

As we have noted, the problem arises because of the vertical integration of the parent Companies, for if the owners of crude-oil were selling it to outsiders, rather than using it themselves, they would, as crude-oil producers, have wanted to expand output to the point that yielded the highest net revenue from crude production. Profits would be distributed to owners

[1] Since British Petroleum was the sole owner of the old Anglo-Iranian Company before the nationalization, yet for political reasons could not be given a majority interest in the consortium, it is probable that the critical percentage of total shares which determines the output of the consortium is such as to ensure considerable weight to BP. BP owns 40 per cent of the consortium, and if the critical percentage were less than 60 per cent, then all others together could possibly determine the output of the consortium without BP, although BP would be entitled to 40 per cent of the output; if this percentage were more than 60 per cent BP alone could determine the maximum output. When we consider the circumstances surrounding the negotiations between the Companies to set up the consortium, it seems improbable that the percentage would have been less than 60 per cent, even though a greater percentage would give BP very much more weight in the decision determining total output than would appear from its 40 per cent shareholding. However, in spite of the fact that BP has abundant oil available elsewhere, it would be surprising if its nominations for the consortium's output were unduly low in view of its political interest in ensuring a level of output satisfactory to the Iranian Government.

Shell and Mobil have been short of 'cost oil', while Gulf and Socal have very large supplies in the Eastern Hemisphere; hence it is probable that Gulf, Socal and perhaps Texaco would be the 'low bidders' for Iranian oil. But low nominations from those Companies would be defeated if the critical percentage were less than 79 per cent, since together they own shares equal to only 21 per cent of the total. Therefore, since BP would not have wanted Iran's output to be held down by these Companies, the critical percentage is probably less than 79 per cent, and more than 60 per cent, and the total output of the Iranian consortium is probably determined by the bids of either Jersey Standard, Iricon, CFP, and perhaps Texaco.

according to their equity, but presumably any owner having better use for its investible funds would dispose of part of its shares rather than restrict the total investment by others in the producing company.

But when part ownership in an affiliate is obtained, not for the profit directly made on sales of the affiliate's output, but to provide raw material for the owners' further activities, then some sort of inducement will be required to persuade any owner to invest in expanding capacity of the affiliate to an extent greater than necessary to meet its own requirements. Such an inducement may be provided if sales of the affiliate's output can be made between the parents at prices which provide a return on the capital invested by the seller that is at once better than any alternative return and high enough to offset any competitive disadvantages to the seller of providing the oil to the buyer. Provided that the necessity of paying such prices does not reduce the amount demanded from the affiliates by any owner, then the arrangement will not result in a total output lower than would have been produced had each Company been entitled to obtain its requirements at cost.

Thus, there are two conditions which must be satisfied before we can safely conclude that joint ownership of a raw-material producing affiliate by vertically integrated parents will not in itself retard the rate of expansion of the affiliate: (1) The parents as a group must be willing to invest in expanding capacity to the full extent necessary to supply all the demands on the affiliate by all owners, and (2) the price that parents wanting extra supplies must pay must not reduce their demands on the affiliate.

If there had been no question about the first condition being satisfied there would have been no need for rules regulating the amount of oil any Company could lift at some price, and Kuwait-type agreements would have been made. Hence some owners must have wanted restrictions (a) because they feared that without them they would be required to invest more than they wanted to, or (b) in order to put some co-owners at a competitive disadvantage, or (c) as a means of regulating total output to prevent surplus supplies, i.e. 'irresponsible and unreasonable over-production', or (d) to protect reserves.

For vertically integrated Companies operating in markets where demand was expanding very rapidly, large investments in refining and marketing were required which made heavy demands on capital funds; moreover, most of the Companies most of the time have strongly preferred self-financing. Given the very large differences between cost of production and posted prices, the 'half-way price' charged the 'overlifters' undoubtedly provided for a long time a very nice return on investment for the 'underlifters'—the sellers among the parent Companies—but there may well have been some reluctance to invest in order to provide oil to co-owners, who were also competitors in product markets. It is clear from the record, for example, that some of the Companies at least, wanted to restrict the amount of cheap oil the French could get from the IPC for fear that the French would use it to cut prices in product markets.[1]

In view of the very large reserves available, it seems unlikely that fear of early depletion of reserves was a major consideration in the regulation of output, but there can be little doubt that the very size and accessibility of these

[1] Federal Trade Commission, p. 76.

reserves made it imperative from the Companies' point of view to do the best they could to create firm controls over the rate of exploitation. The arrangements discussed here are, in effect, the mechanism through which the Companies attempted to adjust increasing supplies to increasing demand at ruling prices.

As to the second condition—the willingness of owners to buy extra oil—if we assume that crude oil was not available elsewhere at lower prices, and that the expansion of an individual Company's output of products was not affected by the incremental cost of crude oil but by other considerations, then the higher price charge to 'overlifters' for crude would not significantly affect the amount taken. In this case, however, it would also follow that the competitive position in product markets of rival parent Companies would not be affected by differences in the costs of crude oil, unless higher costs reduced the rate of investment by reducing retained earnings.

There is no way of knowing how important considerations of this kind have been. It is perhaps likely that the companies with higher average crude-oil costs hesitated to expand in some markets which were only marginally profitable in any event. But there can be little doubt that a Company having to pay higher prices than its competitors for crude oil would feel at a *long-run* competitive disadvantage, and would have every incentive to search for a cheaper source of oil. Thus the offtake arrangements in the Middle East provided for some Companies a very strong incentive to find new oil for themselves.[1] The greater the gap between costs and posted prices the stronger was the incentive, since higher posted prices would mean higher 'half-way prices'. After 1960, however, the situation changed considerably; market prices of crude oil fell while posted prices remained up, and some oil could be bought in the market at prices even less than a half-way price calculated on the basis of posted prices. There was thus some pressure to modify the arrangements for overlifting in some companies.

Given the vertically integrated nature of the parent Companies, there is only one way of avoiding the difficulties discussed here and ensuring that the development of existing producing affiliates will not be retarded by the desire of some owners to profit at the expense of co-owners by raising the cost and restricting the availability of oil to them. This is to allow all owners who want to take oil at 'cost' to do so, providing that the overlifters either pay the interest and depreciation costs on the capital investment to the partners putting it up (as is evidently done in Kuwait), or put up the capital themselves, including the investment in exploration and development attributed (perhaps arbitrarily) to the amount of oil demanded. The former arrangement would preserve the relation between the existing ownership shares and the distribution of capital investment, but would change the distribution of oil to the

[1] In a brief presented in a Civil Action brought against the Company by the US Government in 1953, Socony-Vacuum (now Mobil Oil) made the following statement:

'Socony began to embark on a systematic foreign exploratory effort primarily for two reasons. *First*, because substantially all foreign production was in the hands of large integrated companies which were competitors of Socony and, therefore, could not be depended upon for supplies. *Second*, because a long period of time would elapse between the inception of such an exploratory effort and the development of commercial production. . . .' *Current Antitrust Problems*, p. 879.

parent Companies; the latter would ignore ownership shares in both the distribution of capital investment and the distribution of oil, which would then be determined for each Company by the proportion of actual investment it contracted for, including an allowance, where appropriate, for exploration expenditures.

That procedures of this kind are feasible has been demonstrated by the working of the Gulf–BP agreement in Kuwait. Nor would they reduce the chief alleged advantages to the major Companies of vertical integration—an assured supply of crude oil from numerous sources. One of the consequences would probably be a more flexible pattern of investment and oil supplies among the various Companies,[1] but at the same time competition in product markets among owners might well increase with adverse effects on prices. Moreover, to insist on arrangements which would permit Companies to gain benefits from their joint venture which were not proportionate to their original shareholding, or to promote a change in effective ownership shares through disproportionate investment in new capacity, might appear to deprive some of the original investors of ownership rights.[2] Such arrangements may well be much easier to make when only two Companies are involved, as in Kuwait, than when several interests have to be conciliated, especially if the different Companies have very unequal shares. Moreover, in such arrangements, companies with a small percentage of the total shareholding would be able to secure a gain which was out of proportion to the original risk they undertook.

Joint planning of the rate of supply from the Middle East by the major Companies, together with the internal planning of the vertically integrated firms themselves, was sufficient to ensure that the output of the new Middle Eastern fields would be fitted into world markets in an 'orderly' manner. Most of the output in Venezuela was also in the hands of the major Companies, but 65–75 per cent of it was absorbed in the Western Hemisphere, primarily by the US, and hence the rate of supply was closely related to the development of outlets in that market.[3] Exports of crude and products to Europe, which ranged between 17 per cent and 28 per cent of total Venezuelan exports of oil in the decade 1949–59, were almost entirely in the hands of Shell and Jersey Standard.

It is clear that this 'system' was not a planned cartel. Nevertheless, its stability depended on the continuing ability of the international majors to

[1] Quite apart from the context of this discussion there is, I think, much scope for making more use of the flexibility that could be provided by arrangements which permit a change in ownership shares by altering the proportions in which additional investment is made. This could be a particularly useful device for developing countries that want to attract private foreign investment for a period, but prefer over a longer period to reduce the extent of foreign ownership in their industrial economy.

[2] But strictly legal considerations must be weighed against other considerations when appraising the desirability of change. See, for example, the well-argued criticism by F. R. Parra of the 'legalistic' attitude of the Companies in *Exporting Countries and International Oil* (Organization of Petroleum Exporting Countries, 1964). On the other hand, the oil Companies have more than once showed a willingness to re-negotiate agreements in the interest of the producing countries.

[3] In 1945 over 98 per cent of Venezuelan production was accounted for by Jersey, Shell, Gulf and Mobil.

control the total supplies of crude oil coming on the market, and on the self-restraint of those international groups with supplies of crude oil in excess of their integrated requirements. Both could be threatened if prices were held 'unrealistically' high, and it can be argued that in this respect the companies over-reached themselves. This question will be examined in the following chapter.

JUSTIFICATION FOR THE REGULATION OF OUTPUT

Many have argued that the oil industry is by its very nature so inherently unstable that control is a necessary condition for its effective operation, both because of the irregular and uncertain results of exploration, and because of the existence of economies of scale so large that the industry must be treated as a 'natural monopoly'. Moreover, both the economies of scale and the economies of operating existing plant at full capacity are held to make vertical integration essential. Control over output can be exercised either by governments, as it is in the US through a system of pro-rationing, or largely by private firms, as it has been in the international industry. We shall not be concerned in this section with the relative merits of government and private control; rather we shall examine the general question whether a socially acceptable operation of the industry necessarily requires either government regulation of output or the domination of the industry by firms large enough to exercise a strong control over total output and to keep competition, especially price competition, within very narrow limits.[1]

An Inherently Unstable Industry

The proposition that the oil industry could not operate under competitive conditions has been argued most persuasively by P. H. Frankel in his *Essentials of Petroleum*.[2] The argument can be summarized as follows: because of the uncertain results of exploration, the high overhead costs at all stages of the industry, and a high inelasticity of demand in the short run, the industry is not 'self-adjusting' in the sense that a fall in prices significantly chokes off supply or stimulates demand. The uncertain results of exploration create the likelihood that either too little or too much oil will be discovered in relation to the amount that will be bought at profitable prices; the fact that crude-oil production, transportation, refining, storage, and distribution all require large amounts of capital results in heavy fixed charges and forces competitive producers to produce as much as they can as quickly as possible in an attempt to recoup their sunk costs and to cover at least part of their

[1] We are not concerned with that degree of regulation required to ensure that the exploitation of oil fields is conducted with due regard for good conservation practice. There are a number of circumstances in which practices that are both technically and economically wasteful may be adopted by oil companies in the absence of strong government regulation, notably if more than one competitor operates in the same oil pool, if the period for which concessions are granted is so short that companies have an incentive to exploit the fields at too rapid a rate, or if political uncertainty about the future position of the company also gives it a very short time-horizon.

[2] P. H. Frankel, *Essentials of Petroleum: A Key to Oil Economics*. Dr Frankel briefly restated his argument later in *Oil: The Facts of Life*.

overheads. For these reasons, the industry is subject to continuous crises in the absence of reasonably strong control over supply. Because consumption is not readily expanded when prices fall, prices are easily pushed below costs of production by relatively small surpluses. Thus 'hectic prosperity is followed all too swiftly by complete collapse, and redress can be hoped for only from the efforts of "eveners", adjusters and organizers'.[1] And '. . . for technical reasons alone the formation of paramount oil concerns was inevitable; their role could not be taken over by a welter of smallish firms'.[2]

Such a role could only be fulfilled by the large firms if an important bottleneck of the industry could be brought under their control. For Standard Oil in the US in the last decades of the nineteenth century, the relevant 'bottleneck' was transportation, but for the international industry the key proved to e crude-oil reserves. Vertical integration is held to be necessary because fficient operation requires a continuous and secure flow of supplies, and big rms with large and far-flung markets and heavy investments cannot afford to e dependent on others for their supplies unless, of course, they are securely tied up under long-term contracts. Hence '. . . whatever we may do, the fundamental factors come to the surface; the oil industry to exist at all, calls for concerted effort and, however often a co-operative structure may have been disturbed or broken up, it will soon begin to form again'.[3]

This analysis, which has been widely accepted both in the industry and by most outside observers, rests on the proposition that economies of scale are so enormous in relation to the size of the industry that falling prices would not lead to reasonably appropriate adjustments in output at the margin. In other words, once reserves are discovered, their exploitation is carried out under continuously decreasing costs. As a result prices fall, and remain for long periods, far below the average long-run costs of production, or 'supply price'. The notion that producers are forced to produce 'to recover overheads' as quickly as possible assumes that they have a very short time-horizon and take little account of the future (or else are in continuous financial difficulties).

This thesis is commonly illustrated with reference to the history of the industry in the US, where the 'law of capture' almost forced instability on it by putting tremendous pressure on producers to drain their reserves as fast as possible. Under the 'law of capture' oil belongs to him who brings it to the surface. Since oil migrates underground to areas where pressure is reduced, the drilling of a hole may attract oil from an entire reservoir, the extent of which bears no relation to the property lines drawn on the surface of the earth. Thus there may be many properties lying over a single reservoir, and many producers, each of whom has an incentive to produce as fast as possible to get a maximum share of the oil before it is 'captured' by others. In such circumstances, the conclusion that some centralized regulation of output is essential on both economic and technological grounds is unassailable.

But the very special legal conditions that prevail in the US are not found elsewhere. In the international industry both the legal and political environ-

[1] P. H. Frankel, *Essentials of Petroleum*, p. 67.

[2] Ibid., p. 85.

[3] Ibid., p. 97.

ments were different, and the emergence of oligopoly occurred in a very different way from that in the US. The interplay of international politics, and the terms on which concessions could be obtained in the Middle East, gave the already well-established international Companies powerful advantages. The fact that vertical integration and the dominance of very large firms characterize the history of the industry, both in the US and internationally, provides no evidence at all, nor even a presumption, that the industry could not have operated efficiently under a different type of organization; the most we can presume is that either the large firms found it both possible and profitable deliberately to organize themselves and the industry in the way they did, or that the conditions of competition left little effective alternative.

An Inherently Self-adjusting Industry

Apart from historical experience, which in the present case can only demonstrate the *profitability* of the given response to given circumstances, the essential empirical foundation of the theoretical argument has been powerfully challenged, notably by Professor Adelman.[1] He makes a strong case on technological grounds for the proposition that the crude-oil industry is in general subject to increasing costs (i.e. decreasing returns).[2] He also finds that it is not characterized by an exceptionally high ratio of fixed to variable costs, but rather by an 'extraordinary variability'.[3] Since in general the costs of development and production may be expected to rise fairly quickly as output rises, it will pay producers in competitive conditions to produce only up to the point where the rising marginal costs equals price, and the industry will in fact be 'self-adjusting' in the sense that Frankel denies.[4]

Under effective competition, therefore, output in the crude-oil industry would tend to be regulated by the same forces that regulate output in other industries—rising marginal costs for all producers choking off production

[1] Professor Adelman has set forth his argument and the evidence supporting it in four important studies: *The World Oil Outlook*; *Oil Prices in the Long Run* (1963–75); *Efficiency of Resource Use in Crude Petroleum*; *Oil Production Cost in Four Areas*.

In the following few paragraphs I intend only to summarize Professor Adelman's argument, quoting his conclusions at some length. For the evidence adduced in support, and the details of the argument readers are referred to the works cited.

[2] 'Costs are considered over the . . . life of the field . . . and decisions on investment and production are made on that basis. Over the remaining lifetime of a field or well, marginal cost is a rising function of output: the greater is the output the higher is the cost of additional output. Even neglecting the cost of well workovers, fracturing, waterflooding, etc., the basic fact about oil and gas production is *the production decline curve*. . . . To sum up: looking at development-production programme *ex ante*, we must take account of the fact that both total output and total cost are functions of time. In general, if the expenses per unit of time are fixed (and *a fortiori* if they are increasing, as in fact they are), and if the production per unit of time decreases, *then the production cost per additional unit increases the greater is the total production.*' *The World Oil Outlook*, pp. 42–3.

[3] Ibid., p. 35–6.

[4] 'The crude-oil industry, contrary to common belief, is inherently self-adjusting; more precisely, it has a strong adjustment mechanism for determining the level of output and its division among various sources of supply, by the *price* acting upon the cost which must be incurred to bring up more output. . . . The current spasms of the world oil market express both a strong adjustment mechanism trying to make itself felt and also strong barriers to it.' Ibid., p. 32.

as net revenue per unit of output decreased with expanding output. Both exploration and the development of discovered reserves would be functions of anticipated costs and demand taking account of appropriate rates of discount.[1] There may be fields with reserves so large that oil left in the ground has no present value, and where pressure and other conditions of exploitation are so favourable that operating costs are extremely low. But low operating costs of *developed* capacity do not govern price in rapidly *expanding* industry.[2] In such cases, replacement cost has no significance for costs of production, and price will not be determined by the low *operating* costs of developed capacity but by the costs of *developing* further production from the field.[3] Oilmen contend, however, that some developed fields in the Middle East are so large that very great increases in output are still possible at constant, if not slightly falling, marginal cost.

Strains in the Process of Adjustment

It cannot be denied that the mechanism of adjustment may at times be severely strained, and that in the process of adjustment to changed conditions some producers may be hurt severely. No one can predict how much time or money will be required to expand world reserves by any given amount, and chance, in the form of the discovery of unusually large low-cost reserves, can suddenly alter all previous expectations of the costs of additional output. This 'destabilizing' element plays a central role in Frankel's analysis. But whereas Frankel puts the emphasis on the uncertainties of exploration as the primary cause of alternate glut and scarcity under competitive conditions, Adelman puts the emphasis on the costs of developing discovered reserves as the primary factor preventing the unprofitable flooding of the market that Frankel evidently fears.

The discovery of new reserves does not yet constitute supply on the market, and the additional investment required for their development will not be

[1] '. . . under any given conditions of knowledge, the more exploration is done, the higher, probably, will be the finding cost per unit of what is found. The more a given deposit is developed, the higher will, almost certainly, be the cost of additional development. Finally, from the time a well begins to operate, the greater the output the higher the cost of additional output. When the additional cost of another increment of output goes up above the current price, production ceases; when incremental development cost rises above the anticipated near-term price, development ceases; and much more hazily, when finding cost in any place goes above a longer-term price anticipation, exploration stops.' *The World Oil Outlook*, pp. 46–7.

[2] '. . . it is true that once producing capacity is in operation, the well dug and equipped, operating costs are very low. *But this fact has no importance* in an industry where capacity must be increased 10–15 per cent annually because of growing consumption and natural loss of pressure and capacity. To sell crude oil today at a price which would not cover the cost of replacing capacity when one is actually engaged in digging and equipping new wells is unreasonable and is not done. (The exception is in the US, because of a legal and regulatory system peculiar to it, and which exists nowhere else, so we need not discuss it here.) Therefore the relevant incremental cost is not the very low operating cost but the sum of development plus operating cost. The industry really shows unjustified jittery nerves on the subject of costs because of the mistaken idea that the bare operating cost is an important fact in relation to price.' *Oil Prices in the Long Run* (1963–75), p. 158.

[3] In *Oil Production Costs in Four Areas* (1966), Professor Adelman restated in greater detail the principles discussed here and attempted to work out actual cost figures.

made by sensible producers unless they expect the new increment of output to be demanded at prices sufficient to justify the investment, and then only to the extent that the use of reserves for present output is expected to be more profitable than holding them back in favour of future output. It is assumed that firms in the international oil business plan for a considerable period in advance and do not allow short-term considerations to dominate their actions.

Nevertheless, chance discoveries of new reserves which are very large in relation to the industry as a whole, and which can be developed and produced at a cost lower than that of a large part of existing output (even eliminating replacement cost for some time) may violently shift the industry supply curve far to the right. Such discoveries necessitate substantial readjustments among producers and sources of production which may be unacceptable to governments as well as to the industry, and may lead to some form of public or private intervention to mitigate the effects. Adelman's analysis of the history and nature of exploration shows that this type of change has not been a significant factor for the industry as a whole very often, but it is exactly what happened in the 1930's and 1940's as a result of the discoveries in the Middle East, which Adelman considers a 'huge random disturbance' and unlikely to happen again.

Adjustments to 'disturbances' on this scale are indeed painful to the industry and there is no doubt that the leading producers would make every effort, as they in fact did, to control them in their own interest. Part of Frankel's point is that the market may act so ruthlessly in such circumstances that fear of the result prompts producers (and governments) to attempt regulation. One need not quarrel with this, providing it is recognized that the problem is one of cushioning existing interests against changes which are necessary for long-run stability and which, if delayed too long, may create even more difficult circumstances. Frankel seems to hold that sudden changes in the reserve position have been, and presumably will continue to be, characteristic of the industry. Although I think we must reject his conclusion that if it were not for the dominating influence of very large firms, we would regularly find as a matter of course that decreasing costs, inelastic demand, and extremely short time-horizons on the part of producers, would lead to great gluts and ruinous prices because of fluctuations in reserves, thus discouraging exploration until a present scarcity caused prices to soar, we must agree that *if* there should be frequent 'Middle Easts' the resulting instability would in competitive conditions create serious difficulties for those countries a large part of whose production became suddenly displaced.

Adelman's exposition of the nature of the underlying economics of crude-oil production is little more than an application of a fairly simple type of economic analysis to the oil industry—yet is of the utmost importance. Some may feel that he puts insufficient emphasis on the "dynamic" aspects of industrial change—that is the complications that arise as adjustments work themselves out. This is to some extent true, but the emphasis he lays on the fact that producers react to expectations—and in the oil industry to long-run expectations at that—and not to immediate circumstances, or blindly to 'sunk costs', is of the highest importance. Moreover, a more complicated analysis would not alter the fundamental proposition that crude-oil production should not be regarded as a 'natural monopoly' because of economies of scale, and that therefore the same type of equilibrating forces can be expected in this industry as in others. Hence, given reasonably sensible appraisals of their own interests on the part of producers, violent alterations between glut and scarcity need not occur although prices can be expected to fluctuate with changes in supply and demand. Serious shocks may at times be suffered as supply curves are shifted quickly to the right by discoveries which can be developed and exploited on a large scale at an average total cost much lower than the cost of a significant proportion of existing output, and some mitigation of the effects may be required.

All of this, however, is an analysis of the supply and price of crude oil in conditions where individual producers determine their total output, as well as its geographical distribution, with reference to costs on the one hand and to market prices on the other, and where no producer restricts his production because by doing so he expects to maintain the level of market prices. This has certainly not been a characteristic of the actual conditions of supply in the international petroleum industry, and some readers may be questioning the relevance of the entire discussion. Moreover, to demonstrate that the petroleum industry could operate effectively under competition is not to recommend the desirability of requiring it to do so. Nevertheless, even though the analysis provides no simple prescription for policy, it is important to understand the way in which the 'forces of supply and demand' will necessarily operate under the surface. Moreover, even in present circumstances, the analysis demonstrates that the growth of competition among a larger number of producers would not necessarily lead to marked instability, once the required adjustments to the underlying supply conditions had been made.

Because of the attempt to maintain extensive excess capacity and to keep prices far above long-run costs of finding and developing crude-oil reserves, the industry has been, and still is, subject to strains that are due not to its 'nature' but to policy. The continued restraint exercised by oligopolistic competitors with strong market power effectively withstood the pressures for a time, but so long as it is profitable for firms to develop new capacity in the face of existing excess capacity, it is unlikely that anything short of governmental authority will in the end be capable of effectively preserving the disequilibrium thus engendered. The crude-oil producing countries have recognized this and have in fact called on such authority in setting up the Organization of Petroleum Exporting Countries. But in view of the divergent interests among the producing countries as well as among the consuming countries, and between the producing and consuming countries it seems unlikely that even governments will be effective in holding back the pressures that have been pushing the industry closer to a more stable equilibrium.[1]

The argument for control of supply is strongest if put in terms of the desirability of reducing the shock to whole economies (rather than to individual Companies) of sudden shifts in supply curves as a result of new very large low-cost discoveries. If supply curves are rising in all areas as output

[1] See Chapter VII for a discussion of OPEC.

increases, small shifts in the cost of production between different areas will drive out only the marginal production in the higher cost areas and need not result in any serious disruption. Especially if total consumption is expanding rapidly, the chief consequence may only be a relative stagnation of the higher cost areas. Large shifts are another matter; it is indeed possible, for example, that if the oilfields in Kuwait and Saudi Arabia had been developed after the war as rapidly as price-cost relationships and factor supplies permitted, a sharp loss of exports for Venezuela, and perhaps for others, would have ensued. Control designed to mitigate such effects can be defended on the general ground that the economic consequences of extreme fluctuations to national economies are unacceptable. This is, of course, the primary justification advanced for any commodity control agreement, but the successful implementation of such control for oil would be even more difficult than it is for other commodities.

The supply of oil reserves does not 'fluctuate' around a long-term trend as does the supply of a number of agricultural commodities for which commodity agreements to stabilize prices are deemed desirable because of the vagaries of weather and other natural phenomena. Rather, new large discoveries create long-run shifts in supply curves which cannot be expected to move in such a way that the movements to the right or left cancel out over a reasonable period of time. Hence, attempts to prevent, in contrast to slowing down, appropriate adjustments, including adjustments of prices, would face peculiarly intractable problems. On the other hand, and again in contrast to many agricultural commodities, there is no particular reason to assume that changes in supply that are very large in relation to total output will occur very frequently. An industry as large as the oil industry could, *if it is in reasonable equilibrium*, absorb the impact of quite extensive additions to reserves without cataclysmic consequences for any of the major producing countries.

CONCLUSION

Regardless of the general economic justification for control, the fact remains that the major Companies did attempt, with considerable success for a time, to regulate the output of crude oil. The discovery of much low-cost oil in the Middle East within a very short period caused a very great change in the conditions of supply. The potentially disruptive effects of this, it can plausibly be argued, were averted or at least delayed by the control over the development of the oil which followed as a 'natural' consequence of the combination of the joint ownership of crude-oil producing companies, the vertical integration of the major groups, and the restraint imposed on each group by the oligopolistic nature of the competition in the industry.

It is sometimes argued that the oil Companies deliberately retarded the rate of development of Middle East Oil in order to preserve the value of their Western Hemisphere investments. If seems improbable that this would have been a reason, for the maintenance of the value of sunk investment at the expense of new opportunities which would increase total profits is plainly bad business. Whether the Companies were in competition or were acting as a monopoly would have made no difference in this respect, for in either case they would gain if they cut back their Western Hemisphere production and scrapped assets, to the point where the direct costs of production in the Western Hemisphere did not exceed the total cost of further developing Middle East oil. But in a rapidly growing industry such a result is less likely to be effected by an accelerated scrapping of assets than by a change in the location of new investment, which is, of course, exactly what to a large extent in fact occurred. The rate at which the international Companies made new investments, and the allocation of their investments among the several producing countries, were influenced by several types of consideration.

In the first place, the Companies are so large that each had to consider the political consequences of its actions, not only in the distribution of offtake among the several producing countries but also in the rate of importation into some of the consuming countries. For example, a much larger flow of cheap foreign oil into the US than in fact occurred after 1948–9 could have had political consequences for the international oil Companies that they undoubtedly preferred not to contemplate. It must be remembered that a Committee of the House of Representatives of the US was investigating the desirability of compulsory restrictions on imports of oil into the US as early as 1950. At the same time, the possibilities of European governments acting to protect their coal interests had also to be considered.

Secondly, the size and world-wide scope of the operations of the Companies led them to put great weight on the availability of crude oil from a wide variety of sources, both to provide flexibility of supplies under their own control in the face of changing market conditions, and to increase the security of supplies in the event of unforeseen difficulties in obtaining oil from any given area. In this respect, they had on occasion to consider the national policies of the governments of their parent countries. For it must not be forgotten that in the last analysis the 'international' Companies are restrained by the national policies of their 'native' countries, which may not always be consistent with the interests of the other countries in which the Companies operate.[1]

Finally, we come to the important consideration of the effect of increasing output on prices, and especially on the conditions of competition. The Companies seemed to have been convinced that the aggregate demand for oil products was highly inelastic with respect to price (although not the demand for every product). They had therefore a powerful incentive to do what they could to hold price competition to a minimum. This would in the circumstances require not only that supply be controlled, but that a method be found to ensure as far as possible that all Companies quoted much the same price except in short-run special circumstances. Since the pricing policies adopted were an integral part of the operations of the Companies in world markets, we must now turn to an examination of the way in which prices were determined and the level at which they were maintained. We can then examine some of the broader consequences of the type of control over supply and prices that the majors exercised.

[1] This question is discussed further in Chapters IX and X.

Chapter VI

PRICES AND THE ORGANIZATION OF THE INDUSTRY

In the last analysis attitudes toward the operations of the international petroleum firms are likely to be determined more by the views held about prices than by those about any other single aspect of the industry. Thus, there has been much controversy over oil prices, controversy often characterized more by misunderstanding, confused analysis and inconsistent argument than by clear presentation of the essential issues. Some of the confusion and inconsistency has been the natural result of the special pleading of different vested interests, but much of it has derived from a genuine failure to understand the significance of the role and structure of the international firms in the industry and of prices themselves. Most of the controversy has been centred on the price of crude oil; the issue of whether prices were in some sense 'too' high or 'too' low became thoroughly confused with the question of why they were the one or the other.

In detail, the pricing of crude oil is an intricate business; there are a large number of types and qualities of crude oil and a very large number of joint products of the refinery process. Different crudes often require different treatment in processing and for any desired pattern of output in a given refinery, some crudes will be more appropriate than others. Thus, there will be numerous different f.o.b. prices for crude oils, depending on quality, gravity, impurities, etc. In principle, such differences in prices would always be related either to differences in costs of processing (e.g. the necessity of removing impurities) and to the relative costs to refiners of using the different crudes in producing the desired pattern of refinery output at expected product prices. In practice, and because of the absence of an efficient market mechanism, certain types of price differentials, especially differentials related to specific gravity, which were at one time deemed appropriate to a pattern of demand under the conditions of competition prevailing in a particular area (notably in the US) often came to be applied in circumstances to which they were not appropriate. This type of question will be but briefly touched on, since for the most part our concern will be with much broader aspects of the price of crude oil.

At times, and particularly in some areas, product prices have been the centre of attention with respect either to their structure, i.e. the relative prices of different refined products such as fuel oil, kerosene and gasoline, or to their general level. On some of the most important oil products indirect taxation is very heavy and the tax component of the price paid by consumers may exceed the cost of the product itself. Hence, for an economic analysis of the industry's pricing policies one is not so much concerned with retail prices as with ex-refinery or pre-tax prices.

The term 'product prices' encompasses a wide spectrum of different prices for very different commodities sold in different markets, the only connection between them often being that they are produced from a common raw material. Not only are they joint products with joint costs, but the proportions in which the different products can be obtained from a given barrel of crude are technically limited. New technological developments have significantly increased the ability of refiners to vary the product-mix obtainable from a given crude, but it is not yet possible to produce products in any proportions desired. Again, we shall not be much concerned with the relative prices of products, although for products, as for crude oil, the adaptation of the pattern of relative prices to the changing patterns of demand has not always been notably efficient, again largely because of the way in which world prices were linked to US prices until the late 1950s. The emphasis here will lie on the relation of crude and product prices more generally, and on the level and movements of both, with special reference to the organization of the industry.

SOME GENERAL CONSIDERATIONS

The demand for crude oil is derived from the demand for products, for, apart from the very limited possibilities of burning it directly as a fuel, crude is of little use to any but those who can refine it into products. Within the limits of the technical capacity of existing refineries (or special stock-piling arrangements), the amount and qualities of crude demanded in the short period are directly related to the products refiners expect to be able to sell. In general, the short-run elasticity of demand at prevailing after-tax prices is held to be quite low, although there are significant variations between different products and in different areas. The demand for gasoline, for example, seems to be highly inelastic within most prevailing price ranges, although for certain uses it is in direct competition with diesel oil. Fuel oil is in competition with other fuels in many uses and the amount demanded is likely to be influenced by relative costs. The sales of both products, however, are likely to be more influenced by prices if time is allowed for adaptation of the size of engines or the replacement of existing fuel-burning equipment, etc.

All of this implies that in the short run prices of most products could be raised without much effect on demand, but that such action could be at least partially self-defeating in the longer run. Indeed, governments have long realized that the short-run inelasticity of demand makes some oil products ideal subjects for the raising of revenue through indirect taxation; and so heavy is taxation on some of the more important products, especially gasoline, that a considerable change in prices before tax may be only faintly reflected in prices paid by consumers.[1]

[1] A useful survey of the question of indirect taxation can be found in P. H. Frankel, 'Taxation of Petroleum Products and its Impact on Consumption'. Frankel also argues that taxes on products may so distort the pattern of refinery output that average refinery costs are raised.

The industry tends, therefore, to assume that a general reduction of pre-tax prices would not significantly increase the amount bought in the short run and in the longer run an increase might have an adverse effect. Of course, the elasticity of demand for individual sellers is not the same as it is for the industry as a whole, but when producers and marketers are few, each can clearly see the significance for his own operations of a reduction in market prices which would not bring about an all-round increase in the amount bought. In these circumstances, each seller, with due regard for the response of the total market, may refrain from attempting to increase his own sales by shading his prices, and increases in prices only take place when a price leader, expecting to be followed, decides to act; an 'oligopolistic equilibrium' is then said to prevail. Because such an equilibrium has some of the characteristics of a 'competitive equilibrium' from the point of view of the market position of each of the sellers, economists and industry spokesmen often talk at cross purposes in discussing the 'competitive' nature of the industry, and have great difficulty seeing each other's point of view.

Prices and Competition

The difficulty centres around the meaning of price competition. When an economist states that prices in a particular industry are determined by competition, he means not only that the individual sellers in the industry have little 'market power' and therefore cannot influence the market prices for their products, *but also*—and this is the crux of the matter—that the output of each firm is limited *only* by the unprofitability of producing more at ruling prices: a producer refrains from producing more *only* because it would cost him more to produce (and sell) additional output than he would obtain in sales receipts at going prices. He cannot raise his prices without risking a loss of customers to others, but if he lowers his price he would find it unprofitable to sell as much as before because his marginal costs are rising and at lower prices he would have to reduce his output in order to reduce costs. His sales are limited not by demand but by the rising cost of additional output.[1] In other words, prices are not determined by price competition if individual sellers consider the influence of their own sales on market prices and in consequence restrict output in order not to 'spoil the market'. In such circumstances, it is manifestly not price competition but deliberate control of supply that determines the level of prices.

Nevertheless, when an oligopolistic seller looks at his own market position he may well feel that his situation is very similar to that of a seller operating

[1] It is for this reason that price competition will not survive for long in an industry where marginal costs fall indefinitely. Such industries are known as 'decreasing cost industries' and are treated by economists as 'natural monopolies'. As noted in the previous chapter, it has sometimes been argued that both crude-oil production and refining are subject to decreasing costs in this sense. Obviously, costs may fall in crude-oil production for a period when a field is newly producing; in refining, costs will also fall up to a point, but will rise sharply when throughput begins to strain the facilities. One must, of course, carefully distinguish between the behaviour of costs with given capacity (short-run costs) and costs when expansion of capacity is envisaged (long-run costs). So-called 'historical decreasing costs', which take place over time because of changes in the conditions of production, for example, because of technological changes, are not relevant for the problem discussed above.

under 'price competition' as defined above: if he raises his price and others do not follow, he can expect to lose many of his customers, and if he lowers it and others do the same, he cannot expect to expand his sales. From his point of view he has no 'market power' either, no control over market prices. He therefore takes his selling price as given by the market and feels that his scope for action is severely constrained by 'competition'. The difference between the two situations, of course, lies in the factors determining decisions about output. Under effective competition, expansion of output is deemed unprofitable because of rising *costs*, while under oligopoly, expansion of output is expected to be unprofitable because of lower *prices* consequent on the increased supply and the matching actions of other sellers. Thus output is restricted because of each seller's fear of the effect on market prices if he tries to sell more. The economist distinguishes between 'oligopolistic' competition on the one hand, and monopoly and 'monopolistic' competition on the other. The former term applies when individual sellers restrain their competitive instincts because they expect their competitors to react in self-defence, and the latter when they refrain from attempting to sell more, not because of the expected reaction of rivals, but simply because they expect that increased sales could only be made at lower prices and would not increase total profits.

Clearly no sensible seller embarks on a self-defeating course of action if he can foresee the consequences, and when the number of sellers is few, such foresight is more practicable than it is when the number is large and the actions of each single seller have an imperceptible effect on the total market. Understandably enough, firms resent being called 'monopolists' or even 'oligopolists', when such terms have strongly condemnatory overtones, simply because they do not engage in self-defeating competition. Under oligopolistic competition, however, the fact remains that so long as sellers exercise restraint, prices are not forced down to the point where they are ruled by costs at different levels of output. The greater the number of sellers and the easier it is for new sellers to enter the market, the less stable is such a situation. But the mere fact that the number of sellers in an industry is few does not justify the conclusion that the industry will be characterized by this kind of oligopolistic 'equilibrium' or 'restraint' on competition, for there is always the possibility that one or more of the sellers will take the risk of attempting to enlarge his sales by shading prices, particularly if the details of transactions can be kept secret. Nor can one assume that sellers in such a position will be able, or even attempt, to maintain particularly high prices, since there is always a threat of competition from potential newcomers who may try to enter the industry if prices (and profits) make it unduly attractive for them to do so. Unless the barriers to entry are deemed to be very difficult to overcome, such potential competition can be a powerful restraint on the prices charged by existing sellers.

On the other hand, if firms give a high priority to complete self-financing in an industry where price competition among existing sellers is restrained, they may try to maintain prices at a level which makes entry by newcomers dangerously attractive. For a large firm to keep its relative position in a rapidly growing industry, large amounts of finance may be required and this necessarily implies a large cash flow from existing operations if the finance is to be obtained from internal resources. Unless barriers to entry are very strong, newcomers, and perhaps especially industrial users of the product, may decide it is worth their while to try to break in. One can perhaps argue (with hindsight) that the major oil Companies have failed to give adequate weight to these considerations.

It is often assumed that the statement 'prices are determined by supply and demand' is equivalent to the statement 'prices are determined by competition'. This is manifestly untrue whenever producers restrict supply in order to maintain prices. It may well be that a wide variety of competitive (sales promotion) techniques are used by sellers in order to maintain or increase their respective shares of the market, and competition in the improvement of products or the quality of service may be intense, but prices are not 'determined' by this type of competition. It is quite possible for producers to agree in advance explicitly or tacitly, what prices would maximize net returns in the light of the elasticity of the total market demand for their products, or simply to accept a 'going price' for fear of the consequences of cutting it, and for each of them then to limit his own supply to the amount he could expect to sell at this pre-determined price *even when it would pay to produce more at that price*. Prices would still be 'determined by supply and demand', but not by price competition among sellers.

Whether an individual seller has some 'control' over prices (i.e. whether he can by his own decisions influence them) may be tested by whether or not he refrains from producing (and selling) more because he expects that if he does so prices would fall. And the influence of competition on prices may be tested by the extent to which they tend to approach the long-run costs of production and do not involve discrimination between different markets. In the various controversies over oil prices the issue is not, and really never has been, whether the major oil Companies had some 'control over prices' in this sense. The fact that each of the Companies has always attempted to increase its share of the market as far as possible by means other than price competition is sufficient evidence that supply was not restricted by the unprofitability of selling more at existing prices, but rather by each Company's estimates of the amount it could sell without initiating increased price competition in either crude or products markets.

Some Functions of Crude-Oil Prices

Prices may be required to value goods for a number of purposes other than market transactions, for example for taxation or insurance, or for a variety of internal accounting purposes, such as the calculation of depreciation, inter-departmental costing within a firm, or as a means of imputing profit to particular internal activities. Before discussing any given pricing arrangement it is often necessary to ask what the prices involved are used for. Oil products have by and large been priced for the purpose of sales to final consumers. This is not generally true of crude oil, for crude-oil prices are for most Companies primarily internal (inter-affiliate) transfer prices, although some Companies sell considerable crude in the open market or at special prices on long-term contracts. Until very recently no 'market price' in any meaningful sense could even be said to exist for crude oil, although sales have always been made outside the network of the integrated Companies. For this reason our analysis of the price of crude will have to take as much (and sometimes more) account of the organization of the industry as of the influence of 'market forces'.[1]

Apart from tax considerations, the price of crude oil is of much less importance to an integrated firm than the price of products. Price competition in product markets which forced down product prices would, by reducing total profits, necessarily and directly reduce the value of crude oil to producers. Control over crude oil supplies would confer no monopolistic benefit on integrated firms engaging in unlimited price competition in product markets. It is clear, therefore, that firms with extensive control over the production of crude as well as over refining and marketing would, in the absence of pressures from governments or a differential impact of taxes, have a stronger interest in the prevention of price competition in product markets than in the maintenance of crude-oil prices as such.[2]

In practice, not only are government policies, and the differential impact of taxation of great importance, but neither vertical integration nor monopolistic control are anywhere near complete. Some of the Companies are short of crude, or even of products for their marketing networks, and must purchase some of their requirements; others have more crude than they need, which they want to sell. At all stages of the industry 'independents' are active in varying degrees and their actual (or potential) competition cannot be ignored by the dominant firms. Moreover, the combination of the various elements which determine the behaviour of the firms in the industry changes over time in response to economic incentives and political pressures, and these changes themselves provide inducements to further change. Our problem is to trace and analyse the broad outlines of these interacting considerations.

PRICING BEFORE THE SECOND WORLD WAR

Before the Second World War international trade in crude oil was insignificant; an overwhelming proportion of petroleum exports was in the form of products which had been refined at or near the source of the crude, and the US was an important supplier in almost all markets. Prices of oil products

[1] For example, Shell Transport and Trading reported in the 1961 *Annual Report:*
'When oil crosses international frontiers it is, as far as the Royal Dutch/Shell Group is concerned, almost invariably the subject of sale and purchase, either between two Group companies or between a Group company and some third party. The prices at which these purchases and sales take place are therefore of particular significance not only as one of the major factors influencing the profits or losses, and hence the tax liabilities, of each individual Group company, but also in relation to the balance of payments problems of nearly every country where a Group company is trading.

'Supply pricing both to Group companies and third parties is largely based on the f.o.b. export selling prices for crude oil and products, which are "posted" by Group companies at various export centres. Posted prices for crude oil at the principal points of export outside the US were introduced in the early 1950s. Their purpose was to provide a rational framework which would represent a fair commercial price as between sellers and buyers and would be acceptable to governments as a basis for calculating profits.' P. 15.

[2] However, if there are independent refiners, and therefore some independent demand, an integrated firm, even if it does not sell much crude in the open market, may still be very much concerned with the market price of crude because cheaper crude available to the independents may encourage price competition from them in product markets.

everywhere tended to equal f.o.b. prices from the US plus transport costs, although since the elasticity of demand differed in different markets there were some deviations from these prices, depending on the conditions of competition in particular markets. Moreover, the prices charged to their affiliates by the integrated Companies were not always identical with those charged to independent buyers.[1]

As we saw in Chapter III, rivalry among the major Companies was often intense and price wars were not uncommon. A particularly significant one broke out in India in 1927 between Asiatic Petroleum, a subsidiary of the Royal Dutch/Shell, and Standard Oil of New York. One of the immediate causes of the 'war' was the introduction by Standard of kerosene from Russia in the Indian market. The USSR had expropriated Shell properties in Russia without compensation, and Shell quite naturally resented Standard's action in subsequently buying the 'stolen' oil and using it in competition with Shell in the Indian market. So bitter was the feeling between the Companies at the time that the war rapidly spread to other markets, and was partly responsible for the initiation of one of the best known and most comprehensive of the cartel agreements among the major Companies—the 'As Is' or 'Achnacarry' Agreement of 1928.

With this agreement the 'Big Three'—Shell, Standard, and Anglo-Persian—hoped to limit competition in product markets among themselves and others who joined them, by agreeing on the principles that would govern market shares and prices in all world markets except the US, by laying down the broad outlines of the type of organization of the market through which the objective could be achieved, and by setting out the rules to be followed. The general principles of 'As Is' provided the basis for a number of agreements in local markets, few of which seem to have been entirely successful for long, but which clearly indicated the desire of the Companies to conduct their own operations with due respect for the established positions of all, to discriminate against outsiders and to co-operate as far as possible in supplying world markets without disruptive price competition.[2]

[1] In discussing Jersey Standard's foreign operations in the 1920s, for example, Gibb and Knowlton point out that foreign prices were affected by local supply conditions, including the 'political squabbles that delayed the development of Middle East oil reserves', '. . . agreements between competitors to divide markets', and '. . . Jersey Standard's very strong marketing position in certain areas'. They go on to add: 'Profiting generally from its ability to get better prices abroad than at home, the Jersey Company also directed its price policy at meeting specific competitive and supply situations. Two export price schedules were maintained—published prices and so-called inside prices. Published prices formed the basis for most of the business with non-affiliated companies, while the lower "inside" prices were quoted principally to affiliates . . . To retain the large foreign business done with the Anglo-American Oil Company Limited, and the Vacuum Oil Company—neither of which was an affiliate—the Jersey Company at least occasionally extended to them the privilege of the prices quoted by Jersey to the others, and the parent company sometimes differentiated in these family quotations in order to allow a particularly hard-pressed affiliate to meet a local competitive situation. Prices were also changed in order to swing the American supply point for the foreign trade to whatever loading port was most advantageous in view of the domestic supply and refining situation.' P. 499.

[2] The inter-war attempts at cartel agreements are described in the Federal Trade Commission Report. I do not intend to discuss them here since their details have little bearing on the theme of this study. A useful summary of the motives and expectations of the Companies is

Development of a Pricing "System"

It is not clear how effective the market-sharing aspects of the 'as is' agreements really were, and their importance was undoubtedly different in different markets. It seems probable, however, that the explicit cartel arrangements were less important in restraining price competition among the larger firms than was the pricing system adopted by the major Companies. Apart from sporadic price wars or deviations in response to local and short-period market situations, common pricing procedures had grown up which persisted well into the middle of the Second World War. The system provided in many ways a more secure basis for an 'orderly' adjustment of the industry to the changing conditions of supply than comprehensive market-sharing agreements accompanied by restrictive and somewhat complicated intra-cartel pricing arrangements could have been expected to achieve.

Because the US was the marginal supplier to most world markets products were priced by the addition of transport costs to prices f.o.b. the US Gulf. The prices so calculated were referred to as 'world parity prices'. The system grew out of, and may for some time have been consistent with, the underlying comparative cost conditions of the industry and could not by itself be taken as evidence of monopolistic restrictions in the market, as cartel agreements necessarily must be. Even under highly competitive conditions low-cost oil products from the Persian Gulf would not be sold in nearby markets, such as India, at prices less than those of similar products from the US so long as the latter were required to meet market demand. When, however, common recognized formulae for calculating prices are adopted by all sellers and the system becomes institutionalized, as it were, it turns into a restrictive basing-point system.

A basing-point system is in essence a system in which all sellers, no matter where located, calculate delivered prices by taking generally accepted f.o.b. prices at one or more specified locations ('basing points') and adding standardised freight charges (not actual freight payments) from the 'basing points' to the place to which the commodity is shipped, regardless of the actual origin of the commodity. Such a pricing system is a very effective device for ensuring not only that uniform prices are quoted by all sellers but *also that low-cost producers cannot use their lower costs to expand their share of the market by reducing prices.* The system is restrictive precisely because, when adhered to by all sellers, it prevents the expansion of low-cost production by price competition.[1] Because competition would in fact produce uniform delivered prices in any given market, the oil Companies pointed to the uniformity of "parity prices" as evidence of the existence of competition.[2]

---

contained in a speech by Sir John Cadman, Chairman of Anglo-Persian, delivered in 1932, in which he indicated his confidence that the principle of 'as is' was becoming 'the keystone of co-operation in international petroleum trading outside the US'. *World Petroleum*, December 1932, pp. 20–31.

[1] See Fritz Machlup, *The Basing-Point System*, for a full exposition of the economic effects of a basing-point system.

[2] Here we have another instance of ambiguity in the use of the word 'competitive'. Businessmen often call their prices 'competitive' simply because they are not higher than the

Clearly, however, uniformity of prices provides no empirical criterion for distinguishing competitive from collusive behaviour; we must look at the way in which low-cost producers are enabled to expand their share of the market.

It is not clear when and to what extent the basing-point method of pricing became more or less formally adopted by the Companies, but it is clear that delivered 'parity prices' played an important role in their pricing policies. With the advent of the notion of 'as is' in the Achnacarry Agreement, the prevailing ideas of 'normal' prices would inevitably have had to be included in the definition of a position to be kept 'as is', and this would equally inevitably have required that sellers adhere to prevailing prices. However, it would not necessarily have required the Companies formally to adopt an agreed and uniform system of pricing, or even to agree on a firm schedule of prices for each of their many products in all markets. If they simply refrained as much as possible from upsetting existing conditions, the combination of the historical position of oil exports from the US in world markets and a general desire among the major Companies to observe the status quo, would have created a situation with most of the characteristics of a basing-point system, but without any formal and collusive adherence to such a system by the Companies. A formula for calculating freight charges on a common basis would, however, have had to be devised.

Under effective price competition the production and use of low-cost oil from the Eastern Hemisphere would have increased to the point where no further expansion of output was deemed profitable at prevailing prices. In other words, if to call the price structure and the parity of delivered prices a 'basing-point system' is merely a misleading way of describing a natural historical equilibrium, we should not expect to find attempts to restrain competition in order to maintain the 'parity prices', nor evidence that the expansion of low-cost sources of supply was being constrained to avoid spoiling the market. In fact, we find an abundance of such evidence.

We must first note, however, that the system was not wholly effective, in part, precisely because it provided little scope for the expansion of markets for new sources of supply. And if existing occupants of a market are not prepared voluntarily to make way for new producers a 'price war' is likely to break out. The immediate background of the 1928 Achnacarry ('As Is') Agreement was the price war in India which resulted partly from the difficulties the Companies supplying the Indian market had in coming to agreements that would give acceptable shares of that market to suppliers from the nearby Persian Gulf, while at the same time maintaining the parity system

---

prices of other sellers (if their prices are lower they are called 'very competitive'). Whenever oil prices were challenged, as they were, for example, by the Damle Commission in India in 1961 (see below pp. 224–8), the oil Companies concerned insisted that the prices were 'competitive', and to prove it they blandly pointed out that oil could not be obtained more cheaply from any other source. So long as each Company could say this with reasonable confidence, the critics remained baffled and silent, or else muttered darkly about cartels, for the existence of which they could produce no conclusive evidence. (It is perhaps ironic that after 1960 the Companies pointed to the wide range of prices and terms offered to secure contracts as equal evidence of extensive competition!)

of pricing.[1] The very existence of the Agreement, and the arguments put forward for it, are evidence enough that the observance of any pricing system that may have existed was not sufficiently effective to prevent significant price competition.[2] It should be remembered, however, that economic conditions in the later 1920s and the early 1930s were difficult for many industries, including the oil industry, the problems of which had been aggravated in the US by an unusually large discovery in Texas.

The Agreement itself provides evidence of the desire to restrain competition in order to maintain prices. The 'Preliminary Statement' attached to it stated: 'Excessive competition has resulted in the tremendous overproduction of today, when over the world the shut-in production amounts to approximately 60 per cent of the production actually going into consumption.'[3] And what evidence we have of the costs of production in the Persian Gulf, and of the conservative policies followed by Anglo-Persian in spite of its large reserves of low-cost oil, supports the conclusion that the possibilities of 'over-production', which could only mean 'cheap products', was very much in the forefront when output was planned.[4]

Finally, as an increasing proportion of Eastern Hemisphere markets came to be supplied from the Middle East, the delivered prices of oil products should have fallen in these markets and f.o.b. prices in the Middle East should have fallen relatively to those from the US Gulf, since the marginal costs of additional Middle East Oil were evidently very much lower than those prevailing for the relevant rates of supply in the US, especially after pro-rationing was introduced in that country. Such changes did not take place before the war.

The first decline in delivered prices for oil products from the Middle East took place during the war, and then only under the pressure of governmental authorities who forced the creation of a second 'basing point' in the Persian Gulf for fuel oil, although at the same time they accepted the Companies' demand that f.o.b. prices at the Persian Gulf should equal f.o.b. prices at

[1] In India the delivered 'parity' prices were often treated as a kind of standard minimum price. Local agents were often free to ask more if they could get it and, if necessary, they could even discount this price to retain their outlets, sometimes on their own initiative and sometimes only after reference to the head office. A useful discussion of the way the system operated in the inter-war period in the Indian market can be found in B. Dasgupta, Chapter VI.

[2] Then, as in more recent times, a good deal of the trouble was attributed to Russian exports and the complaints have a familiar ring: 'The abnormal situation created by excessive competitive production in the US, a country which ordinarily dominates the position in the oil world, has been rendered more acute by the larger exports from Russia, where the peculiar conditions seem to prevent—for the time being, at least—the ordinary operation of economic law. It is obvious that this large source of supply, freed temporarily from the various charges and restrictions which affect the production of petroleum in other parts of the world, must be a disquieting factor in a market situation already very much disturbed.' Chairman's statement at the Eighteenth Ordinary General Meeting of the Anglo-Persian Company Ltd., November 2, 1927, p. 4.

[3] Quoted from the Report of the Federal Trade Commission, p. 200.

[4] '. . . it is in the interests of Persia, as it is of this Company, that production should be steadily controlled—that is to say, steadily regulated in conformity with the world's demands; that the reserves of oil underground should not be extravagantly and uneconomically forced to the surface, regardless of the world's requirements. Fortunately, we are not impelled to overproduce. . . .' Chairman's statement, Anglo-Persian Oil Co. (1927), p. 5.

the US Gulf.[1] The adoption of two 'basing points' meant that delivered prices could be calculated on the basis of f.o.b. prices plus a standard freight from *either* the Persian Gulf *or* the US Gulf to any given destination. Clearly there would be an area over which prices based on the Persian Gulf would be less than those based on the US Gulf, and a boundary at which delivered prices from the two areas would be the same.[2]

POST-WAR CHANGES

Although in 1939 nearly three-quarters of the crude oil produced outside North America and the non-communist world was refined in or very near the countries producing it, there was already a discernible movement toward the establishment of refineries in consuming areas. After the war this movement gained momentum; between 1947 and 1951 refining capacity in Western Europe nearly tripled, and by 1951, the second year after the re-opening of their refineries, Japanese capacity had exceeded its pre-war level. Thus the amount of crude oil entering into international trade rose rapidly and the relative importance of trade in products declined. In consequence, the price of crude oil began to take on more significance from the point of view of both importing and exporting countries,[3] although most refineries were owned by the integrated Companies producing the crude oil and the 'price' was still an internal transfer price.[4]

Crude Oil

Between 1943 and 1945 the Persian Gulf emerged as a second basing point for crude as well as for fuel oil. The oil Companies' insistence that f.o.b. prices in the Persian Gulf should be set equal to those in the US Gulf, in spite of the lower costs of the former, has been defended on the ground that demand could be expected to rise rapidly in the US after the war while there was considerable uncertainty as to the speed with which Middle East productive

[1] This issue has aroused much indignation, largely because it touches on the sensitive question of abuse of monopoly in times of national emergency. The indignation has centred around the prices charged during the war and the post-war reconstruction. Private firms are often in a dilemma in such circumstances, since even if they are averse to 'profiteering', i.e. using the national emergency to enlarge their own profits, they at the same time do not want to have their 'post-emergency' position seriously undermined by concessions made during the 'emergency'. The matter is now of historical interest only; the behaviour of the oil Companies over the issue of prices, in particular the refusal of Aramco to supply the US Navy in 1943 at lower prices, and the Companies' threat to withdraw supplies from Western Europe in 1948–9 if the ECA pressed too hard (see Leeman, pp. 146 ff), does not arouse one's admiration, but one can see their point of view.

[2] The detailed working of this system in the oil industry has been extensively analysed elsewhere (see the Report of the Federal Trade Commission, pp. 352 ff, W. Leeman, Chapters 4 and 5, and H. J. Frank, Chapter 1). For this reason I shall not go into further detail here. One weakness of the analysis in these works, and especially that of the Federal Trade Commission, is a failure to make clear the conditions under which the results of the operation of the system, ostensibly consistent with competition, may become restrictive.

[3] See UN Economic Commission for Europe, *The Price of Oil in Western Europe*, p. 4.

[4] Shell International has itself pointed out that in 1950 most of the trade from export terminals outside the US 'was confined to integrated companies' and that there was a 'consequential absence of prices established by normal arm's length trading. . . .' *Current International Oil Pricing Problems* (August 1963), p. 5.

capacity could be expanded. In these circumstances, the short-run marginal cost of expanding Middle East production could have equalled or exceeded marginal costs in the US, thus justifying the oil Companies' position.[1]

Such movements did not in fact take place; with the removal of price controls in the US in 1946 the Companies promptly raised crude-oil prices both in the US and in the Middle East, but Middle East prices were not generally raised as far as those in the US. In 1948 the relative decline of Middle East prices was intensified and the Caribbean emerged clearly as the Western Hemisphere basing point, since the US had become a net importer and Venezuela had been supplying far more oil to Europe than had the US.[2] The relationship between Caribbean-United States Gulf prices and Persian Gulf prices which had been established in the 1948–9 negotiations between the Companies and the Economic Co-operation Administration remained roughly the same until the aftermath of the Korean war. After a slight drop in 1953, the *relative* posted price of Middle East oil rose, only to resume its decline after 1956.[3]

The relative decline in Middle East prices has been the subject of an elaborate analysis involving picturesque theories explaining the location of a 'watershed' where maximum delivered prices from the Persian Gulf and the Western Hemisphere were equal and which 'swept dramatically westward until it reached the Atlantic and then, perhaps, crossed the ocean'.[4] It is true that if the Companies were calculating prices using formulae appropriate to a dual basing-point system, the relative f.o.b. prices at the basing points would determine the boundary or 'watershed' at which delivered prices from both areas would be equal. It is also true that the price movements that in fact occurred up to 1948–9 were in the direction that one would expect if the markets for Middle East oil were being expanded by straightforward price competition. For some this has provided adequate proof that 'real competition' existed, and the 'moving watershed' type of explanation is cast in terms which imply the operation of competitive forces in crude-oil markets, no cause other than 'competition' being adduced to explain the movements.[5]

[1] H. J. Frank, p. 27, seems to accept this point of view.

[2] H. J. Frank has given the most detailed description of the movement of oil prices in this period yet to appear.

[3] The ratio of the average posted price of 34° crude f.o.b. Ras Tanura to the average price of oil of the same gravity f.o.b. the US Gulf in each of the years 1948–61 was as follows:

1948	·76	1952	·63	1956	·66	1960	·57
1949	·67	1953	·61	1957	·58	1961	·54
1950	·63	1954	·66	1958	·62		
1951	·63	1955	·66	1959	·57		

Calculated from the figures given by Charles Issawi and Mohammed Yeganeh, Table 21, p. 68. Formal posting did not take place before 1950 and prices for the earlier years are estimated using published data.

[4] The nature of the theory has been painstakingly set forth by Leeman, from whom the quotation is taken (p. 96). It was earlier foreshadowed by Frankel in *Oil Forum*, November 1948, and elaborated by W. J. Levy in 'The Past, Present and Likely Future Price Structure for the International Oil Trade', 1951.

[5] Issawi, for example, seems to accept the notion that the crude oil was priced in order to make it competitive '. . . it was necessary that the *relative* price of Middle Eastern crude oil should fall so as to enable it to compete with Western Hemisphere oil in Western Europe', p. 66. But he nowhere explains *why* this was necessary, given the inter-affiliate nature of the trade, and one is left with the implication that it was a necessary condition of selling the oil.

The Federal Trade Commission Report, however, found the same circumstances sufficient to sustain the conclusion that the oil companies operated a collusive basing-point system.[1] Adelman holds that the evidence is insufficient to support either the 'collusive' or 'competitive' hypothesis. While not denying that the pricing system was discriminatory in the sense that the companies received different returns from similar sales to different buyers until the price adjustments that took place in 1948–9 when Middle East prices declined in 'the anxious presence of the US Government',[2] he is not convinced that the subsequent price movements are to be explained in terms of a monopolistic policy of the oil Companies acting as a group. If the Companies merely let matters alone and did as little as possible to disturb prices, such changes would have taken place under the circumstances without the necessity of collusive action.[3]

Thus, the same circumstances are used to support widely different conclusions. On the evidence, Adelman's conclusion seems more nearly correct, but neither his, nor any of the other explanations, takes adequate account of the most significant structural characteristic of the industry—the vertical integration of the major groups who controlled virtually all of the oil of the Middle East and around four-fifths of the refining and marketing capacity of Europe, as well as 60 per cent of the refinery capacity (excluding asphalt plants) on the US east coast.[4] For these groups, the market *prices* of crude oil could not have determined the 'competitiveness' of oil from the Middle East in relation to oil from other sources to be used in their own refineries, simply because they were the buyer/refiners as well as the seller/producers of the oil.

The Importance of Integration

It would make no sense whatsoever for an integrated Company to be guided by market prices in the determination of the sources from which it would lift its oil; this decision can sensibly be made only with reference to *costs*. Since most oil moved within the integrated framework of the major firms and there was hardly an effective independent market for crude oil (and no evidence at all of the kind of competitive bidding for independent contracts similar to the evidence that appeared for the years after 1958–9), it seems unlikely that competitive *market* pressures can explain the post-war willingness of the Companies to reduce the relative price of crude oil from the Middle East.[5] The requirements of the internal organization of the vertically in-

[1] Federal Trade Commission, pp. 360 ff.

[2] M. Adelman, 'The World Oil Outlook', p. 81.

[3] Ibid., pp. 85–6. One of the arguments advanced by Adelman for rejecting the collusive hypothesis is that at one time prices were equalized at New York and equalization there instead of London would not maximize profits for the Companies as a group (p. 81). I find this argument unconvincing in view of the fact that the Companies were not really concerned with maximizing profit from crude-oil sales since they were shipping to their own refineries. See the following discussion in the text.

[4] See the analysis in 'The Price of Crude Oil: A Rational Approach'. Paper presented by OPEC to the IVth Arab Petroleum Congress.

[5] Perhaps the first report to bring this out clearly was that of the UN Economic Commission for Europe in its *The Price of Oil in Western Europe*. This Report created considerable annoyance among the oil Companies and it is widely stated that after it the UN was 'warned off' oil. This was reported by J. E. Hartshorn (p. 286), and has been confirmed privately to

tegrated groups provides a much more convincing explanation.

For an integrated Company, the price at which crude oil is transferred from producing to refining affiliates determines the distribution of its total profit between crude production and refining. For any given level and structure of product prices and costs, the higher the transfer price of crude oil, the lower the profits attributed to refining and distribution, but total profits (before tax) are unaffected by the internal (accounting) distribution of profits.[1] The price at which crude is sold to outsiders is, of course, another matter, for on this profits are made for the integrated group. Those Companies owning crude oil that was surplus to their requirements disposed of substantial quantities to other major Companies on very long-term supply contracts, but the prices in these contracts were negotiated special prices; they were not the prices whose movements we have been discussing. Other sales to outsiders, apparently made mostly, but not entirely, at the so-called 'world parity prices', were generally estimated to have been around 10 per cent of total sales in the immediate post-war period. Thus, the chief effect of a relative reduction of Middle East prices was on the prices at which crude oil was transferred from the producing to the refining affiliates of the international oil Companies.

Internal prices have more than an accounting use for the large firm, particularly when the firm is organized as a group of quasi-autonomous affiliates, for they are the prices that are relevant for the calculation of profits by the managers of the affiliates buying as well as selling the commodity concerned. Not only is it likely to be extremely inconvenient from an administrative point of view if internal prices are out of line with the 'market' prices of the same commodity, but in general internal prices should be such as to provide a financial incentive for the managers of affiliates to take the actions desired by their head offices as they endeavour to ensure the profitability of the enterprises they control.[2] Even if the transfer prices of crude oil from producing affiliates to trading affiliates, and finally to refining affiliates, of the same group are of little significance for the consolidated profits of an integrated Company (apart from the incidence of taxation—of which more later),

---

me by a number of UN officials. It is in any event true that since that time no serious analytical discussions of oil economics (as contrasted with oil technology) or pricing have appeared in UN reports.

[1] To the extent that there is outside equity in refining affiliates, the location of profits makes a difference to the integrated company. But shared ownership outside the circle of the major companies does not seem to have been important before the war in the international industry.

[2] 'The delivered supply prices for products at import terminals should be sufficiently below the general level at which sales are made to consumers to permit marketing companies to cover their costs and earn a reasonable return on capital. In the same way crude-oil prices should be sufficiently below product prices to enable refining companies to cover their costs and earn an adequate return on capital. . . . In practice, integrated oil companies arrange their affairs mainly to supply their affiliated companies. . . . Having regard to the long-term continuing basis of inter-affiliate trading, the desire for a measure of stability, and the co-ordination of supplies between different phases of the industry by integrated companies, it is hardly surprising that posted prices for crude oil and products are relatively stable by comparison with many other commodity prices.' Shell International, p. 7.

managers of subsidiaries and affiliates want to make a profit. If they are to be persuaded to take more oil from one source rather than from another because to do so is cheaper from the point of view of the integrated operation, then it will be administratively convenient if the delivered prices they are charged make it profitable for them to do so.[1] Otherwise the integrated Companies will have trouble with the managers of their affiliates who like to report profitable operations to the head office. The desired result can be accomplished either by a reduction of f.o.b. prices or by reducing some other element of the c.i.f. cost, such as freight charges. The choice is likely to be influenced by tax considerations.

As more Middle East crude became available, and especially as European refineries were completed to process it, the interest of the majors in an f.o.b. price of crude from the Persian Gulf that made it profitable from the narrow point of view of their refining affiliates and subsidiaries to use oil from this source, coincided with the interest of the US Economic Co-operation Administration in securing a reduction in the cost of oil imports to Europe from the Persian Gulf. Thus, the post-war 'revolution' in crude-oil prices is more easily explained as a deliberate decision of the major Companies (to be sure, taken under government pressure) than as the consequence of competition among sellers of Middle Eastern crude oil forcing prices down as they endeavoured to extend their marketing areas. It was as *buyers* of crude that the Companies decided when and how far prices should fall, and surely one of the most important factors in their decision was the desirability of using low-cost Middle East crude in their own refining operations in the Eastern Hemisphere, particularly in Europe.

The ECE report on the price of oil made some rough calculations of the profitability of refining in southern as contrasted with north-western Europe, using the formula for pricing which would equalize prices just to the east of Italy of oil originating in the Caribbean and in the Persian Gulf, and found that losses would have been made on refining in north-west Europe.[2] Since the Companies had been doing increasing amounts of refining in that area it would clearly have been advantageous for them to reduce the delivered prices of crude at least to the point at which operations became profitable for their refining affiliates. This would have entailed the relative reduction in prices f.o.b. the Middle East that in fact took place.

None of this is intended to deny that the fall in the relative price of Middle East oil was related to its lower costs; the point is that it was to the interest of the *users* of crude and the *sellers of products* (who happened also to be producers of crude) to adjust the 'price' of Middle East crude to promote the effective functioning of the integrated Companies.[3] Although this is merely

[1] This conclusion is consistent with that of J. E. Hartshorn. After pointing out the ambiguities and inconsistencies in analysing the movements of oil prices with the so-called 'net-back' formulae (which were used to explain the westward sweep of the 'watershed') he wrote, '. . . the companies used these pricing formulae, after all, to "simulate" the conditions that might have applied in a freely competitive market, in order to assist the sensible allocation of resources between "departments" of large international companies in conditions where such competition was largely lacking', p. 152.

[2] UN, Economic Commission for Europe, pp. 30–2.

[3] For this reason I find unconvincing the intricate analysis of the type presented by the

another way of talking about competition between crudes from different sources, the nature of the competition is very different from the market competition implied by the alternative explanations.[1] Furthermore, there is no need at all to look for collusive action by the Companies with respect to the price of crude oil, except perhaps on sales to outsiders.

This way of looking at the price history of the period provides an explanation of the gentleness of the movement of relative prides on the one hand and of its unevenness in detail on the other. Although the gap between prices in the Middle East and the Western Hemisphere widened, it widened slowly, and generally as a result of changes in prices in both areas, with a smaller change in Middle East prices but in the same direction. Had sellers of Middle East crude been reducing prices in competition with sellers from the Western Hemisphere, one would have expected the changes in prices to have been more erratic and more violent, unless the Companies were acting in collusion. For this reason, collusion of a sort is generally assumed; hence, the belief in the 'International Petroleum Cartel'. On the other hand, if the Companies were deliberately acting in accordance with an agreed policy, it is difficult to explain some of the short-term differences in their individual price policies, which were of the kind that one would expect under oligopolistic conditions when Companies are jockeying for position and making adjustments to their prices, perhaps following 'price leaders', to be sure, but with no precisely formulated and agreed policy.

The Importance of the Immediate Past

So long as price competition is weak, any explanation of existing prices in an oligopolistic industry will almost certainly contain a strong historical element. That is to say, prices at any given time will tend to be related more to what they were in the immediate past, than to the underlying supply and demand conditions in the present or expected in the future, since an 'oligopolistic equilibrium' arises out of a simple reluctance on the part of each existing

Federal Trade Commission or by Professor Leeman, which deals with the effects of differing 'netbacks' arising from differing degrees of 'freight absorption' or 'phantom freight charges' on sales in different markets. No wonder the various experts found 'serious anomalies' in their attempts to apply the formulae, and perhaps the businessmen who spoke disparagingly of the 'theoretical netbacks' calculated by 'hopelessly academic' economists spoke more truly than Leeman gives them credit for! (See Leeman, pp. 129 and 122.) Equally unconvincing, however, is the treatment of the oil Companies as profit-maximizing crude-oil sellers.

[1] The Economic Co-operation Administration objected to paying discriminatory prices for oil and products, i.e. prices that gave the companies higher realizations on shipments from the Persian Gulf to Southern Europe than they obtained on similar shipments to North-west Europe. The ECA was successful in securing reductions in crude-oil prices; it failed to obtain similar reductions for products. Leeman, who presents an excellent analysis of the 'ECA case' (Chapter V) asks, 'How can we explain what appears to have been early ECA successes in getting crude prices reduced followed by later failures to obtain reductions in either crude or products?' (p. 147). His answer is that either the position of ECA became weaker as time went on and conditions changed, or that the Companies may have recognized the 'essential justice' of the ECA's position on crude-oil prices. But if some reduction in crude-oil prices was really in the interests of the operations of the integrated Companies, given their increasing use of Middle East oil, whereas reductions in product prices were not, there is no difficulty in explaining the greater willingness of the Companies to reduce crude prices up to a point.

seller to disturb the *status quo*. More or less recognized price 'leaders' tend to emerge and changes take place in a reasonably orderly fashion, unless there are significant differences of opinion among the firms—and such differences may arise especially in times of great uncertainty, and may then lead to some confusion.[1]

There can be little doubt that before the war the international oil Companies did operate a collusive pricing system in a number of markets and that in general the international pricing system was restrictive and highly discriminatory, reflecting prices in the US which were supported by a pro-rationing control over crude-oil production. After the war, the oil Companies resumed their peacetime activities with a level and structure of prices that had been modified during the war as noted above, but which in most respects still largely reflected the pre-war pattern. Almost immediately the further modifications that we have described in the relative prices of crude oil from the Middle East and the Western Hemisphere took place; a formal basing-point system could no longer be said to exist and some of the discriminatory aspects of the pricing structure so far as it affected Western Europe were eliminated, although they seem to have remained east of Suez.

Both during and after the war, such changes were deliberate modifications of an existing price structure; all were *adjustments* of existing levels initiated by price leaders rather than market responses to changing conditions in the sense that fluctuations in cocoa, rubber, or wool prices are market responses (in the absence of effective market stabilisation schemes), or even in the sense that the changes in the market prices of crude oil since 1958 can in many ways be looked at more as market responses than as 'adjustments' deliberately made to a prevailing level.

Any change in any price must, of course, be measured from a given point, but the influence of the past on the magnitude of the change can vary enormously. For example, at one extreme, the price of a seasonal crop in a given year may be determined entirely by the size of the crop in relation to demand and not at all by past prices (except indirectly to the extent that past prices influenced the size of the current crop). Almost at the other extreme, the annual average posted price of crude oil in the US Gulf, where supply was officially regulated, changed in only one of the nine years from 1948 through 1956. Outside the US, the posted prices of crude oil changed more frequently; for example, the annual average posted price f.o.b. Ras Tanura changed in four of these nine years, but except for the drop between 1948 and 1949, in only one year was the change as much as 8 per cent. The average price of Iraq oil f.o.b. Tripoli and Banias did not change at all between 1950 and the Suez crisis.[2]

It is precisely because crude and product prices in the decade before 1957 can be regarded as having been primarily determined by historical circumstances, adjusted minimally to take account of changes in conditions, that

[1] Examples of such confusion are given in H. J. Frank, pp. 29–30, 112–13.

[2] C. Issawi and M. Yeganeh, p. 68, Table 21. Before 1950, when the 'posting' of prices began, the figures are estimated on the basis of published prices for Saudi Arabia. The annual averages do in fact conceal some price movements within each year and give a somewhat exaggerated appearance of stability, but are not seriously misleading.

we can accept the view that no collusion was needed among the Companies to maintain them. The Companies might well have merely respected the historical position, taking little action in view of the uncertain consequences. There are indeed so many reasons, including taxation, and, perhaps, a fear by the Companies that some major consuming countries might react adversely to heavy price-cutting in crude oil that upset either domestic oil producers or the domestic coal industry, why oil prices should have moved as they did and no further,[1] that an explanation in terms of collusion among the Companies is not required.

At the same time the historical relationship between crude and product prices, which placed the greater part of the profits of the integrated Companies in crude-oil production, was no disadvantage to the Companies; quite the contrary, for a higher price of crude oil raised the cost to independent refiners and thus reduced the incentive for newcomers to take up refining. Entry into refining was in the nature of the case easier than entry into crude-oil production in the early post-war period. For the US Companies, the greater the profits in crude-oil production, the greater was the tax subsidy they could claim in the form of a depletion allowance against their taxable income under the US tax laws. The Companies had, in fact, every incentive not to disturb existing price relationships more than was needed to facilitate the increasing use of Middle East oil by their own refineries. Some short-term changes were made, usually related to changes in the US, and at the time of Suez the Companies had no hesitation in raising prices.

Product Prices

Crude-oil 'prices' in the post-war period up to the late 1950s were more related to the internal requirements of the international firms than to conditions in any kind of free market for crude oil. Product prices were another matter, for here the conditions of competition must have had a much greater significance.

The same pressures which during the war forced the Companies to accept a second basing point for bunker fuel in the Persian Gulf and later to extend it to crude oil, led them also to accept it for other oil products. As crude-oil prices f.o.b. the Persian Gulf fell relatively to those in the Caribbean, and as European refinery capacity expanded using lower-cost crude, one might have expected product prices in Europe and in the Persian Gulf to fall relatively to those in the Caribbean if competition had been significant. On the other hand, if the oil Companies were in fact avoiding price competition in their marketing rivalry and were, except in special local and temporary circumstances, calculating their delivered prices on the basis of the same f.o.b. prices, whether from the Caribbean or the Persian Gulf, plus some accepted freight rate, and if independent suppliers[2] were not significant competitors, there would be no force present to cause European ex-refinery prices to fall relatively to product prices f.o.b. the Caribbean. Of course, 'import parity' would be maintained so long as imported products were supplying an

[1] See M. A. Adelman, *World Oil Outlook*, p. 84 f.

[2] A number of independent refiners processed crude on contract for major Companies and hence not all independent refining capacity represented potential 'independent competition'.

important part of the market, but the significant point is that the level of this 'parity' did not fall with the price of crude oil.

In other words, no 'revolution' in product prices accompanied the post-war changes in crude-oil prices, although it is here that the real effect of competition should appear in an integrated industry. Perhaps one of the strongest pieces of evidence that product prices were not 'competitively' determined is the fact that the structure of relative prices maintained by the Companies in Europe reflected, with some modification for fuel oil, the pattern of demand in the US,[1] even though the distribution of demand among the different products was very different in Europe from what it was in the US.

The ECE report, in examining the consequences of a pricing system under which European prices were based on US prices plus transport costs, pointed out that the system made 'the profitability of refining in western Europe sensitive, in a perverse fashion, to price changes in the US. A substantial change in fuel oil prices in the US has only a small effect on the profitability of refining, since fuel oil is so small a proportion of total output. In western Europe, where the proportion is large, the effect will be large also. Conversely, even small changes in motor spirit prices are matters of great importance in the US refineries but are of much less concern in Europe'.[2]

Indeed, so inappropriate were the relative prices of motor spirit and fuel oil that the maintenance of 'import parity' proved untenable in practice and discounts on fuel oil were apparently widely offered. It must be remembered that in the absence of explicit collusion among the Companies, which to be effective in maintaining a rigid and inappropriate *pattern* of relative prices would have required some sort of policing, the implementation of the basing-point system depended on the willingness of all Companies to adhere to the parity prices. If the pattern of prices is seriously out of line with the pattern of demand, some Companies, at least, are likely to be tempted to offer discounts, and competitive pressures will erode the price structure at the edges. In other words, formal and historically determined prices will tend to give way in response to market forces unless governments intervene.

Although it would be wrong to explain the level and structure of prices in terms of price competition in a market ruled by the free play of demand and (uncontrolled) supply, it would be equally wrong to ignore the significance of competition as the driving force of change, slow and imperfect though it might have been. As the 1950s wore on, competition in product markets increased, especially in the rapidly growing markets of Western Europe and

[1] In the *Bulletin de la Communauté européene du charbon et de l'acier*, '*Etudes sur les perspectives enérgétiques à long terme de la Communauté européene*' (Luxembourg, December 1962), it is noted that until around 1958 '. . . *les prix des produits raffinés en Europe occidentale se trouvaient fixés à la parité d'importation ex-Caraïbes ou ex-golfe de Mexique; bien entendu, des circonstances locales, des réglementations de prix, des structures particulieres au niveau de l'industrie de raffinage, etc., affectaient plus ou moins l'application de la règle; mais d'une manière générale, la structure des prix des raffinés en Europe occidentale était largement influencée par celle des Etats-Unis. Ce mode de fixation des prix, qui continue à servir de référence pour les barèmes publies par les grandes compagnies, ne reflète plus toujours les conditions réeles sur de nombreux marchés européens, specialement pour les fuels lourds qui font l'objet d'une pratique étendue de rabais.*' (P. 87.)

[2] UN, *The Price of Oil in Western Europe*, p. 28.

Japan, and by the beginning of the 1960s it was clear that for both crude oil and products a very different picture had emerged from that which had prevailed before and immediately after the war.

EROSION OF THE PRICING SYSTEM

During most of the 1950s, and especially after Suez, the major Companies seem to have used a variety of techniques to grant concealed discounts on the effective c.i.f. price of crude oil in a number of markets where the competition to sell products was particularly keen.[1] The techniques included freight allowances on delivered prices, the re-charter of tankers at favourable rates to f.o.b. buyers, extra quality concessions, 'spiking' to uplift the quality of crudes without extra charge, the long-term financing of crude purchases at favourable rates, and the acceptance of partial payment in soft currencies by the US Companies.[2] By 1958 prices were being openly discounted and the reason clearly was steadily mounting competition in crude as well as in product markets: the cries of pain that competition which is really effective in reducing prices almost invariably invokes from sellers were distinctly to be heard. By 1960 published bids for large crude-oil contracts showed large differences in the prices and terms quoted by rival bidders; no longer could one plausibly speak of a 'world price structure' for crude oil.[3] Between 1958 and 1965 the prices of petroleum products in Europe before taxes fell by nearly 25 per cent, and oil Company officials were using this to point out the benefits of competition to the consumer.[4]

Competition

The kind of price competition that arose was not of the 'price war' variety,

[1] The significance of even small and scattered differences in c.i.f. prices can easily be under-estimated. The Australian Tariff Board, for example, in recommending in 1956 a reduction of protection for refined products, took account of the fact that 'some at least of the Australian refineries have a slight advantage in crude prices. This may be due to the location of refineries, special prices or relationship with suppliers'. It referred to the UN Report on 'The Price of Oil in Western Europe', stating, 'Indeed, for an integrated company, the allocation of profit between crude oil production and refining is merely a matter of internal bookkeeping.' (Report on 'Motor and Aviation Spirits', June 7, 1956, p. 8.)

[2] H. J. Frank has a useful discussion of these discounts as well as of the history of prices since the war.

[3] In 1959 the Australian Tariff Board found wide variations in the f.o.b. prices of crude oil purchased by different refiners: 'It seems clear to the Board that the prices of crude oil vary considerably depending on the arrangements made between the overseas supplying companies and the local refining companies, both of which are often affiliated in some form or other.' Prices per ton of oil varied from £6 2*s* 7*d* to £7 17*s* 5*d* in 1957–8 when £1 per ton difference would make the return on capital increase from 15 per cent to 23 per cent. 'The Board can only conclude that the variations on f.o.b. prices charged for crudes do not appear to be related to any intrinsic quality variations so much as to the supply position and the policies of the oil companies.' (P. 11.) The Parliament of the Commonwealth of Australia, *Tariff Board Report on the Petroleum Refining Industry*, 1959. See also Dr Odell's discussion of Uruguay's refining industry, p. 279 below.

[4] See the speech of W. R. Scott, executive vice-president of Standard Oil (New Jersey) reported in the *Financial Times* (London), October 21, 1965: 'The vigorous competition in the oil industry in the past few years had admirably served Europe's immediate objectives of maximum supplies at minimum costs.'

for it was not the result of an explicit clash of identifiable rivals jockeying for position in the market. It was a much more widespread and deep-seated response to the underlying conditions that we have discussed in earlier chapters. The organization of the industry had been instrumental in delaying the full development of the great low-cost producing areas in the Middle East. The vertically integrated international groups attempted to adjust the rate of output of crude oil to the rate of demand for products through their control over the output of their jointly-owned producing companies on the one hand, and over a large proportion of the world's refining capacity on the other. Instead of the 'price mechanism' of a free market there was planning within the vertically integrated framework of the Companies, but the effectiveness of the planned system was being eroded, partly from within, since the Companies were not in collusion in the regulation of markets and each had an incentive to attempt to increase its market share, and partly by competition from outsiders.

It is not therefore surprising that changes in the conditions affecting supply and prices were associated with changes in the organisation of the industry. In particular, new firms entered the international arena in both crude-oil operations and refining, some integrated, some not; firms already operating internationally on a relatively small scale began expanding their activities; and firms owned or sponsored by governments began to appear in greater numbers in production, refining and marketing. As a result, and in spite of the fact that the international majors still held an overwhelmingly dominant position, the role of prices in the functioning of the industry gained increasing significance and certain of the normal functions of prices, long rusty with disuse, began to reassert themselves.

The pricing of crude oil in such a way as to attribute a large proportion of the industry's total profits to crude-oil production served both the Companies and the producing countries well. After the '50/50' profit-sharing agreements the revenues of the producing countries soared. With this one device the Companies achieved two important objectives: the cheapest method of generating investment funds because of the tax subsidy in the depletion allowance and the offset against their domestic taxes (for the US Companies), and a way of paying handsomely for the goodwill of the crude-producing countries almost entirely at the expense of their own governments' treasuries.

But prices have other functions than to raise funds for sellers and create taxable income for governments; they also have the function of allocating markets among competing sellers by making it possible for consumers to shop around for the cheapest supplies.[1] Where it became unmistakably clear that there were sellers prepared to undercut the ruling posted prices—and not only the USSR—everybody had to look anew at the way prices might be used to capture markets in a booming industry where failure continually to attract new customers could have unfortunate consequences for the long-run position of a firm.

[1] It has been argued that consumers gain from fixed prices precisely because they are saved the trouble of shopping around. This is surely a strange argument for Companies to put forward who believe in 'quality' competition if not in price competition. In any event, it is always coupled with the assertion that prices are 'competitive'.

Russia and the 'Independents'

In the beginning the 'blame' for the fall in prices was placed squarely on the shoulders of the Russians and the 'independents', both of whom were alleged to be guided by irrational considerations from an economic point of view. Not only were the Russians accused of acting largely from political motives (the emotionally conditioned reflexes related to the 'cold war' were freely appealed to), but it was insisted that in any case the competition was 'unfair' since the Russians need take no account of ordinary economic calculations of costs in setting prices. The fact that the Russians charged higher prices to satellite countries of eastern Europe than they did to others was treated as evidence that their prices were determined by non-economic considerations. As to the independents, it was insisted that they were 'under great pressure to start producing and marketing soon after discovery, in order to recover part of their heavy initial outlays'.[1]

None of these allegations stood up for very long in the presence of the facts. Although the Russians may at times have used their offers of discounted oil for political purposes, it certainly was not necessary to invoke the politics of the 'cold war' to explain the appearance of Russian crude on world markets, for it was clear that the country's oil industry had been extensively developed and that it had available large amounts of oil that could profitably be sold on world markets to earn foreign exchange. The hypothesis that the Russians cut prices just enough to break into the market,[2] as would any other seller in a similar position, is consistent with the facts.[3] Moreover, the Western majors were in no position to complain about price discrimination among different regions since geographical price discrimination was by no means unknown in their own policies; there was clear evidence that higher prices were being charged by the majors in those markets where competitive conditions and the elasticity of demand permitted them, and especially in markets where they owned the refineries.

Finally, and above all, it was complete nonsense to argue that lower oil prices were unjustified by costs of production. No evidence was presented that the independents were 'forced' to produce uneconomically in order to recover overheads; this would imply that they were operating under conditions of financial stringency bordering on collapse and had made the grossest of miscalculations in extending their international operations. A perusal of the annual reports of such independents as Marathon and Continental provide no support at all for such an extraordinary interpretation.

Prices and Costs

Professor Adelman has analysed at length the supply price (i.e. the long-run marginal cost) of crude oil.[4] It would not be practical to summarize his re-

[1] See C. Issawi and M. Yeganeh, p. 162, where these various arguments are put forward.

[2] A New York oil man complained to the author about the 'communist technique' of price cutting; apparently he had become so unused to price competition that he found it 'un-American'!

[3] See the evidence with respect to Russian prices in M. A. Adelman, *World Oil Outlook*, p. 95.

[4] M. A. Adelman, *Oil Production Costs in Four Areas*, March 1966.

sults here in view of the important qualifications to any simple statements that would have to be fully explained if misunderstandings were to be avoided. Broadly, however, he came to the conclusion that since the Persian Gulf has by far and away the largest known reserves, the supply price of oil from that area will govern supply price for the world. It will pay oil companies to continue exploring elsewhere only if the chances are good of obtaining new reserves from which oil could be produced at a cost that would be less than the cost of producing more oil from the Gulf. He estimates that at least until 1980 'an abundance of oil' will be available from fields now producing in the Persian Gulf at a cost of between twelve and twenty cents a barrel, which is far below current prices. Because of the large gap between costs and prices Adelman expects a long-run downward pressure on prices.[1]

From the Companies' point of view costs of production include taxes paid to the governments of the producing countries, which amounted to something around eighty cents a barrel in 1966. The question of how far taxes can be expected to set a floor to prices is discussed in the following chapter, but there can be no doubt that the level of taxes is one of the important influences restraining Companies from engaging in even greater price competition than they do, in view of the large supplies of oil available to many of them at a cost still below current prices. As we have seen, for the last ten to fifteen years there has been considerable competition among the Companies in the delivered prices of both crude oil and products, especially in the large markets, such as those of Europe and Japan, but if the price of oil is to fall very much further the taxes per barrel of oil now imposed by the governments of the producing countries must be affected. In other words, individual oil-producing countries that want to expand their own output may try to do so by reducing the tax burden on the Companies producing their oil. This question, too, is discussed in the following chapter.

The Profitability of 'Owned' Crude

That the possession of low-cost sources of crude oil was extremely profitable to the Companies seems incontrovertible. Throughout the period the majors were able to finance from their own funds the construction of an enormous volume of transport facilities, refineries, and distribution networks, and succeeded in maintaining their dominating position in an industry expanding at around 8 per cent per year. Since by far the greater part of their net income in the Eastern Hemisphere was attributed to crude-oil production, this alone is evidence of the value of their low-cost crude.

At the prices attributed to crude oil, the ratio of the income of the producing affiliates in the Middle East to the value of the net assets of these affiliates as estimated by Issawi and Yeganeh reached 141 per cent in 1948 and fell below 100 per cent in only three years between 1950 and 1960, while the ratio of income *after* taxes to net assets in the same period ranged from 57 per cent to 80 per cent—high by any conceivable standards.[2] This 'income' from

[1] P. H. Frankel would seem to take the same view in *Oil: The Facts of Life*.

[2] C. Issawi and M. Yeganeh, pp. 110–13, and Table 23. Issawi and Yeganeh defined 'gross income' as gross receipts minus costs of production, and 'net income' as gross

crude-oil production is little more than an accounting figure to be sure, but it was calculated using an f.o.b. 'price' of crude oil that had to be paid by most outside buyers (until discounting became widespread after 1958) and which also was the 'price' that, together with transport costs, determined the cost of crude-oil imports to most consuming countries. If it can be assumed that refining margins calculated with the same crude prices plus transport costs were at least normally remunerative in most cases, then this imputed value of crude oil can indeed be used to show how very profitable to the integrated Companies were their low-cost supplies. Accounting figures of this type are, of course, very crude measures of the rate of return on an investment since they take no account of the time pattern of receipts and expenditures. Some attempts have been made to estimate the rate of return to the Companies of their crude-oil ventures in the Middle East using a discounted cash flow method of appraisal. The results of these measures, since they give a heavy weight to the long time lag between the original investment expenditures and the receipt of income, show a much lower cash flow rate of return than do conventional accounting measures.[1]

Whether one wishes to call the net value attributed to crude-oil 'profit'[2] or 'economic rent' (and I think the latter term is misleading to the point of outright deception in view of the regulation of the rate of supply)[3] makes no difference to the argument: the value of crude oil was so far above the expected long-run cost of finding and developing more of it that the return from doing so significantly encouraged the development of new supplies, not only by existing producers but also by those users of crude who lacked adequate supplies of owned (i.e. 'cost') crude and had to buy from others. Both Companies and governments clearly showed by their behaviour that at current (and expected) prices they judged it more profitable to explore and develop their own supplies than to buy from the market.

---

income minus payments to governments, but they gave no clear definition of 'costs of production'. In the absence of this, the figures quoted above cannot be taken to represent 'profits', or indeed, as anything more than general indications that the operations in the Middle East were highly remunerative to the Companies.

[1] See, for example, the work of Mikdashi cited previously.

[2] The Companies point out that their crude-oil 'profits' 'had' to finance the downstream investments required for the use of the crude. This only means, however, that the profits attributed to crude cannot be taken to represent the profitability of the integrated operations, a point we have noted before; it in no way alters the conclusion of the argument presented here.

[3] Rent is a return to the owner of a productive asset (e.g. oil lands), over and above the return necessary to call forth the amount produced. Any given total output will consist of oil produced at a range of costs, some at very low average cost and some at higher average cost. In other words, the total supply is made up of the supply of all producers who find it profitable to produce at the market price. If each producer produces all he profitably can at that price, then producers with the lower average costs will be receiving a true economic rent. But if the rate of low-cost output is monopolistically restricted and prices exceed the level necessary to call forth the total output demanded, then all producers may receive a monopoly rent, which should not be confused with true economic rent, for its origin and significance are very different. Economic rent will not in itself induce expansion. Monopoly rent, by contrast, is a standing inducement to 'outsiders' to get around the monopoly and to 'insiders' to cut prices.

CONCLUSION

The blunt fact of the matter was that the institutional arrangements which had for a while kept 'order' in the industry were no longer strong enough to prevent the emergence of 'excess' supplies at existing prices. As a result, for the first time in the history of the international industry a market price for crude oil began to emerge. Before the war there was practically no international trade in crude; for some time after the war very few 'arm's length' transactions at prices that could be regarded as 'market prices' occurred. The emergence of a genuine price for crude would be of far-reaching significance, and as the true nature of the problem became more apparent so did the dilemma of the major Companies. None could afford to stand aloof from the competition in the market place, and yet it was important for Companies with such a large share of the total oil business that all efforts should be made, if not to dam the slide of prices, at least to retard it as much as possible.

The share of the industry accounted for by the major Companies even rose slightly between 1960 and 1966 as can be seen from Table II. They still accounted for 76 per cent of the production of crude oil and three-fifths of the refining throughput outside Canada, the US and the Communist countries. Nevertheless, their position was weaker than it had been, for the structure of tacitly observed conventions that had kept natural competitive instincts in reasonable check for ten or twelve years after the end of the war had been undermined. It was partly for this reason that the high degree of 'concentration' that continued was no longer sufficient to ensure that the major Companies could—or indeed would if they could—act with a common purpose.

More important, however, were the far-reaching and irreversible changes taking place in the industry which would leave the several contending interests in very different positions from those that they had traditionally occupied, whether prices quickly recovered or continued to fall. In particular, the freedom of action of the international oil Companies had become permanently restricted at a number of crucial points where the policies of the governments of both the exporting and importing countries would effectively determine the course of events. The first of these concerned the fiscal relations between Companies and governments in the producing countries and the rise of OPEC; the second was the increasing determination of the governments of a number of the underdeveloped (as well as developed) importing countries to look closely at the terms on which the Companies imported oil and to take steps to reduce the cost of such imports to their balance of payments. These questions are dealt with in the next two chapters.

Chapter VII

THE INTERNATIONAL OIL COMPANIES IN DEVELOPING COUNTRIES I. MAJOR EXPORTERS

Competition and the decline of prices after 1958 created conditions which were to bring about a fundamental and irreversible change in the position of the oil Companies *vis-à-vis* the governments of the crude-oil producing countries. In general, the demands of these countries have centred on the financial returns accruing to the government and on the degree of domestic control over, and participation in, the activities of the industry. In all matters except equity participation in the old-established producing affiliates of the Companies, the governments of these countries have made steady and spectacular gains, particularly since World War II. Exploration and production concessions granted in the early days have been repeatedly re-negotiated, invariably in favour of the countries; where the concessions covered a very large proportion of a country's area, they have been reduced in size; stiffer regulations respecting drilling requirements, reservoir maintenance and similar matters have been introduced; and financial arrangements of all kinds have improved in favour of the countries.[1] Some of these developments have been the direct result of the rapid increase in the quantities of oil produced, but most of them have been obtained by the governments of the producing countries, using a steadily increasing bargaining power to maintain heavy pressure on the Companies.

SOURCES OF INCREASED GOVERNMENTAL BARGAINING POWER

The increased strength of the producing countries in bargaining with the Companies came from a number of directions. First and foremost was the extraordinary value to the Companies of the oil discoveries outside the US in the light of the rapidly rising demand of the industrialized world and the rising costs of production in the US. When the Venezuelan Government began imposing substantially increased taxes in the 1940s, and in 1948 eventually reaching the point at which it claimed 50 per cent of the net profits of the Companies operating there, it met little opposition from them. They feared government intervention, but the oil—so conveniently close to the US market—

[1] These developments are fully described in the works listed in the bibliography. See especially Z. Mikdashi, C. Issawi and M. Yeganeh, J. E. Hartshorn and the publications of OPEC.

the possession of which could be jeopardized by serious conflicts with the government, was well worth the price.[1] And the Companies were willing to concede improved terms in the Middle East when it became clear that governments there would insist on better terms than those offered under the existing concessions. As we have seen earlier, the tax offsets permitted by their home government made this possible at the time at little extra cost to the American Companies. One of the advantages of the '50/50' profit-sharing arrangement was its apparent simplicity and 'fairness', as well as the clear and large increases in payments to the governments of the producing countries.

Secondly, the growing oil revenues led to an increasing dependence of the economies of the producing countries on them and a growing awareness in these countries, not only of the importance of these revenues but also of the very great value of their national resource. This in itself increased the bargaining power of the countries because the great foreign enterprises came to loom larger and larger as alien 'exploiters' in the minds of vocal political minorities. Paradoxically, the greater the need of the governments to placate aggressively nationalist groups, the greater was their bargaining power, for the Companies had, in the last analysis, always to keep in mind the danger to them of more aggressive groups obtaining control of the government. Except when 'extreme' political groups have obtained control over the government, as in Iran under the aegis of Mossadiq, governments have by and large resisted the pressures on them to act toward the Companies in ways which would create serious risks of retaliation. The significance of 'extremist' influence was enhanced by the state of world politics, notably the importance attached in the US and western Europe to maintaining political stability in the Middle East in the light of the 'cold war'.

Thirdly, the activities of the Companies and the payments they made to governments facilitated the development of these countries, and with this development came the growth of political, economic and administrative expertise which progressively decreased the inequality between the governments and the Companies in the bargaining process. Those negotiating on behalf of the governments became increasingly able to isolate and refute the less convincing aspects of the Companies' arguments, and to evaluate how far and in what directions they could most effectively press. It is inconceivable today, for example, that any government would accept the argument put forward by the Anglo-Persian Oil Company in the negotiations over the revision of the D'Arcy Concession in 1933 that a 60-year concession was a 'necessary condition for the sinking of further large capital sums in the installations in Persia'.[2] At any reasonable rate of discount no business firm would make a capital investment for a return on which it would have to wait 60 years. Economists acting on behalf of the oil-producing countries would today make hash of such an argument, not because oil investment is less risky now, but because the present value of anything 60 years hence has little relevance for current investment decisions, *especially* if the risk is high.

[1] See Chapter XI, pp. 292–3, by P. R. Odell below.

[2] 'Statement of the Relevant Facts up to May 1, 1951'. Annex. III of the Memorial Submitted by the Government of the UK and Northern Ireland, p. 51. ICJ Pleadings, Anglo-Iranian Oil Company Case (UK v Iran).

And finally, the formation of the Organization of the Petroleum Exporting Countries in 1960 very much enhanced the bargaining power of the producing countries as a group. It had undoubtedly been intended by the Companies, when posted prices were agreed on as the basis for the calculation of profits on crude-oil production for income tax purposes, that such prices should be adjusted downwards if product prices fell. Hence, as the competitive pressures in world oil markets increased after 1958 the Companies cut posted prices, by some 8 per cent in February 1959 and again, but a lesser percentage, in August 1960.[1] OPEC was formed as an immediate response to the 1960 cut, and from that time on the Companies effectively lost their freedom to alter posted prices unilaterally, even when these became increasingly out of line with realized prices.

Underlying all these developments was the clearly rising competition in the world oil industry, which in some ways increased the bargaining power of the governments of the producing countries, especially with respect to the terms of new concessions. More particularly, however, it brought into prominence the relationship between increased revenues for the producing countries and the rate of supply of crude oil. Thus each of the governments of the producing countries, partly through OPEC, was forced more and more into a consideration of the wider position of the international petroleum industry generally and of the place of its own industry in it; no longer would the governments be willing to view their problem as the relatively simple one of bargaining with the Companies for an increased share of whatever profits the Companies had decided should be attributed to crude oil.

TAXATION AND PRICES

Between 1957 and 1959 production of crude oil in the countries of the Middle East rose by some 30 per cent and the payments by the oil companies to the governments of these countries rose from $1,022 million to $1,293 million, or by some 26·5 per cent; between 1959 and 1961, however, these payments rose by only 13·7 per cent compared with a 22 per cent rise in production.[2] This slower rise in payments is largely attributable to the reductions in posted prices that had taken place in 1959 to 1960. At the time of these reductions there was a widespread belief in the producing countries that the Companies had acted to protect their own profits (which indeed they had), and that they could without difficulty not only raise the prices again but also re-impose their effectiveness in world markets (which they could not). One of OPEC's first stated objectives therefore was to restore posted prices to 1958 levels. Although this proved impossible, the Organization was able to prevent further cuts. Thus the tax on profits attributed to crude oil, which in the old-established Middle East

[1] Some of the American Companies may also have had another reason for wanting to cut the posted prices on which they paid tax to the governments of the producing countries. The greater the gap between posted prices and 'realized' prices, the more the artificial nature of the income tax became evident, and the greater the danger that the US tax authorities would begin to question seriously the propriety of permitting the Companies to offset fully the taxes paid to foreign governments against the Companies' domestic tax liability on their foreign income.

[2] Edward Symonds, *Financing Oil Expansion in the Development Decade*, p. 3.

producing countries (but not in Libya at that time) was 50 per cent of the difference between the (agreed) cost of production and the posted price, became, in the absence of changes in cost, in effect a fixed payment per barrel of oil. With falling 'realized' prices for crude oil and market prices for products such a tax would bear progressively harder on the Companies, for it would absorb an increasing proportion of the net receipts from their integrated operations unless offset by increased efficiency in their operations. The notion of a 'posted price' at which oil was assumed to be actually transferred began to be replaced by the notion of a 'tax reference' price for the valuation of crude oil for tax purposes.

In spite of their inability to raise posted prices, the conviction remained on the part of the producing countries that the Companies could well afford to pay more than they were paying, even with the continued fall in market prices. Payments to the governments were running at well over $2,000 million a year in 1964 but OPEC nevertheless pressed for more. If prices could not be influenced, the costs allowed in the calculation of profits, especially allowances for marketing expenses, and the treatment of royalties as an item to be included in the income tax liability of the Companies might be attacked. The royalty issue was by far and away the most important since marketing allowances had already been drastically whittled away, and some OPEC countries proceeded to open negotiations designed to force the Companies to treat royalties as an expense rather than as part of their profits tax.[1] After lengthy negotiations an agreement was reached in 1964 in which the Companies accepted this demand.[2]

In return for this concession, however, the Companies demanded that the government allow discounts off posted prices for the calculation of taxable profits. The countries conceded this, and thus for the first time conceded in practice that tax prices might be influenced by market prices. Nevertheless, specific discounts were granted only for stated periods and were to be reduced in an arbitrary progression from roughly 8½ per cent in the year in which the agreement was accepted by a government, to 7½ per cent in the second year, and 6½ per cent in the third year, after which the whole question would be reopened for further negotiation.[3] The net result was an offer of an increased payment to the producing countries of an average of some 3½ cents a barrel of oil in the first year rising to 4½ cents in the third year—a substantial increase for most countries.

This agreement marked two very important changes in the relation between oil Companies and the governments. In form it appeared to be a financial arrangement involving concessions by the Companies over the tax treatment of royalties in return for concessions by the governments over the basis of

[1] If royalties are expensed, they are treated as costs and deducted from income before income tax is calculated instead of being treated as part of the 50 per cent government share of income.

[2] For an account of these negotiations from OPEC's point of view, see 'OPEC and the Principle of Negotiation', paper presented to the Fifth Arab Petroleum Congress.

[3] The allowances off posted prices were also designed to correct some of the inappropriate price differentials between different gravities of crude oil that had persisted because of the link between the world price structure and that prevailing in the US market. Moreover, the Companies agreed to consider smaller discounts if market conditions improved.

income tax calculations. But in essence it involved mutual concessions on deeper matters of principle .The Companies openly conceded the right of the governments to have a voice in the determination of the prices used for the calculation of income taxes—a position they had already implicitly conceded since 1960 in accepting their *de facto* inability further to cut posted prices. In return, the governments implicitly accepted, at least for the time, the proposition that their revenues might be dependent, if only indirectly, on the prices prevailing in world markets.[1] By 1967, however, the governments were demanding an end to the discounts, arguing, among other things, that the price of oil was improving.

These mutual 'concessions' brought to its logical conclusion the course of events that began in 1950 when the income of the producing countries was linked directly to the posted prices of crude oil. At that time it was not intended that the governments of these countries should have any voice in the prices at which the Companies sold or transferred the oil, but it would not even then have been difficult to predict the circumstances in which the freedom of the Companies to determine unilaterally the revenues of the producing countries would no longer be acceptable. The price of crude oil was not in fact a true 'price'; it was certainly not determined by the anonymous forces of market competition, and was seen not to be so determined; hence outright bargaining between identifiable groups was inevitable.

It may seem strange that the Companies would agree to increase their payments to the governments of the producing countries at a time when prices were falling and efforts had to be made to reduce costs and increase efficiency in order to keep unit profits from falling also. Perhaps they felt themselves to be in a weak position politically, and they may have feared further expropriations of the type that Iran and Iraq had already effected. United action on the part of the Middle Eastern producers could have had awkward consequences, and at times the negotiations over the agreement took place when Arab political feelings were running high.

Such matters were undoubtedly important. But it is also true that, in part at least, the OPEC agreement could be used as a means of putting pressure on the independent Companies whose price-cutting in international markets was blamed for much of the industry's troubles. In this respect the situation in Libya was crucial, for it was with respect to Libya, where taxation was based on 'realized' prices, that the issue of price-cutting arose in its most acute form.

By freezing posted prices, the oil-producing countries had intended only to protect their own revenues per barrel of oil. But, with given costs of production, their income tax thereby became a fixed tax per barrel of oil and was thus an addition to the marginal costs of all companies subject to the tax.

[1] This point was explicitly made in a note on a subsequent resolution (IX.61 dealing with the control of supply) in which OPEC stated:

'Although most of the Member Countries of the Organization levy taxes on the basis of posted prices rather than actual sales prices, the latter are of great importance . . . for two reasons: first, the weakening of actual sales prices threatens to debilitate further an already unsatisfactory structure of posted prices; secondly, the terms of the supplemental agreements on the expensing of royalties . . . make the elimination of the permitted allowances [off posted prices for tax purposes] dependent on an improvement in market conditions.' OPEC, *Note on Resolution* IX, 61, n.d., p. 4.

The belief grew up, not only in the producing countries but among the Companies as well, that such taxation, uniformly applied, could set a floor to price-cutting in international markets; moreover, the higher the rate of taxation, the more effective should be the restraint on price competition. Competition can seriously disrupt existing market positions and necessarily increases uncertainty; increased taxes merely raise costs and might be expected to affect the operations of large financially strong Companies with diversified and established markets and sources of supply less severely than the operations of smaller less fully integrated Companies.[1] It is likely, therefore, that some of the major Companies saw some advantage in acceding to the demand of OPEC for higher revenues if this would encourage the uniform enforcement of a common level of taxation which would effectively hit the independents, particularly in Libya.

The Problem in Libya

When the first comprehensive Libyan oil legislation was adopted in 1955, the oil Companies were in a favourable bargaining position. The Libyan Government was eager to see the country's petroleum resources explored and developed and was therefore eager to attract as much foreign investment as possible. Before two years were up, however, the Government concluded that the terms of the concessions under this law were too generous to the Companies. In consequence, it insisted on better terms in some of the new agreements negotiated after November 1957; the various allowances (including depletion allowances) were reduced or eliminated altogether, increased bonus payments were provided for, etc. By 1960 it became clear that very large quantities of oil would be available, and the Government began to consider new legislation. This appeared in a Royal Decree of July, 1961.

The development of petroleum legislation in Libya in these ten years reflected the changing knowledge of the petroleum resources of the country. In the face of promising, though almost completely untested prospects at first, it was necessary that the inducements offered to the Companies should be relatively generous in order to persuade them to take the requisite risks. As prospects improved with the growing knowledge of oil resources, the Government became in a better position to demand a higher price for the privilege of concessions to explore and exploit.

There is inevitably an inherent difficulty in this type of situation: terms which may have been necessary to induce the pioneers to take the initial risks may later appear far too generous if the pioneer efforts have been notably successful and the *ex post* profitability of their operation seems high. This difficulty will be intensified if the terms under which the concessionaires operate are more favourable to them than are the terms under which similar concessionaires work in other countries. The fact that the concession terms in Libya were in many ways less favourable to the Government than were those

[1] Price-cutting by independents occurred on two levels: in sales of crude oil and in the sale of products. Some of the oil produced by independents was used in their own integrated operations, and to expand their share of the markets for products they cut product prices. Other oil was sold to outside refiners, and at this level the price-cuts were also substantial. It was hoped that an increase in the effective cost of crude would create difficulties for the independents in both crude and product markets.

obtained by governments elsewhere in the Middle East gave rise to a great deal of local criticism.

The Government, however, refrained from attempting to enforce unilaterally determined changes in the existing concession contracts, but rather adopted legislation designed to create advantages for the companies that were willing to convert their old contracts to the new terms desired by the Government, and disadvantages for those that were not willing to do so. It thus attempted to steer an even course between incurring a loss of international reputation for fair dealing and incurring an unnecessary loss of revenue. In any event, the law itself can be looked on as determining only the minimum terms acceptable to the Government, with more favourable terms being demanded in the actual negotiations for any given concession. Thus, in mid-1965 when new territories were thrown open to concession bidding the Government made clear the kind of improved terms it hoped to obtain.[1]

In spite of the successive alterations of the law and the tighter definition of 'profits' and of the relevant price for tax purposes, the Libyan concession arrangements did not violate the basic '50/50' principle of profit-sharing, nor did any of the provisions up to 1965 penalize Companies if they granted such discounts from posted prices as were necessary to sell the oil.[2] This inevitably put Companies with Libyan oil in a favourable position to supply European markets, and it is not surprising that the Libyan income tax law was seen by the Companies to be an important factor in competitive price-cutting in Europe.[3]

The highest discounts were offered by independent Companies, notably Continental and Marathon, who slashed prices to find markets and who, with Amerada,[4] accounted for about half of Libya's 1964 output. The majors,

[1] It was stated that the Government would favour those bidders who offered 'to base royalty and taxes on posted prices agreed with the Government, and comparable with Middle East prices adjusted for freight; to keep old and new concessions distinct for determining financial obligations; not to merge rent and royalty payments; to submit large contracts for Government approval; to accept detailed Government supervision of work and conservation programmes; to give the Government a share of profits higher than 50 per cent and to expense a high proportion of royalty; to accept that the Government may take half its share of profits in kind; to have market outlets available for disposal of production; to construct refineries or petrochemicals plants; to provide other extra benefits'. *Petroleum Press Service*, July 1965, p. 267.

[2] The Petroleum Law of 1955 had made no provision for the posting of prices and gave no definition of income for tax purposes. The 1961 revisions provided for posted prices but permitted discounts in the form of marketing expenses if the discounts were necessary to sell Libyan crude. See A. A. Q. Kubbah, pp. 78 ff. Mr Kubbah was for some time Petroleum Economist and Assistant Economic Adviser, Ministry of Petroleum in Libya, and he gives a useful discussion of the development of petroleum policy in this period.

[3] 'From a practical point of view, however, there can be little doubt that many of the low-priced sales which tend to depress the market emanate from sources where a realization basis is in force—that is to say, from operators who are paying tax at a lower rate than those of their competitors who are paying taxes calculated by reference to posted prices.' British Petroleum, *Annual Report*, 1962, Chairman's Statement to Stockholders, p. 19. At this time BP did not have Libyan production.

[4] The three companies jointly own the Oasis Oil Company, which was the largest single operator in Libya. Both Continental and Marathon are integrated companies, but neither their marketing nor refining facilities were adequate to absorb all their Libyan oil. Amerada is a crude producer only and sells its oil to Shell on a virtual partnership basis.

notably Esso (Jersey Standard) and Mobil, also discounted heavily on the small proportion of their output sold to third parties; transfers to their own affiliates were made at much lower discounts—and at posted prices to the UK.

As soon as the OPEC agreement was negotiated, Libya was pressed to adopt the new tax regulations and to impose them on the Companies operating there. The independents vigorously objected, pointing out that their rivals, the majors, had been consulted in the making of the agreement but that they had not;[1] they insisted that substantial discounts off posted prices were necessary to sell the oil and that to force them to pay taxes as if the oil had been sold at posted prices would be both unreasonable and a breach of faith by the Government. After some delay, the Government decided in 1965 to adopt the OPEC agreement and to press its concessionaires to 'convert' their concession agreements to the new terms. The Petroleum Law was amended accordingly, and it was also provided that any Company refusing to accept the new terms would not be permitted to obtain any new concessions. On the other hand, the Government offered to refrain from questioning the tax returns back to 1961 of Companies that did accept. The pressures worked, and all the Companies eventually complied with the Government's wishes.[2]

Most of the majors were quick to accept the new law, providing all Companies accepted it. In other circumstances they undoubtedly would have attacked the Government's action vigorously as an act of bad faith, if not an abrogation of contract. In this case, however, it was expected that the higher taxes would create economic difficulties for independent producers,[3] and one or two of the leading majors may actually have encouraged the Government to enact the law.[4] Some commentators have suggested that in view of their support of Libya's action the majors would not in the future be in a strong position to complain of unilateral changes by governments of the conditions under which foreign companies accepted to explore for and develop oil.[5] But

[1] In fact, Continental had signed the 1964 OPEC supplemental agreement as a member of Iricon in the Iranian consortium.

[2] For a judicious summary of the issues and attitudes involved see J. E. Hartshorn, pp. 17 ff.

[3] In February 1967, however, the *Petroleum Press Service* noted that the new arrangements had 'less than the expected effect on prices'. (P. 43.) Indeed, the independents may have tried to sell even more crude than before.

[4] It has been suggested that one reason for such encouragement was that some of the US Companies expect the tax depletion allowance in the US to be withdrawn in the foreseeable future. This would raise the US income tax liability against which they are permitted to offset income taxes paid to foreign governments. If the allowance were withdrawn, it follows that the Companies could pay more taxes to foreign governments without extra costs to themselves than they can now. On the other hand, the US Treasury is known to view with disfavour the practice of allowing the oil Companies an income tax offset for taxes paid to foreign governments which are manifestly not bona-fide income taxes.

[5] It is of course true that when a Company invests in a foreign country it makes allowance for political risk, including the risk that the government might change the terms of the original contract. This risk is probably greater in most of the developing countries than it is in the industrialized countries. The Company will therefore only invest if the short-term returns are considered high enough to justify the risk; in other words, in the initial arrangements this type of risk is already discounted. If, therefore, the risk materializes and the terms are changed to the disadvantage of the Company, it can be argued that the resulting loss to the Company has already been allowed for. However, if this type of argument were to encourage governments consistently to change the terms of their contracts, obviously the risk would become so high that foreign capital would be difficult to attract on any terms.

by 1960 the Companies were caught between competition and falling market prices on the one hand, and a much improved bargaining position of the oil-producing countries collectively on the other. They had little alternative than to attempt to associate, if only indirectly, these countries as a group with the problem, not only of the posted, but also of the market prices of oil. As a group, that is to say, as OPEC, the producing countries are very much concerned with the maintenance of prices; moreover, as a group, they know what needs to be done in order to maintain them. Action, on the other hand, must come from the governments of the individual countries, and therein lie the real difficulties.

Taxes as a Floor to Prices

Payments to the producing countries, though costs to the Companies, are not genuine costs for the industry as a whole since they come out of profits or rents, and can be reduced, without reducing output, by any country that wants to promote an increase in the sale of its own oil by enabling producers to offer it at lower prices. Taxes could only set a firm floor to prices if all governments not only agreed to enforce a common level of taxation, but also succeeded in making their agreement fully effective.[1]

So long as the demand for oil is inelastic at current prices, the imposition of higher taxes, which push up marginal costs for all Companies and inhibit price-cutting, would be to the advantage of the producing countries considered as a group—the higher the taxes (and prices) the greater would be the revenue obtained. But as OPEC itself has recognized, the demand for an *individual* country's exports may be highly elastic, at least in the short run. Hence each country can reasonably consider whether it might gain by giving its own oil exporters a tax incentive to cut prices and increase sales. OPEC. naturally argues that such action would be 'suicidal' for any country since it would drive down prices for all.[2]

This may well be the case, but even while accepting the general argument, the government of each individual producing country must still decide whether it is satisfied with its existing level of revenues and its existing share of the market. So long as revenues can be increased by increasing tax prices without

[1] OPEC at its XIth Conference in April 1966 urged that all governments use agreed posted or 'reference' prices for the purpose of determining the tax liability of the oil Companies, and recommended that no new rights be granted or contracts entered into for exploration or exploitation of new areas unless the income to governments was not less than would be so obtained. This resolution was presumably aimed at certain exceptions to the general arrangements for taxation. For example, companies with which the National Iranian Oil Company was in partnership were excepted as well as the sales of Aramco to non-affiliates (the latter exception was later eliminated). (See *Petroleum Press Service*, July 1966, pp. 242 ff.) Moreover, Venezuela was then still taxing, at least nominally, on 'realized prices'.

[2] 'The demand for crude oil at the source is price-inelastic, that is, the demand for crude varies little with changes in price at the point of export. That is true for demand overall, but it is not true for demand as between different areas. What is true of prices is also true of taxes, and the low-tax area will attract faster development than the high-tax area. But in view of the overall inelasticity of demand, it would indeed be suicidal for any major producing area to encourage development of its own resources over an extended period of time by low taxes at the expense of other areas. This form of competition in taxes would in the long run only serve to drive down Government revenue for all.' *Taxation Economics in Crude Petroleum*, a paper presented at the Fifth Arab Petroleum Congress, p. 16.

affecting output, it is clearly to the advantage of the producing countries to increase them. But this kind of action can only be taken by the countries acting together (or in co-operation with the Companies); it is very difficult for an individual country acting alone to increase its revenues in this manner, unless, like Libya, it had been accepting lower unit revenues than other countries to ensure the rapid development of its oil.

Now if one examines the present circumstances in the several oil-producing countries and their plans for future development, one is forced to conclude that the pressures for greater revenues are likely to increase in most of them.[1] All have made extensive efforts to invest oil revenues in a wide variety of undertakings designed to develop and industrialize their economies, but for a long time to come their foreign exchange receipts will depend heavily on crude oil, and foreign exchange receipts have a special importance for their development programmes. Even Kuwait is finding its large revenues extremely useful as a means of consolidating its position and assisting other countries in the Arab world by investing in development projects in other Arab countries, as well as in its own industrial and welfare plans.

The Problem of Market Shares

With existing tax rates, revenues will continue to rise steadily as the consumption of oil rises, but it is already clear that not all countries are content with the share of the market allotted them. Among the major oil-producing countries, only Venezuela has tried to stand alone in holding out for higher tax prices even at the expense of a slower rate of development and a reduced share of the market, but she alone among these producers has very limited proved oil reserves.[2] The oil reserves of most of the others are so large that they need not in practice worry about the depletion of their 'wasting assets' for many many years to come, and in contrast to Venezuela, some of them, notably Iraq and Iran, are insistently pressing for larger shares of the market.

When a country does not have to invest its own resources in finding and developing oil and when existing reserves are so large that there is no advantage in holding back production today in order to save it for the future, the optimum rate of increase of output is that which maximizes current revenues. The greater the danger that the price of oil will fall, the greater will be the desirability of obtaining maximum output in the present. This danger is increased by successful exploration for either oil or gas, particularly if the discoveries are made in territories that were not previously important pro-

[1] M. Iskander has made just such an examination in a D.Phil. dissertation being prepared for Oxford University.

[2] Venezuela's proved reserves are equal to about sixteen years' consumption at present rates (which is undoubtedly a conservative estimate), and she has for long accepted the proposition that a slower rate of development is preferable to parting with a 'wasting asset' for a reduced return. Not only has she refused further concessions to the foreign oil Companies, but she has restricted the discounts allowed in the sale of oil. As a result, investment in Venezuelan production by the Companies declined drastically and exports increased at a much slower rate than those from the Middle East. Not all of this has been due to Government policy, since Venezuela suffered more than the Middle East from the US import quotas and from sagging freight rates (which reduced the relative delivered cost of Middle East oil in Europe). Moreover, much of Venezuelan oil is more costly to produce. Her position is discussed in detail by Dr Odell, pp. 292–7 below.

ducers and by Companies without extensive market outlets of their own. Similarly, the pressures from both old and new producers for greater output, and the growth of national oil companies in crude-oil producing countries which compete for outlets in world markets, must increase the chances that prices will continue to be under pressure in the near future. Any prospects of a slackening of the rate of increase of demand must have a similar effect.

All this may well increase the demand in some at least of the producing countries for greater output in the present, and one way of obtaining this would be to make an increased offtake financially advantageous to the existing concessionaires. Because the major Companies controlling most of the oil production in the Middle East have interests in a number of different countries, each Company can, within limits, vary the proportion of the total oil that it takes from any individual country according to the relative costs to it of so doing. Thus in 1961 the Middle East accounted for 99 per cent of the oil produced by Jersey Standard from the Middle East and Africa together, by 1966 its share had fallen to 67 per cent. In other words, the increased Libyan offtake by Jersey Standard was largely at the expense of Saudi Arabia and Iraq (and especially Iraq, with whom the Companies were disputing), for there is no reason to believe that if Libyan oil had not been available, the increase in the requirements of the Company would not have been met primarily from increased Middle Eastern production.

Although there are strong political pressures on all producing countries to prevent any one of them from competing against the others by accepting lower revenues per barrel of oil, there is no escaping the dilemma that will arise if pressures to increase the rate of supply continue to force the Companies to engage in price competition whether they like it or not: the producing countries will have to decide either to reduce the role of the oil Companies as producers and sellers of their oil or to permit the Companies to receive a return they will accept for continuing the expansion of their activities. The different countries may well view these alternatives differently, and the outcome will very much depend on the political atmosphere in which negotiations take place.[1]

The governments of the producing countries have to consider not only the rate of output of existing concessionaires, but also the effect of granting new concessions. In the late 1950s competition from non-major Companies for oil concessions began to become important. The discovery of oil in new countries, the development of techniques for offshore drilling, and the power that the producing governments had acquired to reduce the areas held under concession by the older concessionaires, enabled a number of governments to make many desirable areas available for competitive bidding. For reasons discussed in the previous chapter, both the established majors and large numbers of newcomers not only offered increasingly favourable financial terms but even accepted without demur a variety of arrangements for sharing ownership as well as profits with the local governments in order to secure concessions.[2]

[1] The alternatives are discussed in Chapter IX.

[2] For a description of some of the changes in the newer concessions see OPEC, 'Radical Changes in the International Oil Industry During the Past Decade'; and 'From Concessions to Contracts'.

In some countries, where the monopoly enjoyed by the majors had for long been considered prima facie evidence that they were 'exploiting' the countries, independents seeking concessions were not only welcomed but sometimes favoured. By the middle of the 1960s doubts about this policy began to arise; competition began to bring it home to some officials in the oil-producing countries that monopoly profits (and therefore government revenues) would be threatened by market competition and that while the competition of independents for new concessions might be financially lucrative in the short run, in the long run the consequent increase in competition in world markets might bring severe disadvantages. The question began to be openly debated whether new concessions should not be withheld from independent groups, for these were the sellers most active in cutting prices, and whether the producing countries should not help to protect the position of the integrated majors. Although there were signs that in some countries this type of consideration was beginning to influence their policies with respect to the granting of new concessions, the desire for an increased rate of development of their oil resources remained the more powerful influence, and the granting of new concessions continued unabated.

The close relationship between the aggregate rate of supply, prices, and the future of oil revenues was officially recognized by OPEC in 1965 when it agreed to a resolution to establish a voluntary programme for the control of supply.[1] Under this programme, each member country was requested to keep the increase of output over a twelve-month period below a specified target rate. The operation of the programme was not very encouraging for OPEC. It is difficult to see how such a programme can be implemented in practice so long as the Companies, not the governments, are the effective offtakers of oil, and it would not be easy for the governments under existing circumstances effectively to control the Companies' offtake programmes, which are related to Company plans for their international operations. The Companies objected to the scheme and some demanded arbitration on the ground that it interfered with their commercial freedom of action, although the target quotas were apparently based on Company estimates.

Moreover, a number of the member countries of OPEC openly expressed their dissatisfaction with the quota arrangements, and there was little evidence of any effort by them to co-operate significantly in any practical manner. Iraq did not ratify the 1964 OPEC supplemental agreement for a variety of reasons. But it did state its case for a larger share of the market, in spite of the fact that the permitted increase in its oil production when the OPEC programme first started was greater than was in fact attained.[2] The Oil

[1] IXth Conference, July 1965, Resolution IX. 61.

[2] The Director of Iraq National Oil Company's economic department, for example, argued in a paper presented to the Conference of Arab Economists in Baghdad, 1965, that although 'any policy aiming at the planning of production on an international scale is a reasonable policy which is acceptable to all', the Iraq Company realized 'that this should not necessarily mean that Iraq be deprived of a just share of production for the following reasons: 1. The policy of production limitation cannot succeed unless it is implemented on an international scale. 2. The claim that what is not produced is saved for a better tomorrow is a mere guess. 3. The restriction of Iraq's production in the face of the steady increases in world output will inevitably lead to a weakening of its position in world oil markets.' He

Minister of Saudi Arabia expressed his lack of confidence in the production programme and indicated willingness to accept it only if bigger quotas were allotted his country.[1] Late in 1966 Iran demanded and obtained a substantial increase in her rate of production, hinting that strong action would otherwise have to be taken against the oil Companies. Much of the Iranian argument rested on the country's need for oil revenues for its development programmes, and it asserted a basic right to a larger share of the market.

It was in practice difficult for each individual country to accept the proposition that the mutual restraint required to protect all should be strictly applied in its specific case, and in 1966 the whole programme was quietly dropped. In essence OPEC attempted to work a voluntary price-maintenance scheme, and as such the chances of success depended upon the extent to which the interests of the several participants coincided. Since prices can be maintained appreciably above the long-run cost of producing the amount of oil demanded at such prices only if some oil that would otherwise have been put on the market is deliberately held back, there are obvious difficulties when there are fundamental disagreements over how the restraints should be shared, and when at least some of the problem arises from the efforts of newcomers to find a place in the sun.

Nevertheless, the direct intervention of the governments of the producing countries in the affairs of the industry is likely to become greater the less successful are the attempts to stabilize the level of prices in the near future and the greater the consequent danger to oil revenues.[2] A comfortable level of prices makes possible financial bargaining and the achievement of a *modus vivendi* between Companies and governments without any major modifications of the structure of control; but if prices reach a point at which the Companies feel compelled to ask for re-negotiation of the existing financial arrangements, some governments at least are likely to feel equally compelled to insist on re-negotiation of existing concession terms in other respects—notably the vexed question of 'participation', i.e. a share in the ownership of the producing companies.

To some extent joint operations between the international Companies and local companies in new areas (or in relinquished areas) will help to satisfy the demands of governments, but in the last analysis demands will increase for

---

then pointed out that at an 8 per cent discount rate Iraq would have to receive $1·25 in five years' time, or $1·84 in ten years' time, on each barrel of oil the production of which is postponed today, thus foregoing the $·85 it would have yielded to the Government. 'In order to obtain this kind of income the price per barrel should rise from $1·72, for example, to $2·70 and $3·90 respectively. We do not know anyone who could guarantee that oil prices will actually rise to this level in the future.' Izzat Al-Gharbawi, *Review of Arab Petroleum and Economics* (Baghdad), November 1965 (Volume 1, No. 10), pp. 11–12.

[1] *Financial Times*, February 14, 1966.

[2] OPEC has already raised the question whether 'it is truly a proper function of commercial entities to administer prices. The administration of crude-oil prices cannot be indefinitely entrusted to a handful of international companies whose primary motivation is commercial.' 'The Price of Crude Oil: A Rational Approach', p. 8. It can be expected that the doubts of OPEC regarding the propriety of crude-oil price maintenance by the Companies will increase, the *less* effectively are prices maintained.

a share in the ownership of the great established producing companies. This raises fundamental matters of 'principle' for the international Companies as well as for the producing countries. There is a limited convergence of interests between the producing countries and the major oil Companies, but it extends only to the question of prices, and then only to the extent that taxes can be expected to defend the price level without running significantly afoul of what each country considers to be its appropriate share of the total output. As the scope for bargaining over financial gains decreases, however, matters of 'principle' are likely to gain in importance. Even the 1964 OPEC agreement, which did bring financial advantages to the producing countries accepting it, was not ratified by all countries, partly because it raised, for some at least, matters of principle which easily and quickly became entangled in local politics.

MONEY VERSUS 'PRINCIPLES'

Most problems arising in the bargaining between Companies and governments raise two types of issue: those involving economic or financial advantages and those involving principles. The former are more easily negotiable than the latter, for on principles it is more difficult to achieve 'compromise with honour' than on money. Most arguments are, of course, couched in terms of principle (e.g. prices should be raised because the countries are being cheated of the 'true' value of their natural resources), and for some of them it is sometimes difficult to decide which aspect is uppermost in the minds of the negotiators.

The intractable nature of the dispute between Anglo-Iranian and Iran in the early 1950s, in spite of the substantial financial concessions by the oil Company (including a version of '50/50'), was in large part due to the fact that for the Iranians of Mussadiq's persuasion the problem was more one of principle (and politics) than of money. The prolonged negotiations with Iraq over that country's acceptance of the 1964 OPEC supplemental agreement involving the expensing of royalties, arose not so much over money but over principles arising from the dislike by the Companies of recognizing in any form the validity of the 1961 law expropriating their properties,[1] and from the dislike by Iraq of the provisions for compulsory arbitration, which she

[1] A Company that has been expropriated without adequate compensation always faces a problem in deciding the period of time that must decently be allowed to elapse before it 'forgets' the matter, even after it seems unlikely that any restitution will be made. At the time of the Iraqi expropriation the *Petroleum Press Service* insisted on the consequences of the 'moral wrong' done by Iraq: '. . . it can only do harm to Iraq's international standing, for he [Qasim] has put his country in the wrong, morally, in the eyes of those abroad. . . . No reputable concern is likely to make an agreement to spend money in territory only made available by breaking an agreement. By thus condoning the breach, it would be eroding the basis on which its own agreement would rest.' January 1962, p. 7.

In fact, of course, such things are often relatively quickly forgotten when there is no point in keeping them alive, and it is perhaps astonishing how quickly private business, after having been knocked about, will try again when it seems profitable to do so. But in Iraq the Companies feared that if they recognized the expropriation, very valuable oil properties would fall into the hands of others. However, even there the Companies came to terms in 1965.

felt also touched on her sovereignty. The Ministry of Oil in Iraq wanted the IPC partners to join with its national oil company in the exploitation of some of the expropriated areas, and the draft agreement mentioned above provided, among other things, for a joint company to be formed for the purpose. The company was to be 33 per cent Iraq National Oil Co., and 66 per cent IPC shareholders, except Jersey who refused to participate.[1]

In joint operations the Government's share of the profits would be greater than the 50 per cent share obtained as an income tax because it would also receive a share according to its investment. Under similar arrangements in other countries the government receives 75 per cent of the profits. The reserves of some of the expropriated areas in Iraq are believed to be very large, and the Government might well press for the rapid development of the jointly exploited fields at the expense of the existing operations of the IPC on which it receives only tax revenues. At the same time, the Companies might have to handle oil produced. It is not surprising, therefore, that such arrangements appeal more to independent companies trying to establish themselves in production in the Middle East than to the major Companies with established producing operations entirely under their own control.[2]

The Kuwait legislature waited a long time before ratifying the OPEC agreement. Here also some members stated that some of the provisions infringed the country's sovereignty. They, too, disliked compulsory arbitration, they were unwilling to agree not to impose additional taxes without negotiating with the Companies, and they were unwilling to be bound by a 'most favoured company' clause which would require the Government to offer Gulf and BP terms as favourable as those it might want to offer a new company to induce it to take up a concession.[3] On the other hand, these reservations were 'in principle' difficult for the Companies to accept, largely because of the precedents that would be set for other countries.

Among the issues in which principles become important for each side, the most difficult relate to local participation in the ownership of the oil-producing companies, and to the degree of control over the conduct of the affairs of the companies by the local governments, including control over prices. On the question of prices, the countries have made limited progress through OPEC; in new concession agreements, provision for local equity participation, i.e. shared ownership, has become standard, even one of the majors (Shell) accepting a government as partner in a new venture.[4] But no progress whatever has yet been made in inducing the Companies to accept equity participation by the government in any of the old-established producing companies.

[1] It should be emphasized that Iraq's problems with the Companies arising from the expropriation under Law 80 of 1960 overshadowed in practice all questions relating to the Supplemental Agreement.

[2] ENI, for example, for a long time hopefully courted the Iraq Government for an oil contract.

[3] The agreement as finally ratified in 1967, two years after it was first suggested, allows the Government to give special treatment to its own national company.

[4] As suggested above, if the major concessionaires enter partnership arrangements with the government in unexploited areas, the government's interest in promoting offtake from the new areas can come into serious conflict with the Companies' interest in the old-established areas.

For the producing countries the issues involved in 'participation and control' raise principles related to their own concepts of independence and national sovereignty, while for the Companies they impinge on their 'commercial freedom of action' and have far-reaching implications for the organization of their integrated operations as a whole. Moreover, unless equity participation means nothing more than a greater payment to the governments (or nationals) of the producing countries, there can be no doubt that the potential area of discord between governments and Companies would be enormously widened if 'participation' were extended to include the major sources of the Companies' crude oil.

Equity Partnership and Government Control

It has been argued that a share in the ownership of the local producing companies would bring material advantages to the producing countries in addition to the psychological satisfaction of their national aspirations. Not only are greater financial returns expected but also an increased assurance that the local companies would operate more in the interest of the domestic economy as a whole. In particular, some governments feel that the use of profits by the local companies could be more effectively channelled in the national interest.

In some of the new concession agreements the government is entitled to buy shares in the local company after oil is discovered by putting up the appropriate proportion of the costs incurred. In addition to income taxes the government is then entitled to an equity share of the profits. In the NIOC–AGIP[1] agreement (1957), the equity share is 50 per cent; in the Kuwait-Shell agreement (1961) it is 20 per cent. In this type of 'participation' the government avoids the exploration risk, but its further significance depends on its effect on the control of the use of the profits attributed to the local company. If the parties merely intend that the government will be paid the relevant percentage of the profits attributed to the company, the only effect is an increased government 'take'. Alternatively, the government may receive dividends (after taxes) on its holdings at the same rate as does the foreign partner, with the remaining profits retained for reinvestment in the company. The government may then use its share of dividend payments for general expenditure or for reinvestment in other industries, while the private partner may use its share for reinvestment outside the country.[2] Again the only effect is to increase the payment to the government.

On the other hand, if the local company is allowed to retain the funds required to finance its expansion, then the required investment will be shared by the foreign partner and the government in proportion to their respective equity interest, whereas if the government withdrew its entire share of profits, the investment required to expand the local company would be borne entirely by the foreign partner, which would be incompatible with the existing percentages of the shared equity. In neither of these circumstances, however, is there any material gain for the government greater than could have been obtained by a simple revision of income tax arrangements (apart from the

[1] A subsidiary of Ente Nazionale Idrocarburi.

[2] This is, in general, the effect of the British Government's participation in British Petroleum.

fact that the foreign Company bore the entire risk of initial exploration). In both cases the 'profit' attributed to the local company will depend either on the price received by the local company for its oil, or on the 'tax reference price' agreed to by the government. The government would still have an incentive to keep this price as high as possible, and its ability to determine the price would not be changed by virtue of its equity participation.

But if the effect of participation by the government is to put it in a position to control the entire use of the profits arising in the local company or to influence the level of production, real difficulties arise. One of the complaints of the producing countries in the Middle East has been that the international oil Companies use the 'profits' derived from Middle Eastern crude to invest in exploration and production in other competitive areas. The Companies point to their need for 'flexibility' of supply and also to their need for investment in downstream operations before crude oil can be used. The governments recognize the need for downstream investment, but have long insisted that they should have some share of downstream 'profits', which they hold are, in part at least, to be attributed to their own crude.

From the point of view of any given oil-producing country, the most desirable arrangement would presumably exist if large taxable profits were attributed to domestic crude production, and if profit after tax and minimum dividends to the foreign Companies were invested either in the domestic economy, or in downstream operations in the international Companies that moved its own crude and in the profits of which the government also shared. In other words, joint ownership of the major crude-oil producing affiliates of the international Companies could give rise to pressures from governments not only to extend the partnership to all of their vertically integrated activities but also to relate downstream investment directly to particular sources of oil.[1] This would severely hamper, if not destroy, the Companies' cherished 'flexibility' of operations if applied to their major sources of crude oil, while pressures from the government for an increased share of profit would be in no way abated.[2]

The demands of the oil-producing countries have not been formulated in this way, but the situation described is the logical result of their expressed desire for (1) as much revenue as possible while leaving the Companies a 'fair return' on investment; (2) participation in downstream profits; and (3) their dislike of the Companies' investment in competitive sources of oil. On the other hand, the several countries do not all take the same view of their needs and there are differences of opinion within each. OPEC must pursue a course satisfactory to as many of its members as possible and has concentrated so far on the more urgent question of doing something about prices and excess supplies, saying relatively little about the longer-range aspirations of the countries.

[1] This may, in effect, be the consequence of some of the partnership arrangements made by the national oil companies, for example, those in which the foreign partner offers the producing companies guaranteed markets for all or part of their crude oil.

[2] It is for these reasons that the proposals made by M. A. Mughraby in *Permanent Sovereignty over Oil Resources*, Chapter IX, are not appropriately applied to the major concessions of the oil Companies.

In papers submitted to the Arab Petroleum Congresses, however, OPEC has endorsed the principle of 'evolution' in concession agreements, and has stated that the 'governments of OPEC member countries in the Middle East are deeply aware of how outmoded the old 50/50 Agreements in the concessions held by the international majors have become', for these have been 'overshadowed by new Agreements providing for taxes of up to 57 per cent of earnings and full, active partnership for the governments in the various enterprises . . . "Fifty-fifty" provides the government with tax revenues but none of the means by which a true partner can protect his economic interests—and these interests are of vital importance to whole nations.'[1]

Plainly, 'evolution' towards the type of situation outlined above would require the oil Companies to organise their supply lines on a very different basis from that which exists at present. The required reorganization could well be more disruptive of their international integrated operations than alternative arrangements under which they brought all their crude oil on long-term contracts, or else produced it as contractors to the governments of producing countries, with which long-term contracts for offtake could be made.[2] Such issues are not yet raised by the participation provisions of the new concession agreements since the amount of oil involved in relation to the total oil supply of the Companies concerned is relatively small; these provisions seem in any event to be looked on largely as a means of increasing the proportion of profits going to the government and of satisfying national aspirations.[3] But they would be raised if similar participation arrangements were attempted for the chief established producing affiliates—the IPC, KOC, Aramco, or the Iranian Consortium—not necessarily immediately, but certainly before much time had passed. The countries, having got used to a higher level of revenue, would soon search for ways of becoming not only suppliers of the raw material but participants in the manufacture and distribution of oil products in world markets as well.

It is clear, therefore, that even apart from the obvious problems that would arise if managerial decisions became involved in local politics, the question of equity participation by the governments in the major concessions of the international Companies does involve very important principles for the Companies, on which they are unlikely to give way, and certainly only if the government partner is permitted to exercise little or no control. The arrangements in the Iranian Consortium are no exception, but rather illustrate the problem. Although Iran 'nationalized' the oil industry, the operating companies, in Iranian Oil Participants Limited, are not controlled by the

[1] 'Radical Changes in the International Oil Industry During the Past Decade', p. 17.

[2] That this would be the most desirable state of affairs from the producing countries' point of view was suggested by OPEC in its paper to the Fifth Arab Petroleum Congress, March 1965, *From Concessions To Contracts*, see especially pp. 5–6.

[3] This, at least, seems to be the case in practice. In principle, however, government participation is explicitly looked at as a potential means of control: '. . . it provides the host government with an opportunity to play an active part in the exploration and exploitation of its mineral resources; it enables the host government to control oil operations effectively; and it is an important step towards governments' sole management and control over the industry, if that is deemed both practicable and desirable.' OPEC, *From Concessions to Contracts*, p. 14.

Government. The conditions under which they work are carefully specified in the agreement, and the Government is in no better position to intervene directly in their price, offtake or investment arrangements than are the governments of the other producing countries.

In bargaining with the producing countries the Companies rely heavily on their legal position and stress the sanctity of contract. For them unilateral or even the forced acceptance of changes in concession agreements involve legal (moral) principles.[1] The producing countries, on the other hand, have insisted that the original contracts were made between unequal partners, and that with changed political and economic circumstances revisions are morally right, even if they have to be forced on the Companies. It is complained that the very long terms of some of the concession agreements are intolerable because with changing economic and social conditions must come changing property relationships; to freeze property rights or tax arrangements for very long periods is unacceptable.

The economies of the great oil-producing countries are dependent on oil revenues and cannot lightly risk their cessation, but they have been able to modify the terms of their old agreements significantly, even insisting on their right to break contracts in the public interest—which in practice tends to mean when the government thinks that the country's economy is being prejudiced by the existing state of affairs. At the same time, even though the more advanced oil-producing countries could produce oil, and probably even successfully find and develop it, no individual country is prepared to risk losing the market provided by its major concessionaires. The position of the major Companies has been somewhat improved by the growing appreciation in the producing countries of the advantages to them of the integrated distribution channels of the majors, in which the price at which crude is transferred is partly sheltered from the competitive pressures of the market. The difficulty of maintaining prices when there are a large number of producers is increasingly realized. But there is no reason to assume that the steady encroachment by the governments on the freedom of action of the Companies will cease, and it seems highly probable that in the foreseeable future a contractual form of relationship between Companies and governments for the production of crude oil will emerge. In the meantime, to achieve their aspirations to enter the oil industry in their own right, the producing countries have turned to the development of their own national oil companies.

NATIONAL OIL COMPANIES

National oil companies are looked on as the means by which the producing countries can acquire experience in the industry and in world markets, for not only do they operate in domestic markets but some of them have already reached out abroad. The most advanced is the National Iranian Oil Company (NIOC) which distributes products in Iran, arranges exports of gas as well as oil, and, in partnership with foreign companies, is launching out into the production of petrochemicals. It is also building a refinery in India in associa-

[1] As noted above, however, the moral position of the majors has been weakened by their attitude towards the 1965 changes in the Libyan law.

tion with a US independent oil company (Pan American, a subsidiary of Indiana Standard), and will supply crude oil. Companies seeking exploration rights in Iran under the Petroleum Law of 1957 are given preference if they agree to take NIOC as a partner. In 1967, arrangements were made with the Consortium for NIOC to obtain stipulated amounts of royalty oil to be used for marketing in Eastern Europe, largely on a barter basis. It was expected that sales in that area would not affect Consortium sales.

The Kuwait National Petroleum Company (KNPC), established in 1960 with a 60 per cent government and 40 per cent local private ownership, was, at the time of writing, building an export refinery and had just concluded an agreement with a Spanish oil group for a joint exploration venture in areas relinquished by the Kuwait Oil Company (BP and Gulf).[1] Petromin, the Saudi-Arabian Oil Company, has also been exploring partnership arrangements. In Venezuela the Corporacion Venezolana del Petróleo (CVP) has been extending its activities in the domestic market but has not yet reached abroad. Even Libya, a very new producer, is trying to set up its own oil company.

Iraq's National Oil Company (INOC), established in 1965, got off to a slow start because of the prolonged negotiations with the IPC group on problems associated with the expropriation of their properties. At first its prospects looked very promising because the Government (and the national company) succeeded in negotiating a very favourable agreement with the owners of the IPC which would have opened the door wide to exploitation of very valuable oil lands by the INOC in partnership with major Companies on terms which would have given Iraq greater revenues per barrel of oil than those obtained from any other major concession in the Middle East.

The negotiations had been difficult, but the Companies were eager to secure the oil and the Iraqi negotiators were skilful. One of the problems in the negotiations related to a difficulty which has faced many of the underdeveloped countries with small markets that have wanted to engage in crude-oil production, and more particularly in refining—the difficulty of disposing of surplus production that cannot be absorbed by the local market. It is easy enough for a government to reserve the domestic market for national companies when it wants to do so, but a similar reservation cannot be extended to foreign markets, and if the production of oil (or products) exceeds domestic demand the excess must be disposed of. The host countries are given the right in their concession agreements to sell their royalty oil to the foreign Companies at the posted prices, and they have generally preferred to do so, having no markets of their own. In joint ventures, particularly with the integrated Companies, a similar arrangement whereby the foreign partner agrees to take the part of the government's share that it could not profitably sell is obviously convenient for national companies without regular markets. Such provisions are common in the new concession agreements. The Iraq National Oil

[1] To overcome the advantage possessed by the integrated companies of being able to guarantee markets, the Spanish Government offered for fifteen years a guaranteed 25 per cent of Spanish oil imports to the joint Spanish/Kuwait company (KNPC, 51 per cent) for the Spanish share of the oil, or a minimum of 130,000 b/d by 1970.

Company requested a 'buy back' arrangement in its negotiations with the Companies, the price to be half-way between cost and posted prices. Such arrangements may, however, commit Companies to taking more oil than they would like at that price, which is not attractive if posted prices stay up and market prices fall.[1]

The 1965 agreement came under heavy political attack by opponents of the then Government, for it brought back on the scene the IPC owners with whom any seeming accommodation can be presented as a 'victory for imperialism' by those who choose to do so, even though the country itself would have gained. On the other hand, there are many in Iraq who feel that the country must, in its own interest, 'break the monopoly of the IPC'. The Government could not survive the political opposition and the agreement was not ratified. As a result, oil exploration in Iraq continued to be blocked and the industry's development continued to be retarded.

After the Israeli war, negotiations with the IPC group seem to have been broken off, and in August 1967 the Government promulgated a law, which assigned to the INOC most of the area of Iraq, including the territorial waters, for exploration and development. This is virtually a 'no concessions' law, but the INOC is allowed to enter into partnership with foreign Companies for exploration and, presumably, production. The new law is a further step in giving effect to the expropriation under Law 80 of 1961, but it did not preclude the IPC Companies from proposing partnership arrangements with INOC to explore for oil in the unexploited areas of the country. One difficulty, which has deterred other foreign companies from actually making arrangements with the INOC to produce oil, lies in the fact that the former owners of the concession covering the expropriated territories in Iraq threaten action against any company selling oil it obtained from such areas. The major Companies involved evidently believe that they have a case in international law against Iraq for illegal cancellation of their concession contract without compensation. Whether or not the case stood up in the Courts when brought against other companies, the potential trouble the majors could cause might well lead others to prefer alternative ways of obtaining oil.

However, in November, 1967, an agreement was signed by INOC with ERAP, the French State-owned oil company, for the exploration and development of part of the expropriated areas, but excluding the rich North Rumaila field. The terms of the agreement are very much less favourable to Iraq than were those of the abortive 1965 agreement, but it was defended by one of the officials of the INOC as an important step in breaking the IPC monopoly in the country.[2]

Negotiations were in process when this book went to press between the Government and CFP, the French partner in the IPC, with respect to the North Rumaila field and it looked as if Iraq and the IPC group might have embarked on a collision course.

[1] As noted above the draft agreement with the IPC group provided for only a one-third share for the INOC. Presumably the chief reason for this smaller share related to the fact that the larger the equity holding of the national company, the greater may be the amount of oil the foreign partners would have to buy at a half-way price.

[2] For an analysis and criticism of the terms of the agreement, see *Middle East Economic Survey*, Vol. XI, No. 5. December 1 .1967.

In negotiating with Iraq, the major Companies have always to bear in mind that if they grant Iraq greater revenues per barrel of oil under new concession (or contract) agreements than other Governments obtain from them, there will be an immediate demand by the others for the same terms in line with the 'most favoured country' clauses contained in the older agreements. On the other hand, if the great reserves of Iraq are developed by companies (including the national oil company) that do not have adequate markets under their own control, the consequences might be serious for the price of oil.

INOC hopes in due course to enter world oil markets, as do the other national companies. All are aware of the dangers to them of competing with their own established concessionaires and various devices have been adopted to avert this. Hartshorn has pointed out[1] that national oil companies, which he defines as companies acting as agents of governments whether State-owned or not, have also been created in a number of consuming countries, and that deals between such companies and the national companies of producing countries have become a means by which the latter can enter markets without cutting prices. Such companies in consuming countries can guarantee markets to their producing counterparts on preferential terms. The fact remains, however, that oil, no matter how it enters, is an addition to the supply on the market and is in competition with other sources either directly or indirectly (except, perhaps, when it is sold to communist countries). There is no escaping the conclusion that when the national oil companies enter world markets there will inevitably be a conflict of interest between them and their concessionaires which will tend further to depress the price of oil.

CONCLUSION

The producing countries, through OPEC, maintain that their objective is to 'optimize' their revenues while maintaining a 'fair' price for oil. The difficulty, of course, lies in the quoted words; 'optimize' is undoubtedly a euphemism for 'maximize', and what is 'fair' is a matter of opinion, for the producing countries and the Companies are not the only groups concerned with the price of oil.

It is by and large the importing countries which provide the funds out of which not only are most oil operations financed, but all taxes to the producing countries are paid. To the extent that payments to the governments of the oil-producing countries come from funds that would otherwise have been paid as taxes to governments of the countries in which the international oil Companies are incorporated, or that would have been retained for distribution or reinvestment by the Companies themselves, the gain to the producing countries is not at the expense of the importing countries generally; but if taxation by the crude-oil producing countries becomes instrumental in maintaining prices, then the gain of the producing countries will to that extent be at the expense of the importing countries. The importance of this for the underdeveloped importing countries will be dealt with in the following chapter.

[1] In a seminar paper given at Riyadh, April 18, 1967.

Chapter VIII

THE INTERNATIONAL OIL COMPANIES IN DEVELOPING COUNTRIES
II. IMPORTERS AND MINOR PRODUCERS

Outside the major crude-oil exporting countries the developing countries have varied interests in the operations of the international oil Companies. A broad common denominator is provided by their desire to minimize the costs of their oil supplies, usually with special reference to the foreign-exchange cost,[1] and where possible to develop domestic crude-oil supplies. A large number of countries produce some crude oil, a few are self-sufficient and several earn a large amount of their foreign exchange from exports. But even countries that are largely self-sufficient, like Egypt for example, must often import certain qualities of crude oil or products to meet their domestic requirements.

In most—but not all—underdeveloped countries where crude oil is produced, the issue of foreign control has caused political conflict; a number of countries, notably Mexico, Argentina, Brazil, Chile, Syria, Indonesia and for a while both Bolivia and Turkey, have insisted that their resources must be exploited under State ownership or auspices.[2] In many countries, the dislike of a foreign-dominated industry has extended to refining, and distribution is either reserved for State companies or is government controlled, private distributors (if permitted) sometimes being required to take their supplies from a government company, as in Brazil for example.

It is not possible from published statistics to draw up a balance sheet of the impact of the operations of the Companies on any country although in some cases the results were of crucial importance. A description of their economic, social, and cultural projects outside the oil industry, their exploration and investment projects, the efficiency of their local marketing organizations, and the extent of their trading activities, would indicate the wide-ranging nature of their contribution, but would be tedious, and without hard figures inconclusive. It is not, therefore, the purpose of this chapter even to attempt to appraise either quantitatively or qualitatively the significance of the operations of the international Companies for any country or group of

[1] There are, however, important exceptions, for some countries seem so eager to have a refinery that they are prepared to pay very heavily for it. In Lebanon, for example, the terms on which the Companies have agreed to undertake refining in the country and which were accepted by the Government are such that a net loss for the economy may well be incurred as a result of the operations of the industry. See pp. 232–4 below.

[2] Odell's discussion in Chapter XI below deals particularly with the political problem in Latin America.

countries; rather we shall explore some of the problems that the underdeveloped importing countries and the minor producers face in the formulation of their petroleum policies, particularly with reference to the role of the large foreign Companies in the industry.

GENERAL ATTITUDES

The markets of the countries of Asia, Africa, and Latin America are small, but some were considered important enough for the Companies to struggle over long before industrial development began to become important. Until the great expansion of refining capacity in the developing countries after the war, those countries without significant domestic supplies of crude oil had to import their oil products. The prices of products, as we have seen, were set by the suppliers within a pricing system supported by effective control over supplies and open or tacit market-sharing arrangements. There were very few 'independent' suppliers willing to cut prices to sell products.

The structure of prices, and the price/cost relationships established, were of considerable consequence for the underdeveloped importing countries, but before the Second World War many of them were colonies, and others were in no position to appraise, let alone to influence, the effect on their economies of the operations of the Companies. In any event, most of these countries saw little reason to be concerned with the industry except when there seemed some possibility of domestic oil production.

A number of countries where a significant amount of crude oil was produced, early preferred to remain independent of the foreign Companies and others came into serious conflicts with them. As early as 1927 Chile established a State monopoly in crude-oil production. In 1937 Bolivia expropriated Jersey Standard and set up a State company, keeping operations in Government hands for 15 years, after which it decided to accept international help. But even now the State company (YPFB) is the largest producer, although the international Companies also operate in the country; Gulf, for example, even exports. Mexico expropriated the international Companies in 1938. Peru has been in conflict with Jersey Standard's subsidiary International Petroleum since 1922 over the legal status of La Brea-Pariñas oilfields, although it has so far stopped short of nationalization.

Such pre-war examples are relatively few and were confined to Latin America, where political independence was achieved very much earlier than elsewhere in the underdeveloped world. In the countries of Africa and Asia there was little government 'intervention' in the industry, although colonial governments often took steps to prevent firms from other than the mother country from getting a foothold, particularly in production. After the war, with changing economic, social, and political ideas in most countries and the advent of independence in some, the appropriate role to be allotted to the foreign Companies, not only in crude-oil production but also in refining, became more and more widely debated, and clashes between the oil Companies and the countries in which they operated were fairly common, with a few countries engaging in outright nationalization.[1] Few, like Chile have pursued

[1] Retaliation from the home governments of the international Companies is always a

a consistent policy of excluding the foreign oil companies; some, like Argentina, Turkey, and Bolivia, vacillated between exclusion and acceptance; still others, like Ceylon, India, and Pakistan, took no significant action until they were convinced not only that some of the policies of the Companies operated to their disadvantage, but also that there was a practical way of reducing their dependence on them.[1]

In most of Africa, on the other hand, with the notable exception of Egypt (and of course, Algeria, which is not considered here[2]), governments have so far made little attempt to impose their control on the industry in ways acceptable to the major Companies. The wave of refinery construction in African countries did not begin until the early 1960s, and consequently the activities of the Companies consisted of the importation and distribution of products, exploration in a number of areas, and crude-oil production in a few. ENI was among the most enterprising of the international Companies in instigating the building of refineries in the less developed areas of Africa and it early offered the governments a 50 per cent interest in such enterprises in order to obtain the contracts. Thus, government participation in refining became more the

---

consideration when an underdeveloped country contemplates nationalization of foreign properties. Such retaliation may be serious if the country is dependent upon (or hoping for) economic aid in promoting its development. Under the US law, aid is automatically withdrawn from any country nationalizing American property without prompt and adequate compensation. In 1965 political demands that the Peruvian Government expropriate the La Brea-Pariñas properties of the International Petroleum Company (a subsidiary of Jersey Standard) were widespread, although by no means universal, in the country. Among the problems facing Peru's President was the possible reaction of the US: 'Meanwhile the latest developments in this long controversy between the Peruvian Government and IPC are being closely followed by the US Government, which is currently studying Peruvian plans for communications, irrigation, and other projects, for the financing of which Peru is seeking some US $50 million in Alliance for Progress loans. President Belaúnde is aware that any rash unilateral action such as expropriation by the Government would create an unfavourable impression on the US Government and people, and the implications are that President Belaúnde would welcome a settlement acceptable to both his political supporters and IPC.' *Petroleum Press Service*, August 1965, pp. 303–4.

[1] 'Special relationships' established under colonial rule are often used after independence has been gained to continue a preferential position for firms from the ex-colonial power. Sometimes this is part of the continuing economic dependence of the ex-colony on its erstwhile protectors—the relationship of Algeria and France, for example, under which Algeria has gained much. Sometimes it seems to be little more than the continuance of an 'economic imperialism'; it is difficult to see, for example, why US oil firms should have a preferential position in the Philippines. Even in the Congo it is reported that the Belgians pressed for a revision of the refinery agreement between ENI and the Congolese Government (under which the refinery was to be owned in equal shares by the company and the Government) to admit Petrofina (the Belgian company) together with Royal Dutch/Shell and two US companies. The proposal was, apparently, to increase the Government participation to 51 per cent and to force ENI to share the rest with the four other companies. See *The Financial Times* (London) April 28, 1966.

[2] The agreements between France and Algeria with respect to the oil industry are not only complicated and confusing but have been changed frequently. The relationship between these two countries is a very special one, and it does not seem useful to consider it here. The arrangements governing the oil industry are more a function of French Government oil policy and Franco-Algerian political relations than they are of the policies of or towards the large international firms, which is the subject of central interest for this study.

rule than the exception, with distribution as well as crude-oil production remaining largely in the hands of the private Companies.

In Egypt, however, production, refining, and distribution were almost entirely under the control of the international majors, chiefly Royal Dutch/Shell, until the 1952 revolution when the Government began progressively to take over control. At first special support was given to State enterprises and then, after the attack on Suez in 1956, the Government formed a General Petroleum Authority to control imports and exports, production and refining. At the same time, however, the Compagnie Orientale des Pétroles d'Egypte (COPE), which was formed in 1957 and is jointly owned by the Government and by ENI, remained, and continued in both exploration and production. Until 1968 COPE was the most important producing company in the UAR. Exploration was permitted by other foreign independent Companies (including Pan American, a subsidiary of Indiana Standard, and Phillips) under agreements which included the provision that upon the discovery of oil a company would be formed in which the Egyptian State company held a 50 per cent interest. Thus, Pan American's discoveries are being developed by the jointly owned Gulf Suez Petroleum Company (GUPCO). In distribution generally, the Government dominates, although Mobil and Esso have so far been allowed to continue with a small percentage of domestic distribution.

The fundamental question with which all of these various countries have been, and for the most part still are, wrestling, relates to the amount of control that the foreign Companies should be permitted to exercise over the local industry. The international Companies have shown themselves eager to build refineries even in small and relatively uneconomic markets whenever this was necessary to hold them for their own crude-oil production. The cost of importing crude oil is less than the cost of importing the products, and a considerable saving of foreign exchange on oil imports can be effected if crude is domestically refined, especially if foreign Companies finance the refining investment. Similarly, the desire of countries to develop crude-oil supplies where this seemed possible needs no explanation in view of the growing importance of petroleum and the cost of imports; and again the international firms have shown themselves willing to explore and develop crude-oil resources wherever geological conditions seemed favourable and the laws governing their operations were liberal.

In view of the very large capital investment required in both refining and production, and of the risk involved in exploration, it seemed obvious not only to the international oil industry and its spokesmen, but also to Western governments and to international agencies like the International Bank, that the poor countries should call on the willing Western capital both to develop their resources and to build their refineries. There can be no doubt that the international Companies, because of their capital resources, their ability to bear risks, their advanced technology and managerial expertise, and their control of markets (which enables them to guarantee outlets for crude and to deal with surplus refinery products), have much to offer the poor countries in their attempts to develop their domestic industry at all levels. On the other hand, foreign private capital may be expensive, and foreign control may have undesired side-effects in refining and distribution as well as in crude-oil production, although the latter has probably been given more attention than the former. In the next two sections of this chapter we shall examine some important aspects of these questions, first with respect to refining and distribution and then with respect to exploration and production.

REFINING AND DISTRIBUTION

Refining capacity outside the US and Canada and the Communist countries rose from 93 million tons (1·8 million b/d) in 1939 to 974 million tons (19 million b/d) in 1966, at which time 26 per cent of the total was in Latin America, Africa and Asia (excluding Japan and Australasia).[1] Much of the increase in the capacity in the developing countries was in response to direct pressure from their governments. At the same time, because the Companies needed to keep or expand their crude-oil outlets they made available relatively cheap and abundant supplies of capital for the construction of refineries in these countries, even when the possible market was too small to support a refinery of economic size.[2] The refineries in the Central American Republics provide an extreme example of this problem. The demand for domestic refineries has been widely regarded as the consequence of economic nationalism and an economically irrational desire for 'prestige projects'. Although there is some truth in this, the governments of the developing countries were not the only ones to intervene in order to encourage, or enforce, domestic refining.[3]

When the international Companies built or financed refineries in importing countries they became responsible for the supply of crude oil to these refineries. Few countries saw any disadvantage in this as long as crude seemed to be imported at prices similar to those paid by other countries and being charged by alternative suppliers. When this situation changed, new problems arose which stemmed directly from the vertical integration of the Companies. Similar problems also arose for a country importing refined products where the distribution facilities were owned by the integrated Companies. Perhaps the best way of making clear the nature of the problems posed for underdeveloped importing countries by the organization and policies of the international Companies is to discuss the experiences of India, Ceylon, Pakistan and Lebanon in the early 1960s, for these provide excellent illustrations of the issues involved.

India

As early as 1928 the President of the Tariff Board in India had urged that a refining industry be established in the country based on imported crude oil,

[1] W. L. Newton, in a paper presented to the international petroleum seminar at the School of Oriental and African Studies, University of London, March 1967.

[2] An interesting example of the way in which research and innovation respond to changing external circumstances is provided by the changes in refinery economics as political circumstances force the building of smaller refineries. Faced with this demand, the companies quickly developed technology which significantly reduced the cost of smaller scale refining.

[3] As early as 1948, for example, the UK officially announced its policy of preferring domestic refining to save foreign exchange, and in its White Paper on *Fuel Policy* (Cmnd. 2798, HMSO 1965) it was made clear that the Government expected all refining to take place in the country as soon as practicable.

but it was not until 1954 that action was taken. The international Companies were at first reluctant to establish refineries, insisting that domestic refineries in India could not compete with imported products. They calculated that an annual loss of Rs.20 million would be incurred unless domestic prices were increased by at least 10 per cent.[1] India, on the other hand, was concerned not only to reduce import costs, but also to obtain greater security of supply. Some 70 per cent of Indian imports of products came from Abadan, and after the Iranian crisis in the early 1950s, when supplies were severely curtailed in spite of the Companies' efforts to maintain them, the Government's case was seen to be strong.

The Companies finally agreed to establish refineries to process their imported crude oil, but the Refinery Agreements under which they operated provided for guaranteed duty protection for 10 years 'as it was made out at the time of entering into the Agreements that a reasonable return on capital investment would not be available otherwise'.[2] Domestic (pre-tax) prices were to be based on 'import parity'. The Government agreed not to nationalize the refineries for 30 years and to exempt the companies from some of the more onerous provisions of the Industrial (Development and Regulation) Act. Crude oil was to be imported at 'world market prices' and paid for in sterling or dollars; the source of supply was to be at the Companies' discretion. Three refineries were built between 1954 and 1957 by Burmah Shell, Stanvac,[3] and Caltex;[4] thus, all the refineries were wholly-owned subsidiaries of international major Companies.

Although the direct cost of oil imports to the Indian economy was reduced with the establishment of refineries financed by the international Companies, the refining companies were entitled to remit their profits and to remit payment for crude oil in hard currencies; hence the Government had a legitimate interest in their cost and pricing policies. Internal prices to consumers are influenced more by the rate of government taxation than by duty protection.

At the same time the Government wanted to develop domestically controlled production and refining. The State-owned Indian Refineries Ltd. was created in 1958 to operate new refineries that were being planned; in 1959 the Indian Oil Company was formed to undertake marketing, and the Oil and Natural Gas Commission was established to carry out crude-oil exploration and production. The Indian plans intended the developing petroleum industry to become part of the 'public sector', i.e. State controlled. But quite apart from this, the Government was becoming increasingly dissatisfied during the later 1950s with the pricing policies of the refining subsidiaries of the international majors, who still controlled all of the country's refining capacity. This dissatisfaction was brought to a head in 1960 when Russia offered the country crude oil on very much better terms than those at which the major Companies were importing to India. Moreover, India knew that the same

[1] B. Dasgupta, *Oil Prices and the Indian Market, 1886–1964*, pp. 106–7.
[2] Government of India, *Report of the Oil Price Enquiry Committee* (Damle Report), 1963, p. 3. The Companies gave up their duty protection in stages after 1956.
[3] A jointly-owned affiliate of Jersey Standard and Mobil Oil.
[4] Jointly owned by Standard of California and Texaco.

Companies were charging Japan lower prices than they were charging her.[1] A Government Committee on oil prices (the Damle Committee) was appointed in that year and reported in 1961.

The Damle Committee was very critical of the policies of the refining companies and was not at all convinced that the companies had done the best they could to get their crude oil at the lowest possible prices:

> 'The oil companies consider it a coincidence that their offer of discount on crude oil was made at about the same period when the offer of Russian crude oil for processing at the coastal refineries in India was being discussed with them by the Government. The oil companies did not accept this offer on account of the freedom allowed to them in the Refinery Agreements in regard to choice of source of supply of crude oil. One cannot fail to recognize that the competition generated by this offer influenced their suppliers in their final decision to allow discount at the current rates. On account of the traditional nature of the discounts, it should normally have been possible for the oil companies to successfully bargain for discounts off posted prices with their suppliers long before the offer of the Russian crude oil was made to them.'[2]

In general the Committee confirmed the widespread suspicion that the local subsidiaries of the international Companies were not acting in a straightforward commercially independent fashion:

> 'The Refinery Agreements give freedom to the oil companies to make their own arrangements for import of crude oil. . . . We fail to see why the companies cannot deal directly with the original producers/suppliers of crude oil, or furnish precise information regarding the names of such original producers/suppliers. The oil companies (viz. Burmah Shell, SVOC and Caltex) operating the coastal refineries in India, with their world-wide connections, should normally be capable of dealing directly with the Consortium operating in Iran in which their principals have interest, in respect of supplies obtained through that source, and also with the original producer/suppliers at other places.'[3]

It is not, therefore, surprising that the Indian Government should have decided that it needed its own domestically owned refineries which would enable the country to take advantage of the cheapest sources of crude oil available. The policies of the foreign Companies reinforced the prevailing Government view that important industries of this type should be Government controlled. Although the establishment of the refineries was a gain to India compared with what the position would have been had she been importing products at posted prices, the fact that the refineries were affiliates of foreign Companies and their oil was imported from affiliates of the same Companies significantly reduced this gain, especially when oil became available at lower

[1] An open break in prices of crude oil in inter-affiliate sales occurred in January 1959 when Stanvac gave a 10 per cent discount to its affiliates in the Japanese market in response to the pressures from the Government for lower-priced crude.

[2] Damle Report, p. 22.

[3] Ibid., p. 21.

prices than were granted to the Indian refineries.[1] Partly for this reason, the Government refused to allow the privately owned refineries to expand their capacity unless they agreed to give up the Refinery Agreements of 1951 and 1953 and import crude at prices comparable to those offered by the Russians. The Government also refused to let them participate in the building of new refineries unless they agreed to a majority shareholding for the Government. The Companies declined to do either of these and the Government went as far as it could consistent with the Refinery Agreements to deny the established majors any further share in the Indian market.[2] The Government proceeded to expand the activities of the national oil company that had been set up in 1959.

At the same time, the Government was not averse to partnership agreements with foreign Companies nor did it renegue on the agreements previously made with the Companies. Nevertheless, the Damle Commission found that discounts available on crude oil were not reflected in the prices of products from the refining affiliates of the majors, and that these affiliates did not obtain the discounts known to be available in world markets for their bulk imports of deficit products. It therefore recommended that in the allocation of foreign exchange to the refineries for their crude-oil imports, discounts available in world markets should be taken into account. Similarly for the import of bulk refined products: 'As discounts are admittedly available in crude oil, there is every reason to believe that the discounts should also be available on refined products' and the Committee recommended that 'it would be quite appropriate to base the landed cost of products on imports at cheaper rates, if available, from sources other than those, which have been normally utilized by the major oil companies provided continuous supply of equivalent quality is guaranteed'.[3]

The Damle Committee's expressed criticisms of the refining companies for

[1] B. Dasgupta in his study of oil prices and the Indian market found that between 1956 and 1960 Burmah Shell alone accounted for 60 per cent of the total outflow of foreign dividends from all sources. Remittances were at the rate of Rs. 80 million a year for the oil Companies compared with Rs. 30 million for all foreign manufacturing firms in India. Had a discount of 20 per cent been allowed between 1955 and 1961, Rs. 473 million would have been saved. Thus his conclusion was that the foreign refineries did help India's foreign exchange problem in the short run but were expensive in a rather short longer run. See also the statement of R. S. Bhatt, Executive Director of the Indian Centre, that 'against a net inflow of Rs. 48·3 crores of foreign capital into petroleum during 1956–60, the net outflow from petroleum was Rs. 41·2 crores out of a total outflow of Rs. 71·3 crores', *The Economic Weekly*, February 16, 1963, p. 317.

[2] By 1965, however, Burmah Shell indicated a willingness to accept a Government majority shareholding in new ventures, to purchase crude at the 'most competitive prices' in world markets, and even to agree that the Indian Oil Company (State owned) should distribute the refinery products, subject to certain safeguards. See Dasgupta, pp. 117 ff.

[3] Damle Report, p. 91. 'The stipulation in the Refinery Agreements for determination of the prices of products on "import parity" is not intended to preclude the availability of supplies in the Indian market of products from sources other than those of the major oil companies.' Ibid.

It cannot be inferred, however, that because discounts were available on crude oil they would also be available on all products, as the Damle Committee evidently inferred. The relation between supply and demand at any one time is different for different products, and it may be that discounts can be had for gasoline, for example, but not for kerosene or gas oil. This seems to have been the situation in the Far East at this time.

not pressing harder for discounts, and its inability to understand 'why the companies could not deal with the original producers/suppliers of crude oil or furnish information regarding the names of such original producers/suppliers',[1] arose from its mistaken attempt to treat the local affiliates as independent companies. To do otherwise would have required an investigation into the structure and general pricing policies of the international Companies. These may or may not have been clearly understood by the Committee or by the Government, but the Government did understand that foreign control of the country's refining capacity prevented India from shopping around for her crude-oil supplies.

The Government was also dissatisfied with the Companies' exploratory activities, for the amount being spent in India was not large, although a subsidiary of Burmah Oil had in fact discovered the Nahorkatiya oilfield,[2] and it decided to embark upon its own exploration programme. A number of oilfields were discovered at considerable cost, some with the help of the Russians, and in 1966 the Government re-designed its pricing policy to give protection to indigenous crude.[3]

As a result of the restrictions placed on the expansion of the activities of private companies, the completion of the new Government refineries (which brought half the country's refining capacity under Government control), the leading role allocated to the Indian Oil Company in distribution, the prohibition of oil imports except through IOC, and the expansion of local domestic crude supplies, the Indian Government had by 1966 effectively destroyed the monopoly position of the international Companies and built up its own domestically controlled industry. It cannot be pretended that the Government's policies were based entirely on cold economic calculations, but the attempt of the oil Companies to hold their international line in the Indian market while giving way elsewhere offered considerable political provocation.

The position of the Companies is understandable, for they were struggling to maintain an international price structure in the face of Russian willingness to cut prices to obtain markets and of growing pressures on prices in world markets generally. As international Companies they could not be expected to offer special treatment anywhere unless they had to, since special treatment in one case can easily call forth other special cases. They were desperately concerned to establish the argument that Russian prices were 'political' and that any other discounts going were related to 'distress sales'. In other words, in 'captive markets', such as the Indian market was when the majors controlled

[1] Damle Report., p. 90.

[2] See Dasgupta, pp. 115–16. There can be little doubt that in earlier years the foreign companies did not give a high priority to India's oil development. Burmah Oil, for example, had a concession in the Assam oilfield, but did relatively little exploration—apparently just enough to keep the concession rights. With reference to the investment in Assam, Burmah Oil's Chairman stated in 1933 that the 'investment was thoroughly justified on marketing considerations alone, since in other unregulated competitive hands . . . it would have cost us much more in the shape of "cut" prices than we paid for it. . . .' Quoted in *Petroleum Times*, June 10, 1933, p. 599. After the war, however, exploration activities were greatly increased.

[3] Another investigation of oil prices was carried out in 1964 by the Talukdar Committee and the Government subsequently revised its pricing system. *Report of the Working Group on Oil Prices*, 1965. See *Petroleum Press Service*, April 1966, pp. 128–9, for a discussion of the new policies.

its refinery capacity, and where in consequence there was no competition among the suppliers of crude oil, only pressure from the Government could be expected to force reductions in the transfer prices of the Companies. In its power to allocate foreign exchange the Government had an effective weapon, which it could use without violating its agreements with the Companies so long as the 'world market prices' at which the Companies were required to buy under the Agreements were indeterminate.

Reductions in the cost of crude-oil imports could not, however, entirely solve the problem, for internal ex-refinery prices were based on 'import parity' under the Agreements. This meant that the Companies could sell their products at prices 'not higher' than equivalent imported products could be made available. Thus, the Government's interest in world product prices extended beyond the question of imports since domestic prices based on them determined remittable profits, which would be higher, the higher the assumed import parity prices for products. Again, therefore, the Government laid down discounts on the posted prices of products on the basis of which the Companies were required to calculate their prices for the domestic market. Retail prices remained unchanged, the Government imposing a duty on products which collected the difference from the Companies.

Ceylon

Before 1962 the internal distribution of petroleum products was exclusively in the hands of three international Companies: Caltex, Stanvac, and Shell. These Companies owned all of the distribution facilities in the country. Products were imported at posted prices from the Persian Gulf. In 1960 the Government requested the Companies to reduce their import prices. This was refused, and in 1961 the Ceylon Petroleum Corporation, a Government-owned enterprise, was established and given the right to take over some of the properties of the international Companies. Arrangements were made by CPC to import products from Russia at prices lower than those being charged by the international Companies. In view of the availability of cheaper products, the Government imposed maximum c.i.f. prices for petroleum products which were only slightly above the prices offered by the Russians. The Companies refused to accept these prices, arguing, as they had in India, that the prices were 'unrealistic' and would force the abandonment of 'normal commercial pricing'. They also pointed out that 'prices charged in Ceylon are the prices prevailing in the market area of which Ceylon is only a minor part'.[1]

As might be expected, the whole issue was highly charged politically, and there is no need here to recount the details of the subsequent controversy. In the beginning, the CPC took over only selected parts of the Companies' properties, arguing that it would be wasteful to construct new facilities on the Island to enable the State Company to enter the business of distributing

[1] These quotations are taken from *News Letter* produced monthly by the Public Relations Department of Esso, undated but probably about the middle of 1963. The oil Companies' case is widely discussed in the trade press of the period. Ceylon's case is to be found in the Parliamentary Debates of the Ceylonese House of Representatives in April/May 1961 and August 1963, when the Bill to establish the Ceylon Petroleum Corporation was debated. *Petroleum Intelligence Weekly* reproduced a statement of the Ceylonese official view on November 4, 1963.

petroleum products. The foreign Companies and their governments protested the expropriation and demanded compensation at 'fair market value'. No agreement was reached on compensation and in February 1963 the US Government suspended aid to Ceylon in accordance with the 'Hickenlooper' amendment to the US aid legislation. The Companies continued to refuse to accept the Government's maximum prices for their imports, and in 1964 the Government took over the rest of the Companies' facilities with the exception of bunkering and aviation refuelling, which served international customers.

The Companies had much less incentive to come to an agreement on prices with Ceylon than they had had with India, since the Ceylonese market was a minor one and the investment at stake was much less. At the same time the Government was less astute in handling the problem than the Indian Government had been. But the issue was essentially the same: an insistent attempt on the part of the international Companies to maintain their international price structure in those markets where they had a complete monopoly while giving way elsewhere.[1] The cost to the Ceylonese Government of expropriating the Companies' properties at 'fair market prices' would have been very high; at the same time, the country was in fact being forced to pay prices for imports higher than those at which alternative supplies were available simply because the distribution facilities were in the hands of foreign Companies who were not interested in importing at the lowest available prices. The arguments put to the Ceylonese Government against a reduction in prices were thoroughly spurious, as events have since shown; the only justification that the Companies could give for their prices was that they were 'commercial'.

It is not necessary to condemn or approve the way in which the Ceylonese Government handled the affair to recognize the difficulty of the underlying issue. To be sure, the Companies were in an awkward position in view of their international interests, and of course they had a strong case in international law over the issue of compensation. But they were treading on very sensitive toes and they never seem to have removed their heavy boots, refusing to see the broader implications of their position. For example, they refused to lend, lease or sell tankage or other facilities to the Government in order to enable the CPC to get a start. If they were in fact unwilling to 'meet competition' on prices, it is difficult to see how they could have expected the Ceylonese Government to accept their position, given the prevalence of lower prices in world markets at that time.[2]

Pakistan

Pakistan approached the problem of oil prices much more cautiously than had either India or Ceylon. The attitude of her Government toward both private and foreign enterprise was perhaps less critical, but she was equally concerned

[1] The Ceylonese, as had the Indians, found no lack of illustrations of discounts in other markets.

[2] After the defeat of the Government of Mrs Bandaranaike, which had carried on the petroleum policies begun under the previous Bandaranaike Government, the new Government of Ceylon, faced with serious foreign exchange and economic difficulties, began negotiations with the companies. In June 1965 a compensation agreement was signed under which the companies were to receive Rs. 55 million, about half the amount they had originally suggested as the 'fair market value'. US economic aid was subsequently reinstated.

to avoid excessive dependence on a small group of Companies. She accepted Russian assistance in exploration, and has insisted on local participation in the equity of both production and refining companies. Until the end of 1962 almost all supplies of products were imported, there being only one small refinery using domestic crude, and apparently posted prices were paid for oil imports.

In 1962 a large local refinery went on stream at Karangi near Karachi, owned 40 per cent by local capital and the rest by Esso, Shell, Caltex, and Burmah Oil, who supplied the crude oil. The refinery agreement required the refinery to import crude at 'competitive' prices, but it also provided that if the Government was offered crude at a lower price, the foreign partners in the refinery had an option to meet the lower price. Government spokesmen complained that the provision was of little benefit, since independent suppliers were loath to offer crude when they knew that their offer would be used merely to force lower prices on existing suppliers. On the other hand, suppliers of crude to the refinery were obliged to buy back those products that were surplus to the country's requirements, and this reduced their incentive to compete for the right to supply the crude oil. Burmah Oil, though supplying crude at posted prices, also bought back products at posted prices and insisted that the net foreign exchange cost to the refinery was therefore no higher than it would have been had both transactions taken place at discounted prices. In 1965, however, crude oil was offered to the refinery through a local Pakistani interest (Abdul Jalil, Chairman of the Amin group owning Pakistan National Oil) at prices reported to be substantially lower than those of the supplying Companies.[1] The Government required the refinery to take this crude over the objections of the Companies, and in August 1965 advised the foreign Companies to reduce the prices of their products by similar amounts. The Companies then also reduced crude prices, but no longer bought back products at a premium.

A second refinery in Chittagong is owned 70 per cent by Pakistani capital (Government and private) and by Burmah Oil, but the Pakistan Government is entitled to supply half of the refinery's crude. As to marketing, the Pakistani National Oil, a local private company which was created in 1962 to market petroleum products, was in 1965 granted the sole right to import products into East Pakistan. The company was reported to have been importing products at prices much below Abadan c.i.f. prices, and crude oil at discounts that compared with those given to independent Italian and German refiners in Europe. The international marketing companies Burmah Shell, Burmah Oil, Caltex and Esso were told that they would receive no more licences to import products if PNO could fulfil its commitments. But arrangements have been made for a joint marketing company in which BOC has a 49 per cent interest to import deficit products into East Pakistan, and also for Castrol, a BOC affiliate, and PNO to manufacture and market lubricants in Pakistan.

Thus, Pakistan also brought its refining and marketing industry very largely under Government or local control without disturbing the existing rights of the foreign Companies, largely by ensuring that expansion of the industry was carried out with local participation. But for some time Pakistan clearly paid higher prices for her crude oil and products than did a number of countries

[1] The oil was said to have been obtained indirectly from Compagnie Française des Pétroles.

with more freedom of action. In the expansion of the industry, the role of the major Companies has been closely controlled.[1]

Lebanon

Not all countries, however, have been as vigilant in attempting to obtain better terms from the oil Companies as conditions changed as have the three countries discussed above. As late as the end of 1967 the refineries in the Lebanon, for example, were still paying posted prices for crude oil and operating under agreements that were in some respects extraordinarily unfavourable to the Lebanese economy. Lebanon is a small market, consuming not more than 7 million barrels of products annually (around 20,000 b/d). In spite of this, it has two refineries with crude runs of about 17,000 b/d each, both of which were established under agreements with the Government and are subsidiaries of international major Companies. The first, established during the war, is owned by the Iraq Petroleum Company and receives its crude oil from the IPC pipeline. The second, Medreco, was established in 1955 and is incorporated in the US. It is owned by three US Companies—Standard of California and Texaco (through their jointly-owned subsidiary, Caltex) and Mobil Oil, and receives its crude from Aramco, in which its owners have a 70 per cent interest. These refineries supply the local market and sell their surplus products, almost entirely fuel oil, to their parents for sale abroad. Although the refineries are small, they are not below the minimum economic size, but their agreements with the Government include complete protection of the local market against imports, exemption from taxes and fees of all kinds, as well as other provisions advantageous to them.

The refinery agreements were widely criticized in the country, especially after it was realized that crude oil was available in world markets at prices very much lower than posted prices. To illustrate the nature of the problems involved, we shall consider the position of the Medreco refinery.[2] Under the original concession agreement in 1945 the refinery was exempt from all taxes. Subsequently a law was passed (in 1956) imposing income taxes on all concessionary companies unless they signed new agreements with the Government. Accordingly, the company sought a new agreement and in 1965—ten years later—a 'Supplemental Agreement'[3] was negotiated in which the existing practices of the refinery would have been written into the law. The agreement was not ratified by Parliament because of political opposition.

Under the proposed Supplemental Agreement, the company was permitted to continue to import crude oil at posted prices and to price its products in accordance with a formula that guaranteed it sufficient revenue to cover its operating and management expenses, the purchase of crude oil, depreciation of its installations over fourteen years, and an annual profit of 7½ per cent on its total assets before depreciation, and including the cost of land and its working capital. Although no taxes, fees, duties, or fiscal charges of any kind were to be levied on the refinery, a very small payment of 7½ cents a barrel was

[1] I am not here concerned with the position of the foreign Companies in the search for and production of crude oil and natural gas, which is significant.

[2] The IPC refinery is in a similar position.

[3] Decree No. 3508, November 17, 1965.

to be made on products delivered for internal consumption, in effect in lieu of discounts on crude oil. In view of the fact that something like one-third of the refinery's output was sold outside Lebanon, this was a very small 'discount' indeed, for in 1965 at least 35 cents a barrel on all crude purchased was widely granted elsewhere. The Supplemental Agreement was attacked on a number of points, the three most important of which related to the provision for the import of crude oil at posted prices, the exemption from all taxes, and the method of calculating the rate of profit.

As we have already seen, few markets other than 'captive' markets were still paying posted prices in the middle of the 1960s. Lebanon is indeed a captive market, for the international majors supply all its oil and products, but it should presumably have been able to make a strong case for a re-negotiation of crude prices in view of the evident discrimination by the Companies in requiring it to continue to pay posted prices for its imports. On the other hand, the refinery sold its surplus fuel oil to its parents at the Lebanese ex-refinery price, which was about $1·00 a barrel higher than the price at which oil could be disposed of in world markets. Thus, the parents subsidized the refinery in the prices at which they bought the fuel oil. We have noted earlier, however, that it is to the advantage of an internationally integrated Company to show its profits, wherever possible, in countries where effective taxation is lowest. Had the fuel oil been sold in foreign markets at a profit, presumably these profits would have been taxed, while in Lebanon no taxes at all were paid by the refining affiliates. It is, therefore, reasonable to assume that the tax position provides at least a partial explanation of the willingness of the parent Companies to subsidize their refining affiliates, for their 'losses' on the sale of fuel oil can presumably provide tax offsets for profits made elsewhere. It is also possible, however, that the refining company might have been criticized if it had sold to its parents at prices lower than the ex-refinery prices charged the Lebanese marketers and its affiliated distribution companies in the Lebanon. It is possible (though unlikely) that the subsidy was greater than the 'over-charge' to the refinery on crude oil arising from the failure of its parents to grant it appropriate discounts, and that the Lebanese economy gained when the 'over-charge' and subsidy are considered together. But if the geographical distribution of the profits of the international Companies was to the financial advantage of the Companies, then presumably they would have been willing to grant any necessary discounts on crude, yet still retain the subsidy so long as they could claim exemption from income taxes in the Lebanon.

Exemption from income taxes (as well as from other taxes) was presumably granted to the refineries in the light of the limitation imposed on their prices and overall rates of return. The Supplemental Agreement permitted Medreco, for example, to earn a maximum of 7½ per cent on its assets. But in calculating this rate of return, *net* profits after depreciation were related to *gross* fixed assets, that is, assets at original cost, plus land and working capital. Thus, the ratio of profits to assets, which expresses the rate of return, was to be calculated with a most unusual accounting formula that deflates the numerator (by deducting depreciation charges) in relation to the denominator (where they were not deducted) to produce the 7½ per cent rate of 'profit' permitted.

Normally in such arrangement one might find net profit related to net assets, or gross profit related to gross assets, but to relate net profit to gross assets is surely not a usual way of calculating a rate of return. The effect is, of course, to produce a rate very much lower than would normally be accepted in accounting practice.

The intention of the Government in making the refinery agreements was apparently to regulate the refineries as a kind of public utility, granting them a monopoly in domestic markets but regulating their prices in order to ensure a 'reasonable' rate of profit. Clearly, however, the Government could not have had a very clear understanding of the problems inherent in any attempt to treat the subsidiaries of international firms as if they were independent entities and to regulate rates of return without reference to the transfer prices applied by such firms in their inter-affiliate relationships. Moreover, in view of the reputation of the Lebanese as shrewd businessmen, it is not easy to understand why the particular formula for the regulation of prices and profit rates was accepted by the Government. It is difficult to escape the conclusion that the companies took advantage of the political situation and of their monopoly on the other hand, and of the lack of expert knowledge of the industry among the Government negotiators on the other, to charge prices in the Lebanese market considerably higher than would have yielded a properly calculated 7½ per cent rate of return and one which made allowance for the discounts on crude oil prices that prevailed in world markets.

The Supplemental Agreement discussed above was not ratified; critics from Parliament and from outside made the Government aware of the anomalies it contained and negotiations for a new Agreement were resumed, in the light of the attacks of the critics. At the end of 1967, two years after the Agreement had been proposed and rejected, negotiations were still going on with no result reported. Lebanon continued to pay posted prices, exempt its refineries from taxes and accept the formula for determining ex-refinery prices.

Conclusion

On the grounds that any outside observer must consider weighty, three of the countries discussed came to the conclusion that it would be to the advantage of its economy if a greater degree of independence from the international oil Companies could be attained. So long as their refining capacity and distribution networks were owned by the vertically integrated Companies they could be looked on as 'captive markets' by the Companies. This created no overt problem when there was little or no price competition in world oil markets, since greater freedom in the choice of suppliers is of little value if all suppliers offer the same terms. In such a case, importing countries are at a disadvantage, not because of the vertical integration of the supplying Companies, but rather because of the monopolistic organization of oil markets. With the advent of price competition in the larger markets, such as Japan and Western Europe, and of Russian offers to sell more cheaply than the international Companies, all three countries were spurred to action.[1]

[1] It might be noted in passing that the Japanese Government has been concerned about the same type of problem, for in Japan not only were many refineries partly owned by the majors, but many were built with loans to which were attached crude supply contracts. In

In this brief review I have not tried to appraise the extent to which the specific actions taken were wise from an economic point of view; internal political considerations undoubtedly carried too much weight at a number of points and it may well be that alternative methods of obtaining the desired results were inadequately explored. In any event, such an appraisal would require a detailed analysis of the specific circumstances, and for this adequate data are not publicly available.

It is clear, however, that countries in the position of India, Ceylon or Pakistan, in the circumstances of the international oil industry after the early 1950s, had a choice of three broad courses of action if they were not to be exploited in a very real sense by the international Companies struggling to maintain their prices except where competition forced them to offer discounts: (1) they could force out the Companies or severely restrict their activities, especially their expansion; (2) they could set up private or State local companies to compete with the foreign Companies; or (3) they could bring heavy pressure on the Companies to reduce their prices and improve the terms of their supply contracts. The governments of these poor countries were in a weak position to engage in all-out competition with the foreign Companies; nor were they in a strong position to bring pressure on the Companies except through governmental threats to restrict their activities. Hence, it was almost inevitable that the first course of action would be used to support the other two. At the same time, the usefulness of the foreign Companies in some activities was recognized, particularly in India and Pakistan, and this, together with their need for foreign capital, led to the establishment of a variety of joint ventures which enabled the countries better to protect their own interests, while at the same time making use of foreign capital and enterprise.[1]

EXPLORATION AND PRODUCTION

The establishment of domestic crude-oil production is looked on by most countries largely as a means of saving (and, hopefully, earning) foreign exchange, and partly as a means of ensuring against essential supplies being cut off should external disturbances interfere with the availability of imports. The industry is not very important as an employer of labour, or as a 'leading sector' directly stimulating the growth of the rest of the economy. Domestic-

December 1965, it was reported that 70 per cent of Japanese crude purchases were tied to loans (*Financial Times*, December 12, 1965). Japan was in a better position than other Asian countries to get discounts from the Companies in view of the size of its market, but the Government also intervened directly with foreign exchange allocations, and the requirement that refineries take some Arabian crude from the Japanese Company.

[1] The oil Companies eventually became prepared to go to considerable lengths to meet the desire of the importing countries for increased local control and participation, even making arrangements under which the refineries constructed would revert to local ownership after a specific period. As Edward Symonds has pointed out, this was quite an innovation in the field of direct investment when it began. 'It is . . . noteworthy that, in many of these ventures, full ownership of the refinery or other asset will pass to the host country after a specified number of years. According to the hitherto accepted norms of direct investment abroad, the foreign asset remained indefinitely in the hands of the investor.' *Financing Oil Expansion in the Development Decade*, p. 5.

ally produced crude oil might, of course, reduce energy costs to other industries, but taxation and the widespread prevalence of 'import parity' pricing by governments as well as by oil Companies, and of protection for the domestic industry, forces one to conclude that 'cheap energy' for consumers is not one of the major reasons why governments encourage the development of domestic crude production. The cost of imports and the hope of exports are the more compelling reasons, especially if geological indications are such that a government is convinced there is crude oil to be found for the searching.

The international oil Companies have shown themselves willing to explore for and develop oil reserves under petroleum legislation which incorporates fairly detailed regulations with respect to speed and extent of drilling, relinquishment of non-producing areas, conservation of resources, employment of nationals, financial arrangements assuring substantial revenues to governments, etc. In spite of this there is a clear and widespread reluctance on the part of many an underdeveloped country to take full (and sometimes any) advantage of the willingness of the foreign firms to devote their own resources to the search for oil. This reluctance usually has a strong political and ideological content, but there is also often a strong feeling that the country would benefit economically if it could develop and run the industry itself instead of being dependent on foreign firms, which after all would expect a handsome reward for their services.

The actual production of crude oil from proved fields is not, by and large, financially or technically beyond the capacity of the more advanced underdeveloped countries to conduct without outside help, provided that the industry does not become the football of politics, and is not starved of funds and personnel because of the conflicting pressures of numerous vested interests.[1] But oil deposits are wasting assets, and unless reserves are enormously large, they must continually be replaced. Exploration, the most uncertain aspect of the business, must be a continuing activity unavoidably associated with production, and this raises peculiarly difficult problems, for there is no clear relationship between the resources devoted to exploration and the amount of oil discovered. The arguments about the advantages to a country of accepting foreign investment are somewhat different for this industry than they are for other industries, partly because of the great risks involved in exploration.

The Nature of the Risk

The 'output' of exploration is essentially knowledge about the underlying geological characteristics of the explored area. The 'proving', or establishment, of oil reserves, is nothing more nor less than the process of acquiring knowledge about the location, size and characteristics of deposits of oil than can be pro-

[1] It can, of course, be argued that administrative incompetence, corruption and political interference are so very likely in an underdeveloped country that government enterprises will almost necessarily be extremely costly. If one accepts this position, it follows that the country would gain if all investment was left to foreigners which local private enterprise could not be expected to undertake, whether because of the risk involved, the amount of finance required, or the length of time before a commercial return could be expected. This general argument need not be considered here, however, since it involves nothing peculiar to the petroleum industry. Nevertheless, if one were analysing the prospects for a government oil entity in any *given* country, such matters would require careful appraisal.

duced economically with known technology and at prevailing or expected price/cost relationships. Many wells may be drilled, only a few of which may prove to be commercially productive. Although knowledge is the 'output', the purpose of exploration is clearly not knowledge alone; the considerable expenditure incurred to obtain it is undertaken only by those who hope to be able to make profitable use of the information acquired. For oil companies, whether they be private or government, the really profitable information is that obtained by drilling which tells them with certainty where commercially exploitable hydrocarbon reserves exist. When this knowledge is obtained exploration is deemed 'successful'.[1]

The risk that is said to be inherent in exploration is the risk that the knowledge obtained after the expenditure of large sums of money will be of little economic value. Such a result can occur *either* because nature has not in fact provided very much by way of oil deposits, *or* because the information obtained from man's exploratory efforts is insufficiently comprehensive. Thus in any given period of time, success in the search for oil in any country is not entirely dependent on the existence of oil in place and the size, distribution and character of the oil-bearing formations. These are not known with any precision before drilling is undertaken, and the problem is to drill in the right places and at the requisite depths; just what any particular drilling rig will find is still very much a matter of chance.

Since the prospects of finding any oil that is in fact in place are very much influenced by chance, the likelihood of finding it will be greater, the greater the size of the exploration effort (that is, number of seismic teams, drilling rigs, etc.) devoted to it. Nevertheless, judgment and technical skill are also important, and the more experienced are the explorers and the more advanced their technology, the less, *ceteris paribus*, is likely to be the time and effort required to find any oil that is to be found.

In general, there can be little doubt that the larger international Companies have advantages over any underdeveloped country with respect to the size of the effort they can mount and the technology they can use, and their chances of discovering a country's potential oil reserves are correspondingly greater. Not only can more drilling be done but the interpretation of information as exploration proceeds, the selection of drilling sites, etc., is likely to be more effective. Although the amount of resources any one Company will devote to exploration in a particular area will be limited by the risk of failure that it is willing to accept, if a large number of Companies come in, a very wide search indeed can be obtained in which the resources risked by any one Company will be very much less than the total at risk for all Companies together. The risk of failure facing each Company is not reduced because of the larger number of Companies, but the risk that no company will find available oil is much reduced.

Within limits, the greater the number of Companies exploring, the greater may be the chances of success because of the diversity of judgments that can

[1] Of course, information that oil is not to be found in a particular place is also valuable in that it reduces the area of uncertainty and increases the stock of knowledge about underground formations. In general, however, this is not what is meant by 'success' in exploration, although general geological knowledge of this kind is often saleable.

be brought to bear in the selection of exploration programmes. In other words, with a number of companies making different evaluations of the prospects, and developing different exploration programmes accordingly, it is argued that the chances of at least some of them being right, and therefore the chances of successfully finding available oil more quickly, are greater than would be the case if the judgment of only one group had been involved. This argument has been used to support the contention that the 'centralized judgment' of a State monopoly is likely to be less effective than the 'diversification of judgment' obtained from a multiplicity of private efforts.[1] It can be readily seen that this argument is a variant of the argument that the more extensive the search, the greater will be the chances of success, for even if a number of exploring groups operate on different theories about the location of oil, and consequently have different sets of priorities in choosing where to explore, a comprehensive search would in principle encompass them all. But if the search must be limited in any given period, and if the theory on which a firm operates systematically excludes certain possibilities while there is no way of knowing before the event which is the correct theory, then the more firms there are exploring and using different theories, the more likely is success.

On the other hand, some kinds of information are most useful when they are obtained and analysed over as wide an area as possible. For example, seismic information obtained by a single exploration company covering a large geographical area may well lead to a more efficient selection of drilling sites than would have been the case if a number of companies had each obtained information over the same area but in smaller blocks, and each had analysed it without reference to the rest. In practice, when a number of companies are exploring in an area there is often considerable exchange of information.

Although a larger exploration programme in a given period reduces the risk that oil will not be found in that period because exploration was insufficient, the risk of failure arising from the possibility that the geology of the country does not contain commercially exploitable oil deposits is in no way affected. Hence an extensive search is no guarantee of success, and in this respect the greater the effort, the greater the total amount of resources put at risk. These characteristics of oil exploration have obvious financial implications which may create special difficulties for a poor country if it undertakes an extensive programme on its own.

The Financial Risk

Any country wanting to establish a domestic crude-oil industry must face the

[1] This point is given considerable prominence in the Report by W. J. Levy Inc. to the International Bank for Reconstruction and Development on *The Search for Oil In Developing Countries: A Problem of Scarce Resources and its Implications for State and Private Enterprise* (1960). See especially Chapter III, p. 3. However, there is no particular reason why State operations cannot be conducted by several exploration companies. The French have done this, as the Levy Report points out, but underdeveloped countries may not have sufficient resources. This Report, hereafter cited as the 'Levy Report', presents the arguments in favour of the developing countries using the international oil Companies to develop their crude-oil production. I am grateful to Mr Levy for giving me his criticisms of this chapter, which have saved me from a number of errors. As will be seen, however, he cannot be held responsible for the conclusions arrived at here.

fact that the first task involves a very expensive search for information which may or may not turn out to be economically profitable. Moreover, if oil is discovered in commercial quantities, the proving and development of any fields that are found will also be expensive, and may take a number of years. Even if production is eventually established, therefore, the period of time that may elapse, which is likely to be lengthened by any limitations on the exploratory effort, may well be considerable. In the meantime, the country's imports of crude oil, which may be a significant proportion of its total imports, will continue to increase as the economy develops in other directions.

The risk of loss involved in exploration is, for an economy as a whole, simply the risk that little or no increase in national income will be created by the real resources poured into it. The cost of failure will be the opportunity cost of the real inputs used. The notion of 'risk' includes both the chance of loss and the significance of the amount lost; the more valuable to an economy is a given resource, the greater the risk to the economy of losing a large amount of it. The analysis of both types of risk is usually cast in terms of financial risk—the risk of not recovering the money invested with an adequate return. For obvious practical reasons the question must be dealt with in this way in the first instance, with whatever adjustments are required to make allowance for the failure of prices to reflect social costs and benefits. Exploration must be financed, in the sense that money must be supplied to pay for the workers, equipment and other resources required, in advance of any return, just as any investment must be financed. Much of the finance for an underdeveloped country will, however, have to be in the form of foreign exchange to pay for imports of equipment or to secure the services of foreign experts of various kinds, and the opportunity cost of foreign exchange is likely to be high.

Thus, the nature of the risk in oil exploration poses a dilemma for the country: the risk of failure with a modest amount of expenditure is greater than with a large amount, but the risk to the economy of losing a large amount rises as the amount rises. Apart from any other financial difficulties, this 'increasing risk' will place severe limits on the size of the exploration effort that the country can reasonably support in a given period, whether it be undertaken by the government or by private enterprise, yet the smaller the effort the less are the chances of finding what oil there is to be found.

A developing country might well be able to afford the same amount of exploratory effort over a longer period as foreign companies would be willing to undertake in a shorter period. Even if we assume that the same (successful) results were achieved in both cases, the time lag in itself would add to the cost of the local effort because of the greater outlays required for imports over the longer period and the delay in the increase of the national income. It cannot, of course, be assumed that private foreign companies will be willing to come in on a scale large enough to conduct a really intensive search. In India, for example, foreign companies held exploration concessions for many years and yet very little exploration was done. It was only when the Government took an active hand in the matter that the rate of discovery in India appreciably quickened.[1]

[1] B. Dasgupta, pp. 163–6.

Government Financing

In developing countries, the government often plays a very active part in all aspects of the petroleum industry, including exploration and production. Sometimes this is the result of ideological objections to private enterprise in the exploitation of the country's natural resources (and of even greater objections to foreign private enterprise); sometimes the planning authorities fear that private firms will not operate in the national interest or according to 'the Plan'; sometimes domestic private capital is not forthcoming and it is feared that if foreign capital were accepted, the exploitation (in the technical sense) of the country's petroleum resources by foreign firms would result in the exploitation (in the pejorative sense) of the domestic economy; the international petroleum firms are very widely feared and mistrusted.

Because domestic private capital is rarely available for activities as risky as is exploration for oil, most developing countries must either rely on government-sponsored enterprises or accept foreign private capital if they want to establish the industry at all. For this reason, the Levy Report discussed the question almost entirely in terms of 'government monopoly' on the one hand and 'private enterprise' on the other, where 'private enterprise' was the synonym chosen for foreign enterprise. This may be the practical dichotomy, but a discussion cast in such terms runs the danger of over-emphasizing the budgetary problems of government and of giving insufficient attention to an analysis of the cost of foreign investment to the domestic economy.

Nevertheless, special problems do surround the financing of State enterprises whenever the government itself takes a hand in making funds available to them. Oil exploration is usually considered to be particularly inappropriate for government financing in view of the high risk involved and the other urgent and pressing demands on the government budget.[1] To be sure, the nature of the activity may create special budgetary and political problems. Losses of government enterprises may be widely known and may cause political controversy which could react unfavourably on the industry. Apart from this, however, any loss incurred because of the risks taken is a loss to the economy as a whole, whether incurred by a public or a private (local) enterprise, if the resources devoted to fruitless exploration could have been used for some other purpose.

The more relevant economic problem relating particularly to State enterprises arises from the fact that a State enterprise may have privileged access to funds because the government gives it outright subsidies, a special legal position in the market, or special powers to raise money. If such sources of funds are important, there is a tendency to assume that the government enterprise is inefficient or 'uneconomic'. Whether or not such an assumption is justified is not easily tested, since to do so a full-scale cost-benefit analysis would be required. A list of the kinds and amounts of subsidy is of little help—except to enlighten the government authorities if they had not realized the full extent of government support (which is indeed possible).[2] On the other hand, rightly or

[1] '. . . the promise of exploration can be a siren's lure. If undertaken by government, it must inevitably be at the expense of other critical projects. Funds allocated for education, transportation, or a power plant would be securely invested. Funds allocated for the search for oil would be risked on uncertain ventures.' Levy Report, pp. I–9.

[2] 'While the patrimony of a State enterprise is an obviously identifiable charge against the

wrongly, no similar analysis is deemed necessary for the appraisal of the desirability of a private undertaking regardless of the degree of monopoly the firm may have, providing that it meets the 'market test', i.e. makes profits.

The Levy Report abundantly illustrates the argument that the financial requirements for the continuing exploration and development needed to support increasing output, even when the oil industry is well established, 'typically exceed the financial resources of a State enterprise'. The report describes in great detail the types of government subsidy and the various forms of indirect 'subscriptions of public capital' that are often used: high internal prices, tax exemptions of one kind or another, tariff or quota protection, excise taxes handed back to the company, low-interest loans from the government, etc.[1] In the financing of Petrobras (the Brazilian State company), for example, the 'infusion of social capital was substantial, both through direct Government subscription of tax receipts and reinvestment of dividends, and indirectly through favourable pricing for Petrobras refinery output'.[2]

Such financial assistance is quite correctly treated as 'public support' of the State company, and there is no doubt that most State oil companies are helped in many ways by their governments. On the other hand, 'indirect subsidization' is hardly confined to State enterprises and 'concealed subscriptions of public capital' (e.g. through high prices) are characteristic of any monopoly, public or private. Failure clearly to recognize all costs may well seriously disrupt economic judgments, but again such a failure occurs at least as much in private enterprise (which rarely consider social costs) as in government enterprises. Moreover, even the international oil Companies demand tariff protection in a number of countries, as well as tax subsidies from their home governments (notably large depletion allowances), and they openly defend their price levels with reference to their need to retain funds to sustain the further investment required to permit them to maintain their existing position in the industry.

It may well be generally true, as the Levy Report says of Petrobras, that '. . . it is not just "commercial" profits that have constituted the company's internal flow of funds, but also a measure of support provided through higher prices paid by consumers and/or government revenues that are foregone'.[3]

government's oil operations, indirect assistance tends to be concealed and is less frequently counted as cost. Thus, it becomes difficult for the nation to evaluate the results of government oil operations against the resources committed.' Ibid., pp. I–10.

[1] '. . . where government seeks to maintain oil operations exclusively through a State entity, social costs are likely to run considerably higher than the nominal patrimony. In the review of government operations further along in this chapter, indirect subsidization is a common experience. It may take the form of tax exemption for the government oil entity, or even tax delinquency where payments to the State are specified. In either event, the government, in effect, is supporting oil operations out of current revenue. Government loans to its State oil entity invariably carry lower interest charges than the entity, or even the government itself, would have to pay in the nation's money market. Domestic prices may be deliberately built up to buoy the government oil entity's earnings so that consumers are in effect providing indirect support to State operations.' Chapter III, p. 5.

[2] Chapter III, p. 31.

[3] Chapter III, p. 18.

Since, however, this can equally be said about almost any of the international private Companies it is not clear what significance should be attached to it except to call attention to the considerable financial requirements of the industry, and the possibility that profitable operations may require domestic prices higher even than those charged by the international Companies for imported products.

The fact that a State company may have difficulty in raising sufficient funds from its sales or from private investors and therefore may require government assistance, usually concealed and indirect, is not sufficient evidence to support the conclusion that the economy loses from its operations. It is not always appropriate to look only at the economic desirability of a given activity taken by itself; it is often necessary to look at it in a broader context. For example, a firm appraises its various activities from the point of view of the enterprise as a whole and does not necessarily insist that every one of its activities taken alone must show a profit; rather the contribution of each activity to the firm's total profits is examined. For a country, it is appropriate to examine each activity from the point of view of the economy as a whole; in many circumstances 'private profitability' (i.e. of the firm) may provide the simplest and the most practicable measure of 'social profitability', but it is usually only a rough measure. 'External economies' play a significant role in economic analysis, not least in the economics of development, and one should always consider whether an activity requiring 'subsidization' may nevertheless be worth undertaking.

If the firm as a whole is the economic unit whose performance is being judged, then the profits made in one direction will be used to support activities in others whenever it is considered desirable from the point of view of the enterprise as a whole to do so. The oil Companies sometimes complain that State concerns have a 'built-in' competitive advantage *vis-à-vis* private companies because they can call on public funds to cover their losses. At the same time, however, they point out that one of the advantages that they have as large integrated Companies is that they can cover losses in one country or activity from their profits made elsewhere. This is partly what is meant by the 'financial strength' of a Company, and smaller companies without diversified sources of finance find themselves at a competitive disadvantage *vis-à-vis* the large Companies which can draw on sources of finance from their other operations.

If the whole economy is the unit under consideration, then finance made available from one activity in the economy may appropriately be used to support another, unrelated, activity if this is considered desirable from the point of view of the economy as a whole. Private companies may under these circumstances find themselves at a competitive disadvantage *vis-à-vis* State companies in exactly the same way as would a small local company *vis-à-vis* a large international firm which was prepared to take losses in order more firmly to establish itself in the local market. In both cases any competitor without the additional sources of finance will be at a disadvantage, and will understandably consider the competition to be 'unfair'.

This being said, however, it must be recognized that a State company may have less incentive to evaluate conservatively its prospects and thus may more easily run into financial difficulties. Moreover, if the funds available to it are dependent on political decisions, there is a serious danger that it will be unable to create a reasonably sound financial structure and that it will be continually short of funds. Petrobras and Pemex (the Mexican State company) as well as Yacimientos Petroliferos Argentinos (YPF—the Argentine State company) all suffered in this way,[1] and a careful study of the Levy Report's discussion of their problems would repay any government planning to establish its own national company. The administrative and political difficulties of government are clearly relevant to the finance of government enterprises.

Risk of loss is inherent in any type of activity—even expenditure on education and roads is by no means 'securely invested' and can be wasted—but the risk is very great indeed in oil exploration; particularly if a large programme is contemplated. For this reason, there may be special advantages in accepting foreign financing on suitable terms.[2]

Foreign Financing

When foreign Companies undertake exploration, the financial risk of failure is borne entirely by the foreigners. If they fail to find oil, there is no loss to the country in which they operate; on the contrary, their expenditures will have added to the foreign exchange receipts of the country and the information obtained from exploratory efforts can be an important addition to the knowledge of the country about its own resources. A similar failure by domestically owned companies, though adding to knowledge, would entail a loss to the country to the extent that the materials and services used were purchased abroad or had alternative employment within the country.

Nevertheless, it is only partly true that a country avoids risk by accepting foreign enterprise. Foreign Companies do not, by and large, invest without hope of return commensurate with the risk undertaken, and the return is expected in the form of crude oil or of remittable profits, or both. In promoting a local company to establish the industry, a government risks having to bear the cost of failure or of having to wait a long time for success, both of which might be avoided if foreign concessions were granted. But if foreign concessionaires discover oil that might just as well have been discovered by a local company, the government takes the risk that the economy may have to pay very dearly for its oil supplies, a risk that is only avoided if domestic companies hold the concessions. Hence with either choice a risk is taken. Private foreign investment may cost an economy little in the short run; it may be very expensive in the longer run, and the problems of making the appropriate appraisal may be extremely difficult. One can easily accept the conclusion that the nature of exploration gives rise to such an unusually high risk that a very great strain is likely to be put on the economy's resources. Whether the risk is worth taking obviously depends on the circumstances of individual countries;

[1] See p. 285 below.

[2] Some countries, notably Iran and Kuwait, have themselves carried out, on occasion, preliminary seismic surveys of likely areas, for when results indicate favourable prospects for discovery of oil the interest of foreign companies may be stimulated, thus increasing competition among them for concessions. In addition, the Government may be able to sell the information it acquires for a good price.

from an analysis of the past experience of different countries a great deal can be learned about the difficulties, but no generally applicable conclusion can be drawn.

The almost unanimous view of Western advisers seems to be that the developing countries should not divert their scarce resources, particularly foreign-exchange resources, to oil exploration when there are so many 'urgent needs in other areas'. This theme runs through every discussion of the subject, and not least through the Levy Report, although in the latter some 'national effort' is nevertheless advocated when it can be afforded. The argument is not that the oil industry is an inappropriate industry to develop, but rather that it is more appropriately developed by foreign capital. Certainly, if the initial exploration expenses are borne by foreigners, and if domestic crude production becomes more quickly established than it otherwise would have been, the immediate gain to the domestic economy may be very great indeed. But if production remains entirely in the hands of foreign firms, profit remittances and other charges could within a very short time begin to absorb a high proportion of the country's foreign exchange receipts. Although foreign investment may indeed make a great contribution to a country's development, the terms on which it is accepted should be carefully examined, particularly in industries where monopoly elements are likely to be significant.

The terms on which foreign companies are willing to undertake investment in oil exploration are negotiable, and the more favourable the geological prospects are the more favourable to the host government will be the terms that can be obtained. The Levy Report emphasizes and illustrates the diversity of the types of arrangements in any particular case 'depending on historical, political, and economic circumstances'. It also points out that 'private enterprise is able to operate in many different environments, under terms that are consistent with the policies of the respective governments'.[1] Some of the arrangements adopted in the developing countries are, potentially at least, much less costly to the host countries than are others.

Egypt, for example, has attained self-sufficiency in crude-oil production and the prospect of considerable exports with a very mixed regime. The State company engages in both exploration and production, but foreign companies are also active. When oil is discovered in commercial quantities by one of the foreign companies, a development company is formed in which a State enterprise has a 50 per cent interest and the State then pays its share of the costs incurred. Three-quarters of the profits of the joint company is to be paid to the Government in respect of royalties, rents, fees, equity interest, and taxes.[2] The whole industry is effectively controlled by the Government.

This type of arrangement has so far been made largely with non-major Companies, for until recently the major international petroleum firms have not been willing to enter into partnership with foreign governments, partly perhaps because of the precedent that might be set for their great Middle Eastern concessions, but largely because such arrangements, not being necessary, were

[1] Chapter IV, p. 32.

[2] This 75 per cent is not directly comparable with the 50/50 arrangements in the Middle East since profits are calculated on 'realized prices'.

simply not 'normal' practice.[1] Hence, ENI and the US 'independents' have tended to blaze the trail which some of the majors are now beginning to accept that they must follow if they are to retain a strong foothold in the developing countries.

Such arrangements may require considerable capital investment on the part of the government, and the question to be analysed is how far this investment is likely to be justified in the light of the alternative cost of the otherwise higher foreign-exchange remittances by foreign Companies. There is danger in emphasizing the very short-run needs of developing countries and ignoring the medium- and long-term cost of foreign capital. It is true that development is 'urgent', but it may unavoidably take a long time, and there is as yet no adequate ground for asserting that the *really* 'critical years' are 'now' rather than later, when strains may be heavier.[2] Hence, it may be necessary to guard against permanent foreign control if maximum benefit from foreign investment in a particular industry is to be obtained. In practice, crude-oil concession agreements run for limited periods, but often this period is very long. Any period over ten to fifteen years after production begins should be scrutinized very closely. Indeed, the principle involved is much the same as that of the patent law: a specified period during which the discoverer may recover the costs of his research and make a profit, after which, the discovery reverts to the public domain.

A number of importing countries have also looked at the possibility of obtaining their own sources of crude oil abroad. The high cost of establishing domestic crude-oil production by their own efforts, and the political as well as economic problems that may be raised if foreign firms are permitted to do so, have induced serious consideration of the possibility of entering into joint ventures with national oil companies in countries where crude-oil production is well-established already. India bid for a concession in Iran, and it has been proposed in Turkey that the Government should seek a partnership arrangement with some national oil company in one of the major producing countries of the Middle East.

Success in such a venture would enable an importing country to obtain an assured supply of crude oil while by-passing the international Companies en-

[1] '. . . these ventures are setting a novel pattern in international finance. Until now, the practice in most manufacturing and many extractive ventures abroad has been for the investor to insist on control, preferably through 100 per cent ownership.' E. Symonds, *Financing Oil Expansion in the Development Decade*, p. 5. It is true, of course, that before the Second World War few developing countries had either the financial or technical resources to support a partner's share of such arrangements. That they have now become more common is partly the result of the economic development of these countries itself. Nevertheless, the Western companies have been slow to adapt to changed circumstances, especially in the oil industry. However, it might be noted that as early as 1911 an oil-producing company had been formed in Egypt which was held jointly by Shell, BP, and the Egyptian Government, the latter owning 10 per cent of the shares. This company, Consolidated Oil Fields, operated until 1964, at which time it was nationalized.

[2] 'By permitting private [read "foreign"] initiative in oil development, . . . India would know that where exploration ends in failure, the loss would be taken by the company. Where there is success, the Government realizes an immediate gain—through its share in the income from production, through tax receipts, and in foreign exchange savings. Thus, private investment could be the means for stepped-up oil exploration *now, in India's critical years*.' Levy Report, Chapter IV, p. 30. (Emphasis supplied.)

tirely, and might be acceptable to exporting countries desiring to develop their own national companies. Nevertheless, even if the capital investment were less than that required to develop their own domestic supplies, it would still be considerable and there would be no guarantee that the amount and quality of the crude oil would be sufficient to justify the expense. Crude-oil exploration, even in the established producing countries of the Middle East, is still a gamble; for a particular country the essence of the issue is simply whether it is able and willing directly to accept the risk involved by devoting its own resources to exploration, or indirectly to pay for it either by purchasing oil from the international Companies or by allowing them to make the investment in the development of domestic oil supplies.

CONCLUSION

In this chapter we have concentrated on the problems raised for developing countries by foreign investment in this industry. Inevitably, therefore, the important contribution to the economy that foreign investors can and have made has received little attention. The benefits to the developing countries are well known: an increased supply of capital, especially in the direct form of foreign exchange, and of technical and managerial skills, both of which lead to a more rapid development of their industry and more effective use of their natural resources. But the cost of these benefits to an economy depends very much on the conditions under which they are supplied.

The proposition that the contribution to the national income of a country that is made by foreign investment will always be worth its cost must, among other things, assume: (1) that the conditions under which the investment takes place do not create a monopolistic distortion at the expense of the receiving country in the patterns of trade, and in the prices of goods, either through cartel arrangements or through ownership ties linking the affiliates of international firms; and (2) that initially high profits earned by innovators or pioneers taking unusual risks will not persist for an 'unduly' long time. (It is, of course, impossible to measure the content of 'unduly', and some rule of thumb is perhaps required, following the precedent of the patent system.) The problems of the oil Companies arise very largely from the fact that these conditions are thought not to have been satisfied by important groups in the countries concerned.

In refining, foreign ownership associated with vertical integration has reduced the ability of a country's refineries to import crude from the cheapest supplier; in crude-oil production the long period for which concession arrangements were usually expected to run, and the monopolistic international position of the major Companies, gave support to nationalistic sentiments and political slogans in many countries, the governments of which often refused foreign investment entirely or insisted on State participation in or control over the relevant enterprises. Such attitudes may have had the support of chauvinistic and autarkic ideas, but it was also sensible to approach the foreign investor very cautiously. At the same time, the growing competition in the international industry and the entry of independent companies, both State and private, into the bidding for concessions where prospects were reasonably favourable has widened the choices available to countries wanting to accelerate the development of their resources. It was indeed significant that the Government of Egypt, in spite of extremely critical attitudes toward both private and foreign investment, could induce foreign Companies to explore for oil at their own risk and subsequently to join with the State in the development of the oil discovered. On such terms foreign investment is likely to be well worth the cost to the receiving country.

Chapter IX

THE DIRECTION OF CHANGE

Because of its far-reaching consequences in so many aspects of modern economic and political life, the Second World War is commonly treated as a kind of watershed marking the boundary of two distinct eras, labelled for convenience, 'pre-war' and 'post-war'. With some justification one can find a similar watershed in the history of the international petroleum industry, for the war with its immediate aftermath was associated with very marked changes in the structure and operations of the industry. But in spite of this temporal coincidence, it would be wrong to look on the war as a significant *cause* of these changes, for its importance for the long-run development of the industry relates more to the pace of events than to their type; some developments were retarded and others hastened, but the type and direction of change had other causes.

In addition to the enormous upsurge of consumption and a shift in the major sources of supply to the Eastern Hemisphere after the war the most important developments in the industry concerned prices, taxes, other aspects of government intervention, and the position of the major companies in particular countries and in the industry generally.

POST-WAR CHANGES

The pre-war pricing system was modified during the war itself with the emergence of a second basing point in the Persian Gulf, and after the war the rapid expansion of low-cost Middle East supplies brought a fall in the prices of Middle Eastern crude relative to those for oil from the Western Hemisphere and for all practical purposes undermined the pre-war system. All during the 1950s there was a fair amount of covert price competition among the Companies; by the end of the decade this had turned into overt competition in both crude oil and product markets, and market prices came under steady competitive pressure. When the '50/50' profit-sharing arrangements in the Middle East were made in 1950 and 1951, the oil Companies began to announce publicly for the first time the prices at which crude oil was to be transferred to affiliates or to be sold to third parties (excluding special long-term contracts), as well as valued for tax purposes. By the end of the decade, these 'posted prices' had become almost completely divorced from the prices at which oil actually moved. They had turned into 'tax reference' prices which had to be agreed on with the oil-producing countries. Thus it became possible for the taxation of the producing countries effectively to reach even beyond the profits that the Companies attributed to their crude production.

Although the flow of revenues to the crude-oil producing countries suddenly accelerated in 1950, and approached $3,000 million by 1966, the governments of these countries continued to press for more, and in addition to press for changes in the terms of the older concession agreements and stiffer terms in the new ones in order to widen their control over the exploitation of their petroleum resources generally. These changes did not always come peacefully. In the Middle East, both Iran and Iraq dealt severe blows to their long-familiar concessionaires, while in Latin America and the Far East a number of countries, notably Argentina, Burma, and Indonesia, took actions which seriously damaged the position of the Companies. All countries seemed more suspicious of their existing large concessionaires than of new ones and, if foreign Companies were at all welcome, the so-called 'independents' were often given preference over the 'majors'. Independents, for their part, sought eagerly for concessions, and the number of new producers in Asia and Africa increased significantly.

In consuming countries, too, the major international Companies not infrequently had difficulties because their policies were deemed inconsistent with the national interest of the countries in which they operated. The Ceylonese Government went so far as to expropriate their properties, while in other countries, particularly in Asia and Latin America, their position in the domestic refining industry became closely circumscribed.

None of these changes can be attributed to the war. Most, if not all, of them were the product directly or indirectly of two concurrent developments: (a) the successful response of the industry to the continued strong economic incentives to explore for oil and to develop, refine, and distribute the oil found; and (*b*) the successful exercise by the governments of both the exporting and importing countries of their growing bargaining power.

The war delayed the full effects of the successful discoveries in the Middle East, but it hastened the growth of the economic and political strength of the countries of Asia and Africa. Both of these developments had begun before the war and would have occurred in any case, bringing with them consequential changes in the petroleum industry; neither has as yet run its course. The search for oil goes on, engaging more private companies, and even governments, than ever before. In the crude-oil exporting countries neither the pressures on the Companies for increased revenues nor the interests of governments in a greater measure of control over the domestic industry show signs of abating. But the governments, aware of what they do gain and might lose, tend only to press, they dare not—and indeed for the most part do not even want to—take drastic or unilateral action to assume control over the companies operating in their territories.

ATTITUDES TOWARDS THE COMPANIES IN PRODUCING COUNTRIES

There are a number of reasons why the oil Companies should have been regarded with great suspicion from almost the beginning of their activities in the modern industry, in developed no less than in the underdeveloped countries.

In the first place, as producers they deal with an unrenewable natural resource, and only in the last 25 years or so have they cared very much about the prevention of unnecessary waste, the destruction of the countryside, or the pollution of great bodies of water; they began to do so only under pressure from governments. To be sure, there was much technical ignorance and in this, as in other matters, individual Companies often had little choice but to conform to the competitive system of which they were a part, and only government could modify the ill effects of the system. Even today, however, the Companies operating in foreign countries are widely suspected of doing less than they should to conserve the natural resource, but the domestic governments enforce rigid codes of behaviour in this respect and it seems probable that most of this suspicion is now unwarranted. The Companies do make great efforts to conform to local views. Nevertheless, it remains true that foreigners having control of any country's valuable natural resources risk being widely feared and disliked, no matter how careful they are, and no matter how much the country may gain financially from their operations.[1]

Secondly, the Companies are very large and individually, as well as collectively, they possess great economic power of several kinds, including the power to influence prices and to regulate output. In these circumstances, the charge of monopoly is unavoidable, if not entirely merited, even from those whose livelihood depends on a share of monopoly revenues, and even more from those at whose expense the monopoly revenues are gained. Thirdly, in some countries at least, a very small handful of foreign firms have it in their power by their own decision severely to damage the national economy. That they have this power gives them a strong bargaining position *vis-à-vis* the government; the fact that they can use it only *in extremis* because of the enormous damage to themselves its use can bring, alleviates only a little the prevailing resentment.

Finally, throughout the developing world, large sections of the public are without any knowledge of the industry, but are easily persuaded that the oil Companies are outstanding examples of Western imperialism.[2] The case can be plausibly presented. One aspect of imperialism is often the acquisition by foreigners of a significant degree of control over the natural resources of another country; the history of imperialism, political or economic, is character-

[1] It can hardly be denied that the Companies have on occasion used their legal rights in a way which, to the outside observer at least, appears unreasonable. For example, in Iraq most natural gas associated with oil production is simply burned off—used for nothing—yet according to Nadim Al-Pachachi, Iraqi Minister of National Economy, in 1958 the oil Companies 'agreed to supply the Government with natural gas from their oilfields for use in the chemical and petrochemical industries at a reduced price *provided the products which use natural gas as a raw material were sold on local markets only*'. (Italics mine.)

This prohibition of potential exports was rejected by the Government, and the Companies were only induced to abandon the restriction when the Government threatened to force them to stop flaring the gas and to reinject it or otherwise use it. See Nadim Al-Pachachi, *Iraqi Oil Policy: August* 1954–*December* 1957. (Baghdad. Translated by the Research and Translation Office, Beirut, 1958.) It is true that no real harm was done Iraq at that time since she had little use for the gas either, but this does not affect the principle.

[2] David Hirst, in *Oil and Public Opinion in the Middle East*, has given a short survey of the various attitudes towards the oil Companies which characterize public opinion in the Arab world.

ized by rivalries among foreign powers or private groups in the struggle for control—and such rivalries have shaped the history of the oil industry; it is characterized by the rise of an assertive counter movement from those controlled—and this can be seen in the politics of the oil industry.

But 'control' is a vague concept; it can mean many things ranging from the complete freedom of foreign owners to make all decisions, including decisions about the disposal of earnings and assets, to a simple monopoly or monopsony. Moreover, one cannot legitimately assume that control in any of these senses necessarily damages the domestic economy, either absolutely or in the light of possible alternatives; indeed, quite the opposite may result, for an economy may gain substantially from foreign management as well as from foreign investment, as the crude-oil producing countries undoubtedly have. But an economic gain is not always sufficient compensation for foreign control in either the 'popular' or 'official' mind, and for this reason extensive, easily identified, foreign ownership or control of a country's major economic assets tends to cause trouble, even where both sides benefit economically from it.

Thus, the charge that the Companies are great imperialists is easy to understand in psychological terms. The fact that the major decisions regarding the disposal of a country's most valuable natural resource are made by foreigners engenders a sense among the people that they are being 'exploited', that is, being used for someone else's ends to their own detriment. Paradoxically, in the underdeveloped countries the intensity of a people's feeling that it is being exploited may tend to vary inversely with the degree of actual exploitation (however defined), for the less developed is a country and its people, the easier it is for foreigners to exploit them, but the less they will be aware of it; while the more developed is the (still undeveloped) country, the more difficult it will be to exploit them, but the more sensitive they may be to the possibility of exploitation. In any event, effective decolonization requires the removal of a psychological feeling of exploitation and inferiority and the substitution for it of the feeling that foreigners are in the country to the benefit of *both* themselves *and* the country.

The question must, therefore, be asked whether the major Companies can, short of voluntarily abdicating their dominating position in each of the producing countries, do more than only slightly mitigate the underlying fear, resentment, and active animosity with which they are widely viewed. They have worked hard at the process of psychological decolonization, but nowhere are the people convinced, although some of the governments may be. Since a few foreign Companies control by far the greater part of each country's major resource, the country will tend to look at them as having an especially privileged position. They can be accepted on equal terms only when they do not appear to be overwhelmingly and unequally powerful. If this is true, it follows that if the feeling of 'exploitation' is to be removed completely, the essential remaining step is likely to be the relinquishment by foreigners of the ownership of their great oil-producing affiliates in each of the countries, and of the managerial control and responsibility that goes with it. This is not seriously envisaged as yet by any but those with extreme views. The reason is obvious: the crude-oil producing countries and the oil Companies have grown into a kind of symbiotic relationship, defined by the Oxford dictionary as 'an association of two different organisms which live attached to each other, or one as the tenant of the other, and contribute to each other's support'. The severence of this relationship, so far at least, would have been very painful. No matter how vociferous the demand for 'nationalization', no government has seen a way of going about it that would not lead to the destruction of the host along with the alleged parasite. Even the violent feelings aroused against Britain, and especially the US, during and after the 1967 war with Israel, did not move any of the countries to attempt an expropriation of the British and American oil-producing companies.

INTERDEPENDENCE OF COMPANY AND PRODUCING COUNTRY

The mutual support characterizing the relationship between the Companies and their host countries has been very important to both. The Companies not only gained rights to search for oil over very large areas, in some cases covering entire countries, but they were granted *exclusive* rights, which meant that potential competitors were kept out. (The general acceptance of the obligation to relinquish territories is very recent). Having found oil, the Companies then had the right to exploit their discoveries in the light of their general market position and their integrated requirements. In turn, the producing countries were able to have their potentially valuable resources developed, for they could not have done it themselves, and, as the 'pioneer risk' to the Companies diminished, they were able to obtain increasing foreign exchange revenues, at no investment cost to themselves, which they were free to use as they saw fit.

Ownership of a very large proportion of the world's crude-oil supplies outside the US and the Communist countries was not only extremely profitable in itself, but was also a decisive factor in the struggle of the major Companies to maintain and strengthen their position in product markets. Hence the maintenance in good order of their concession rights in the oil-producing countries was of very great importance to them, and they were prepared to go a long way to meet the demands of their host governments for changes in their original agreements and for more revenues. At the same time, because the major Companies controlled a very high proportion of the world oil market, the producing countries were almost completely dependent on them for the continuance of their oil revenues, for there was no significant alternative outlet for their oil. Hence, they could not afford to sever the established relationship of dependence. The crude oil industry of the producing countries was indeed under foreign 'control', but the Companies, because they were as dependent on the oil-producing countries as these countries were on them, and moreover were in competition with each other, were not free to exercise control in their own interest entirely as they saw fit.

The governments feared that the Companies' interests would conflict with their own and they therefore imposed any number of constraints on the actions of their foreign concessionaires. These included a large variety of regulations with respect to the operation of the industry, ranging from drilling and conservation requirements to regulations regarding employment and training;

in particular, they included restraints on the valuation of output and the disposal of revenues attributed to it. When looked at as a whole, the 'interference' with the freedom of action of the oil Companies from their host governments has reached very considerable proportions, and in this sense even the affiliates of foreign firms are not wholly foreign controlled—rather control is shared with the local governments.

Although the interdependence between the great crude-oil exporting countries and the existing major oil Companies must remain as long as the former produce most of the world's oil and the latter refine and sell it, there are many signs that the present institutional framework in which it is expressed will not long continue. The essential characteristic of this framework is the ownership of the great crude-oil producing companies by the major international firms. Some producing countries do not like it, but as long as the oil Companies insist on their concession rights there is little the countries can do but continue to press for more money and more control over prices and rates of offtake, and even of operations, while at the same time offering new concessions to Companies other than the majors and trying to develop their own national companies.

As to Companies, however, the more they are required to do things that they would not otherwise have done, that is, the more they must accept higher costs, reduced net revenues, or impaired managerial independence in order to satisfy the demands of their host governments, the *less* advantageous to them is their investment in crude-oil production. After all, the continued investment in their crude-oil producing affiliates is presumably undertaken because it is considered profitable, and it is worth examining the consequences if this profitability becomes seriously undermined.

EROSION OF THE ADVANTAGES OF INTEGRATION

The advantages to an oil Company of extending its integration to the production of crude oil are, broadly speaking, of two kinds: (a) oil may be obtainable at a cost less than the price at which it could be bought, and (b) the rate of supply may be more certain and more easily controlled both as to amount and type of crude, thus facilitating close planning of production, transportation, storage and processing and reducing costs which unnecessary fluctuations or uncertainty might otherwise be expected to bring about. In other words, costs can be lowered through continuous and close adjustment of the supply of crude of all sorts to the desired amount, composition, and geographical distribution of the refinery output of the internationally integrated group. If crude oil could be bought as cheaply as it could be produced, and if each Company could feel confident that it could contract for and in fact receive at market prices the amount and qualities of oil it wanted when it wanted them (with unexpected short-term surpluses or deficits catered for by brokers or some sort of spot market), there would be no reason for any firm to undertake crude-oil production simply because the firm was engaged in the refining and sale of products and needed crude as a raw material. It might, of course, be profitable to produce crude oil for sale or for the firm's own use, but the profitability would not depend on the fact of integration through ownership.

Thus, *taken by itself*, and apart from any tax considerations, vertical integration into crude-oil production is profitable because markets are imperfect, and the imperfections that promote integration are in turn intensified by the very existence of widespread integration. If the integration of the major firms gives them preferential access to crude-oil supplies, then all firms will be forced to adopt the same structure to ensure their own supplies.[1]

As we have seen from the history of the international industry outlined in Chapter III, the importance to a firm of owning crude oil has been very closely connected with the nature of the competitive relationships in international markets for crude oil and products. When large reserves of oil were discovered, the major Companies made arrangements among themselves to develop the reserves in accordance with their integrated needs, while at the same time they attempted to maintain a system of pricing that would effectively prevent price competition in sales to outsiders. As a result substantial profit was attributed to crude oil. For a while at least, and especially in the Middle East, a good deal of this profit could be looked on as a genuine economic rent attributable to low-cost fields. The rest could be looked on either as a rent properly attributable to the high degree of control over crude oil, or simply as an accounting attribution to crude oil of rents arising in 'downstream' operations and more properly attributable to such things as refinery location, brand names, distribution, sites, etc.

Since the end of the war, events have been steadily undermining the various advantages that the international Companies have derived from the ownership of crude-oil producing companies in the major exporting countries. The factors responsible for this erosion are likely to continue to be important, with the result that a new form of relationship between government and Companies is likely to arise in the foreseeable future—a relationship which may be readily accepted, if not sought, by the oil Companies themselves.

In the first place, the cost advantages of integration have been substantially reduced by the increasing taxation of profits attributed to crude-oil production. For some time the inter-affiliate transfer price of crude oil, the price at which arm's length market sales were made, and the price for tax purposes were roughly the same, but in the later 1950s these prices began to diverge and by the early 1960s they had become very different, with 'tax prices' higher than the other two. In such circumstances, it becomes theoretically possible for the governments of the crude-oil producing countries to insist on tax prices that attribute *all* the profit of the integrated business to crude-oil production. (This presumably is the implication of the claim that 'downstream losses' were being made after 1958 by some oil Companies). Moreover, the economists of OPEC are fully aware of the general principle that a tax that falls strictly on rents will not reduce output,[2] and in a paper presented by OPEC to the Fifth

[1] De Chazeau and Kahn came to this same conclusion in their study of vertical integration in the US. The importance of the US tax discrimination in favour of profits attributable to crude-oil production can hardly be underestimated as an inducement to vertical integration.

[2] It seems to be true, however, that the officials of OPEC would not expect any kind of tax to reduce output. A proper income tax would not do so because it does not affect costs, while the present 'income tax', which has turned into a flat tax per unit of output to the extent that neither prices nor costs are allowed to vary, is deemed not to affect output even though it raises marginal costs because demand is assumed to be almost totally inelastic!

Arab Petroleum Congress in 1965, proposals were made for a system of taxation that would give the governments a greater share of the true economic rent arising from the difference in the net value of output in different fields.[1]

OPEC does, of course, insist that a 'fair' or 'reasonable' profit must remain for the Companies, for it has never indicated a desire to push them out entirely; indeed, appreciation of mutual benefit is often expressed. Nevertheless, the trend of policy is clear—maximum revenue consistent with peaceful bargaining; the general method is also clear—the enforcement of 'tax reference' prices, which may have little relation to the market price for crude oil. It seems highly probable that if there are still 'excess profits' to be taxed away, the oil-producing countries will get them.[2]

Secondly, the advantage to a Company of producing its own crude oil will be less, the more the governments of the producing countries attempt to obtain some direct influence over managerial decisions.

Although nationals have been given seats on the Boards of Directors of the major oil-producing subsidiaries, it must be admitted that very little interference with the economic plans and managerial decisions of the Companies has taken place so far, other than pressures to maintain prices, increase rates of offtake, employ nationals, etc. Occasionally there has been some pressure in recent years on some companies to 'diversify' the markets in which they sell their crude oil. Nevertheless, the threat of more direct interference is real and could become pressing at any time, while neither flexibility nor security of supply is guaranteed by ownership alone, as has been adequately demonstrated several times since 1950. In addition, the fact of integration has led governments in some of the importing countries very significantly to interfere, not only with the managerial decisions of refining and distributing affiliates operating in their countries, but also with the ability of the affiliates to take advantage of opportunities for expansion.

If the trends described above—increased taxes, government intervention, and falling market prices—continue, the oil Companies may well find it desirable to explore means of selling their major producing affiliates to the governments of the oil-producing countries and substituting some form of contractual relationship.[3] In view of the alternative uses of their funds both within and outside the industry, a point could well be reached when such a move would actually improve the economic position of the Companies; even as things are at present such 'voluntary nationalization' would remove some awkward political problems that arise directly from their vertical integration across national frontiers, including the difficulties with the importing countries that arise when the price at which oil is imported is an internal transfer

[1] OPEC, *Taxation Economics in Crude Production.*

[2] The American Companies could become very vulnerable if the US Government should decide that it would not recognize the payments made on the basis of 'tax reference' prices as proper income taxes which could be offset against the tax liability of the Companies to the US Government.

[3] The analysis of this section refers only to the ownership by the Companies of the major crude-oil producing affiliates in the Middle East and Venezuela, such as IPC, Aramco, KOC, Creole, Mene Grande or Shell Venezuela. The problems related to lesser exploration and producing ventures by foreign companies or in partnership with nationals are very different and such activities would probably continue.

price, and difficult to defend with reference to objective market criteria.

In Venezuela, as a result of revisions made in 1958, the major concessions terminate in 1983. In the meantime the Government has been raising taxes on the Companies, has been restricting their exploratory and other activities and enlarging the scope of the operations of CVP (the State oil company) to prepare for the day when CVP will take over. Even here, however, such pressures might lead to an earlier 'transfer of power'.

But there still remains a major difficulty if the majors should relinquish the considerable control that they now exercise over the rate of supply of crude oil from the Middle East: what might be expected to happen to the price of crude?

THE PROBLEM OF THE PRICE OF CRUDE

As we saw in Chapter VII, a high degree of control over crude-oil supplies by a group of integrated firms will yield them little advantage if they are engaged in intense competition in product markets, for a fall in product prices will necessarily and directly reduce the value of their crude oil. Thus, as a corollary to the advantages of integration into crude-oil production for the *individual* company—'cost oil' and flexibility and security in planning—it is to the advantage of a *group* of integrated companies to avoid price competition with each other in product markets, and to use their control over crude oil to raise the costs of outsiders and thus discourage potential outside competition as well. We have seen that the major Companies were in the past very much aware of this point, and, more recently, it seems likely that both the oil-producing countries and the major Companies hoped that price competition in product markets would be discouraged when Libya was induced to impose taxes with reference to posted prices in 1965. One of the primary objectives, as we have seen, was to raise the costs of independents and thus retard their price-cutting.[1]

If the governments of the oil-producing countries were themselves responsible for decisions with respect to price and offtake, the question might well be raised whether some might not attempt to increase their share of the total market by reducing their own prices, and thereby bring about a faster and even further fall in crude-oil prices. The lower prices of crude would benefit all buyers, but if at the same time the entry of newcomers were eased and competition in product markets were encouraged, with a consequent fall in product prices, the established majors might well lose more on the swings than they had gained on the roundabouts. From the point of view of the major Companies, therefore, an important part of the problem of appraising the desirability of giving up the ownership of their great crude-producing affiliates would be the effect on their competitive position in product markets, and for those Companies that now benefit from a depletion allowance, its effect on their profits after taxes if they could no longer claim that tax deduction.

In analysing this question, one must first consider whether the ability of the Companies to maintain 'orderly' competition in product markets has been due to monopolistic advantages in refining and distribution, or to their ex-

[1] See pp. 203–6 above.

tensive control over the supply of crude oil, or to a combination of both. The ability of newcomers to establish themselves successfully in product markets is hindered by the heavy cost of overcoming the numerous advantages possessed by the present large occupants. These advantages include such things as well-known and accepted brand names, patents on important improvements in products or processes, the occupation of scarce distribution sites, economies of scale in particular markets which leave little room for new entrants, or the importance of an extensive network of selling outlets.

If the advantages of the large Companies were entirely attributable to their entrenched position in product markets (which means that monopoly rents should be attributed to assets such as those listed in the preceding paragraph rather than to crude oil), then clearly they would benefit from lower crude-oil prices, since the position of potential newcomers would be little improved by the access to cheaper crude oil. It would follow that if taxes were raised to the point at which they not only skimmed off the economic rent properly attributable to low-cost crude production, but also cut into rent attributable to 'downstream' assets, e.g. brand names or specially favourable distribution sites, the Companies would do well to sell their producing operations providing that crude-oil prices were not expected to *rise* because of it. The wisdom of such a step would be reinforced if the governments interfered with management and if the position of the Companies in the importing countries was impaired because of their integration into crude production.

If, on the other hand, the advantage of the major Companies was primarily attributable to their ability to regulate the supply of crude oil, then there seems little to choose between a position in which the government imposed maximum taxes on all rents and monopoly profits, and one in which profits were eliminated because a fall in the price of crude promoted the development of competition in product markets. Of course consumers benefit from the latter, but this is not the present issue.

In practice, both sources of monopoly power have been important. When independent competitors obtained access to cheaper crude oil in the 1960s, the effect was felt in product markets as well as in the markets for crude; but at the same time the direct entry of newcomers was limited, for in a number of markets the cost of breaking through the barriers protecting the position of the existing occupants presented a formidable obstacle. Experience does seem to support the view, however, that if crude oil is widely available at prices which make refining and distribution especially profitable at current product prices, competition from newcomers spreads quickly, and that if the supply of crude is restricted and its price maintained, as in the US, product prices and 'downstream' profits are both raised. Thus, monopoly rent earned by the major firms is properly attributed to the ownership of crude-oil reserves. In these circumstances, the major Companies would have a great interest in preventing the emergence of really strong competition in crude-oil markets. It follows, therefore, that if vertical integration into crude-oil production seemed no longer profitable policy for the several reasons we have discussed, the Companies would still have to examine carefully the nature of the competitive pressures that would arise if their great producing affiliates became national Companies.

The possibility that in the absence of strong concessionaires extensive competition might break out among the oil-producing countries is not to be dismissed lightly. It might be objected that at present no crude oil producing country seems willing to cut its 'tax take' in order to encourage its concessionaires to lift more of its oil, and that there is, therefore, no reason to assume that the countries would be more willing to compete if they were themselves the sellers of crude. Indeed, some would expect a fairly strong cartel to arise in the absence of the moderating influence of the Companies. In such circumstances, the price of oil might even be raised; it certainly should not be expected to fall.

If, before the change in ownership, taxation by the producing countries had reached a level which wiped out the profits attributed to crude at current *realized* prices, then the net income to the oil-producing countries per barrel of oil sold would not rise if they took over the producing companies, unless realized prices could be raised. Indeed, if the purchase of the producing companies were financed by a charge on current sales or by the offer of crude at special prices to the former owners, net revenue per barrel would even decline. Total revenues would probably continue to rise, but perhaps less than output; such a consequence of 'nationalization' might come as a severe shock to the producing countries and be politically embarrassing to their governments, especially if the underlying circumstances had not been realized before nationalization. The political advantages of gaining control of the domestic industry and eliminating the foreign 'monopolist-feudalists' would be great, but the economic price might well bring severe disillusionment.

Attempts to raise the existing level of prices by concerted action might therefore be made. These would probably require the planning of offtake in each country in relation to some overall output plan—that is, the acceptance of production quotas. One could speculate a long time on the possible course of events, but it would be bold indeed to predict that the producing countries could successfully reconcile their separate interest for long. The claims of the various national oil companies to the right to bid for important oil contracts in growing markets, coupled with the pressures many importing countries would continue to exert to get oil as cheaply as possible, would be formidable obstacles to the prevention of a considerable amount of price competition. And the domestic political pressures against cutting prices to sell more oil would not be as great as they are now against cutting taxes to the foreign companies.

The major Companies would remain by far the largest buyers of crude oil. Competition among producers in selling crude with consequent pressures on prices would ease the entry of independent competitors and facilitate price competition in product markets. From this point of view, the Companies would want to reduce competition in selling crude, and might through their own contracts attempt to provide incentives to the national producers not to cut prices to others; for example, they might insist on contracts that provided for re-negotiation of the terms if cheaper prices were offered to other buyers. On the other hand, even if increased competition did not occur in product markets and product prices remained unchanged, those Companies for which the depletion allowance had substantially reduced their consolidated tax liability would find that their profits after tax would be less after nationalization than before. This might be so even if they obtained their crude on contract at an unchanged pre-tax cost apart from the depletion allowance, for by buying instead of producing crude oil they would lose the US tax subsidy. Obviously, some reduction in the cost of crude without any increase in competition in product markets would be necessary if the financial positions of such Companies were not to be worsened.

From all this we can conclude that the willingness of the majors to divest themselves of their great producing affiliates will be very much influenced by what happens to product prices. If market prices continue to fall and the Companies are not able to moderate competition through their existing control over supplies of crude, or should the US Treasury cease to accept posted prices as appropriate for the calculation of income taxes, then the day would come nearer when the major Companies would be willing to sell out, not only because they would expect to obtain cheaper oil supplies from doing so, but also because the chief remaining advantages of owning the greater part of the world's crude-oil supplies will have disappeared.[1]

THE CONTINUED ROLE OF FOREIGN COMPANIES IN EXPLORATION AND PRODUCTION

The above discussion relates only to a small group of oil-producing companies (including Iranian Oil Participants Ltd.) in the Middle East and Venezuela. These few companies account for such a very large proportion of the industry in each of the host countries, and in the international industry generally, that their inclusion within the vertically integrated structure of the major Companies in itself fundamentally shapes the character of the crude-oil market. Foreign ownership of crude-oil production on a smaller scale, even when it involves vertical integration, does not create the same problems. Many governments will undoubtedly continue to attempt to induce foreign Companies, including the majors, to explore for oil in their territories and subsequently to produce it on mutually agreed terms. As we have seen, the newer contracts often provide for joint operations between foreign and domestic companies, and the provisions for offtake, prices and taxes are of many kinds. In participating in the search for and development of new crude supplies, foreign Companies may well obtain some of their crude oil on very favourable terms indeed, but this need raise none of the difficulties for the function-

[1] I have not attempted to discuss how a change of ownership could be brought about. There are many possibilities, and it is fruitless to speculate on what might come out of negotiations between Companies and governments where the will to effect a change existed on both sides. In the first place, all of the Companies would have to be involved, for no one of them could afford to attempt it alone; secondly, the governments would have to be prepared to pay for the assets acquired. There probably would have to be a transition period, which should doubtless be fairly short to avoid confused conflicts of interest. The primary requisite, however, is that the Companies, perhaps as a result of steady government pressure, should act voluntarily, for only in this way could a confrontation disastrous to both sides be avoided and leave the Companies' further prospects in crude-oil operations on a lesser scale unimpaired in the countries. In some countries, however, the willingness of the Companies to sell out might exceed that of the governments to buy.

ing of crude-oil markets that are raised by the present degree of monopolistic integration.

Expectations that consumption will continue to expand at a very rapid rate, together with the ever-present prospect that new low-cost sources of crude oil remain only to be discovered, should continue to draw foreign investors. On the other hand, if the competitive development of existing known fields is intensified and the price of oil comes under greater pressure, a period of several years may elapse before exploration in new areas is pushed with much vigour. But if, as Adelman's evidence indicates, replacement cost for the great crude-producing companies in the Middle East is now, and is likely to remain for some years, close to zero, this is no economic loss for the present.

CONFLICTS OF INTEREST

One thing must be made clear: it is neither correct nor analytically useful to postulate an overall 'harmony of interests' among the various groups concerned with the international petroleum industry. It goes without saying, of course, that producers and marketers want to produce and sell their products, and that users and consumers want to buy them; to this extent there is a coincidence of interests. But this does not get us very far, for the crucial questions turn directly on the terms on which exchanges are undertaken and on the tax benefits from the industry obtained by the several governments of the countries in which operations are carried on. Here there are clear divergencies of interests. Each individual Company wants to obtain as much revenue as possible from its sales—or, if this statement is considered too strong, at least enough to permit it to finance from its own earnings the expansion required to maintain, and preferably to improve, its share of the market from period to period; the importing countries with neither coal, domestic oil supplies, nor extensive foreign investment in oil Companies want to pay as little as possible for their energy; and all governments want greater tax receipts (which necessarily reduces the net earnings of the Companies) as well as the maximum contribution by the industry to national income. To a very large extent these different interests are inversely related to each other: a gain for one being at the expense of one or more of the others. It is true that everybody's revenues can increase while prices go down, if output and consumption are expanding, but for any given rate of output the interests of the various groups are not in harmony. At the same time, there are conflicts of interest among the several oil Companies or groups of Companies, among oil-producing countries, and even among the net importing countries.

It is appropriate to treat the economic relations between the oil Companies and the developing countries largely in terms of 'bargaining' and a balance of power, rather than in terms of ordinary trade and commercial competition, because the conditions in which trade takes place, as well as the nature of the fiscal and other financial arrangements, are to a large extent determined by negotiation between the international Companies and governments—including governments of importing countries. The terms of agreements arrived at are frequently re-negotiated when changes take place in external circumstances or in the effective bargaining position of the governments. In this the oil industry is not, of course, unique, but the type of interest a government is almost bound to take in the exploitation of its major natural resources, the extent of vertical integration in the international industry together with the predominance of a few very large firms whose internal transfer prices need not automatically respond to competitive pressures, and the special importance of the role of energy in economic affairs, all combine to place the oil industry in a particularly vulnerable position in importing, no less than in exporting, countries and in developed as well as underdeveloped countries. In a number of the former, such as France and Japan, the government has intervened extensively to regulate various aspects of the industry. But for the most part, the governments of the developed countries have not made the same types of bargain with the international oil firms as have the undeveloped countries in their desire to obtain foreign exchange as exporters, and, as importers or would-be producers, both to reduce their foreign payments and to attract foreign capital for exploration and investment. In the developed countries the conditions associated with investment are very different from those in most underdeveloped countries: their governments look at the oil industry in a much wider context of trade, investment, and domestic energy and industrial policies, and they often have a very different attitude towards both private investment and foreign enterprise.

Companies as Neutral 'Buffers'

It has often been suggested that one of the important roles of the oil Companies is to act as 'buffers' or intermediaries between the governments of the exporting and of the importing countries, taking on themselves, as 'neutral' international organizations, the task of maintaining a balance among the several conflicting interests.[1] Some even express the belief that the Companies can 'take oil out of politics'.[2] Others have suggested that 'the interest of both producing and consuming countries are reconciled by the competitive nature of the industry . . .',[3] and that the 'pricing mechanism has efficiently co-ordinated economic forces and reconciled divergent interests in the past, and should be allowed to continue to perform these important functions in the future'.[4]

These two approaches are, in a sense, at opposite poles. The first recognizes the power and authority of the oil Companies and, in effect, proposes that these should be used in the 'general interest'; the second denies power to the oil Companies, stressing the extent to which the impersonal forces of competition limit their freedom of action. It should be clear from the discussion in

[1] See J. E. Hartshorn, Chapter XXIII, 'The Business In Between', for an able presentation of this point of view.

[2] 'Not the least service provided by privately-owned oil companies, to producer and consumer alike, is their ability to take oil out of politics.' *Petroleum Press Service*, September 1965, p. 324. Statements of this type must indeed sound curious to Iranians, Iraqis, Ceylonese, Cubans, Indonesians, and even Indians!

[3] Quoted from speech by L. E. J. Brouwer, Managing Director of Royal Dutch/Shell, to the Iraqi Economic Association, May 26, 1964. (Mimeo.)

[4] Shell International Petroleum Company, *Current International Oil Pricing Problems*, August 1963, p. 16.

earlier chapters that neither approach takes sufficient account of the characteristics of the industry. The 'pricing mechanism' does not set the offtake rules which determine how much crude oil each of the joint owners of the oil-producing affiliates in the Middle East is allowed to lift in any period of time and the terms on which it can lift it; it does not determine from which company some of the world's largest refineries will buy their oil, nor is it allowed to determine, for many countries, their choice of supplier; it does not determine, although it may influence, the price at which crude oil and products are sold between affiliates; it does not determine, although it may influence, the geographical location of taxable profits; etc. Competition of a kind has always existed in the industry, and even price competition became fairly widespread in the late 1950s and the 1960s, but one can understand why exporters are loath to believe that it serves their interest, and the benefit to importers has, particularly in the developing countries, often been obtained only after strong government representations.[1]

Nor can the Companies, under present circumstances at least, be conceived of as neutral arbiters. Not only do they stand to gain or lose from any bargain struck and must therefore be considered interested parties, but they are not in fact neutral in their attitudes to different countries since they must respond to the relative strength of political pressures, the relative risks of political instability, and the relative rates of taxation in different countries. The home governments of international firms not infrequently attempt to force the firms to act in the assumed national interest of the home country at the expense of the other countries in which they operate, while 'host' governments may also intervene in their international operations for political or economic purposes.

CONCLUSION

The organization of the international petroleum Companies and their role in the determination of prices and output and in the development of the industry generally was, as we have seen, the outcome of a complex historical process in which each private, profit-seeking firm responded to geological, technological, political, and economic conditions, as well as to a competitive environment created by the responses of other firms, in an endeavour to protect its own position in the industry and promote its long-run growth. The whole structure was not created as the outcome of far-sighted planning ('who would have had the imagination, nay the temerity, to do so?').[2] It was the outcome of the competition of big capitalist oligopolies, with their roots in the free-for-all business attitudes of the nineteenth century and very much oriented towards the requirements of the industrialized West.

[1] Moreover, it is misleading to insist that competition 'reconciles' conflicting interests; all it does is to put severe limits on the extent to which any buyer or seller can influence the market in his own favour. The impersonality of the discipline imposed by truly competitive markets has one great advantage—it makes it impossible to make any particular firm or group of firms a scapegoat on which to blame the troubles of the industry. On the other hand, the dissatisfied can always blame the government—that always convenient scapegoat of last resort.

[2] P. H. Frankel, *Oil—The Facts of Life*, p. 10.

After the Second World War came the so-called 'emergence' of large numbers of countries in Asia and Africa, independent, for a variety of reasons growing stronger in their political bargaining power, acquiring an increasing indigenous supply of knowledgeable civil servants and politicians, and imbued with a strong and somewhat touchy nationalism and spirit of independence. The governments of these countries are increasingly preoccupied with the problems of alleviating the poverty of their peoples and of promoting the economic development of their countries, although progress in these directions has been severely retarded at times by internal political controversies, external military ventures, conflicting views over the type of economic system desired, the measures most likely to achieve results, and confused notions of what constitutes economic 'independence'. All of these have been complicated in some areas by ill-advised intervention from outside powers.

Nevertheless, the changed position of the underdeveloped countries, whether they were oil exporters or importers, brought about significant changes in the environment in which the oil Companies operated, changes which coincided with, and partly reinforced, the other economic developments affecting the industry that we have examined. The Companies were in many respects extremely slow to adjust to the new environment, on occasion at a serious cost to themselves. At times, however, emotionally charged antagonism and internal politics in their host countries made serious negotiations almost impossible, leaving the Companies very little room for manoeuvre. But it is not surprising that they were on the defensive, and that they should fight rearguard actions to delay changes adverse to their own interests; after all, they were not the only groups in the West making serious strategic mistakes in the process of retreating.

But the difficulties that the international oil Companies have faced, and are still facing, in the underdeveloped countries are not entirely related to the conflicts arising because these countries are developing. Similar difficulties have arisen in developed countries, and may well become more pressing and widespread as time goes on. The deeper root of the problem is simply that international firms, including the oil Companies, have not yet found a way of operating in the modern world which would make them generally acceptable as truly international institutions. The officials of most international firms may not as yet even attempt an international outlook, and would consider any suggestion that they might do so to be inconsistent with their national allegiance. This certainly seems to be true of some of the international oil Companies. But in other, more far-seeing firms (also to be found in the oil industry), officials are grappling with the types of problems that we have been analysing in this study, and are acutely aware that in the long run the international acceptability of world-wide enterprise will depend on how successfully they are solved. It is not at all clear, however, that the resolution of the basic conflict is within the control of the firms themselves: this we take up in the next chapter.

Chapter X

THE FUTURE OF THE LARGE INTERNATIONAL FIRM

It is widely asserted in business circles that the twentieth century is the 'century of international business'.[1] This presumably implies that international firms can be expected to play an increasingly dominant role in international economic relationships. These firms are of many nationalities—British, Dutch, Swiss, some other European, and now even Japanese—but most of them, and nearly all of the very largest, are American. In these circumstances, any increasing dominance of international firms will tend to be associated with an increasing dominance of American economic influence. This in itself creates special problems and tends to confuse the fundamental economic issues. To some extent, the Americans are aware of this, for in the US there is much discussion in the universities and the schools of business, in the business journals, and in the speeches delivered by executives of international firms, on the role and responsibilities of 'multinational' companies.[2] For the most part the discussions deal with the appropriate relationship between the 'centre', or head office, and the local subsidiaries in foreign countries, including the conflict between the interests of the local subsidiaries (and sometimes of the local economy) and the maximization of the 'global' profits of the firm as a whole.

The administrative controls of international firms are supranational, confined by no frontiers, yet the policies and executive actions of their head offices often have considerable impact on individual national economies in many parts of the world. Since every government has a responsibility for its own national economy, it is bound to watch the activities of international firms within its own jurisdiction and to intervene when it considers intervention necessary in the interest of the national economy. The firms are expected to be 'good citizens' of each of the countries in which they operate, and they must satisfy many increasingly exacting governments. In the international petroleum industry they have often failed in this task and have suffered increasing restrictions on their freedom of action in consequence. But the oil Companies are not unique from this point of view. Firms in other industries have had similar problems in a number of countries. There are two primary

[1] See, for example, the speech of John J. Powers, Jr., President and Chief Executive Officer of Chas. Pfizer & Co., Inc., before the President's Dinner of the Student Association of the Graduate School of Business, Stanford University, April 13, 1967.

[2] I should perhaps remind the reader that international firms are defined as firms that manufacture in a number of countries. The definition does not include firms that merely export and distribute goods in foreign countries.

sources of difficulty. The first relates to the economic or commercial policies of the firms, including their accounting arrangements, and the second to the type of government intervention in their affairs.

COMMERCIAL POLICIES

Most of the large international firms are privately owned. Because they are private organizations, they are responsible only to 'shareholders' in some undefined sense although, like private individuals generally, they are subject to the laws and regulations of the countries in which they operate, and to prevailing codes of conduct. Also because they are private enterprises and function in accordance with the rules and procedures of private-enterprise capitalism, they are sensitive to the judgments of financial markets, especially in their home countries, and must maintain an acceptable level of profitability.

Foreign investment by such firms is not simply investment in foreign enterprises which are expected to be profitable taken by themselves. It involves the establishment of manufacturing subsidiaries in foreign countries which are, in principle, managed in such a way as to maximize their contribution to the consolidated profit and loss account of the firm as a whole. To this end, as we have seen, the prices charged by the subsidiaries may be adjusted, restrictions may be put on the amount and direction of their exports, charges for services or intermediate products transferred from the parent or between subsidiaries may be used as a means of transferring funds internationally, special subsidiaries may be set up in places chosen as 'tax havens', and other devices may be adopted to minimize income taxes.

Thus, because of the competitive financial environment, the pressures on firms to maintain, and, if possible, to improve their rates of profit, and the accepted attitudes respecting the appropriate functions of competitive private enterprises, conflicts arise between the overall financial interest of the firm and the economic interest of the individual countries in which it operates. A firm's primary need is to make profits; tax avoidance is legitimate, and if governments impose taxes, restrict profit remittances, or otherwise attempt to control profits, international firms may be expected to seek to minimize the adverse effects on their integrated operations. Moreover, in the interests of managerial efficiency international firms will often resist the local interference in their affairs that might arise in some countries if local equity capital were accepted in their subsidiaries, or if local people were given influential managerial post. American firms, in particular, are often thought to have a strongly 'nationalist' approach, preaching the ideology, and endeavouring to promote the interest, of their home country or parent company, insisting on 100 per cent ownership of their foreign subsidiaries, and resisting any real dilution of American managerial control. The international firms of some other countries, for example, some of the Swiss firms, seem no less nationally oriented, at least so far as their financial affairs are concerned.

On the other hand, many firms of all nationalities, including the American, make a special point of promoting 'joint ventures', of welcoming and training local management, and of spending large sums on development projects in local economies.

Nevertheless, the very wide differences in the policies and outlook of different firms make it impossible for any host government readily to assume that the international firms operating within their borders will not take advantage of their position to discriminate against the local economy, either to enhance the consolidated after-tax profits of the international group, or to further the interest of their own country, or to facilitate the maintenance of the undiluted control of foreign management. In such circumstances, many governments do not look on private foreign investment as a simple commercial transaction not requiring close government supervision, and the demands by the firms for freedom from government interference are not heeded.

COMMERCIAL POLICIES AND THE 'PUBLIC INTEREST'

Some observers have conceived of the role of international firms as politically neutral buffers or mediators in the economic conflicts of interests between different countries, basing their decisions primarily on economic and commercial considerations, but at the same time balancing fairly the varying interests of those affected by their operations. The notion that business firms should not be exclusively concerned with maximizing their own profit but should consider the wider interests of those affected by their operations is not new. Over thirty years ago Berle and Means, in their examination of the role of the modern corporation in modern life, wrote '. . . great corporations should develop into a purely neutral technocracy, balancing a variety of claims by various groups in the community and assigning to each a portion of the income stream on the basis of public policy rather than private cupidity'.[1] The issues they raised in their study for economic theory and policy are still unresolved; the role of the very large privately-owned companies in a modern economy is still ill defined and controversial, and also subject to great confusion of thought because the 'models' of the 'firm' implicit in the prevailing theory of a private, competitive economic system, to which both the big firms and their critics seem to cling, are inadequate, even for analytical purposes, let alone as foundations for making judgments about public policy.[2]

Nevertheless, this 'neutral technocracy' solution is often implicit in the statements of conservative 'free-enterprise' business organizations such as Standard Oil (New Jersey). Statements to the effect that 'board members are keenly aware of their responsibilities to shareholders, customers, employees, government, and the general public, and of the need to discharge these responsibilities harmoniously',[3] acknowledge the necessity for the corporation itself to balance the claims of various groups, although they by no means imply that the balance will be unaffected by the financial interests of the corporation. Within rather narrow limits, some of the international oil Companies do go some way to justify the statement of Berle and Means that '. . . business practice

[1] A. A. Berle, Jr., and G. C. Means, *The Modern Corporation and Private Property*, p. 356.

[2] Professor Kenneth Galbraith has more recently dealt with similar issues in numerous books and articles, but for a variety of reasons his views have not had widespread acceptance among his fellow economists.

[3] Standard Oil Company (New Jersey), *A Brief History*, p. 31.

is increasingly assuming the aspect of economic statesmanship'.[1]

Even if this is accepted, however, we are still a long way from the central problem posed by the international firm. Few would quarrel with the proposition that firms should act responsibly toward those affected by their actions: and this is as important for small firms in small communities as it is for large firms in the world economy. But no matter how responsibly firms exercise 'economic statesmanship' according to their own lights and endeavour to act in the 'public interest', they are 'non-accountable' in the sense that they have themselves to define the international 'public interest' to be considered when national interests conflict or government pressures push in opposite directions. There is no higher authority to which they are accountable, and which is capable of ratifying, modifying, or rejecting their conception of the international neutrality or the international public interest they may attempt to observe. In domestic affairs the national government has such authority, but there is no corresponding authority in international affairs.

In many circumstances, however, a large firm, national or international, is in a position to take a longer view than are most political leaders of government. Politicians tend, above all, to try to stay in power—as indeed they must do if they are to be effective leaders of their countries. But as a result they give great weight to the immediate, and as a consequence a government may often take a very short-run view of its actions. By contrast, large corporations, as continuing organizations, often take a very long-range view of events. Although they cannot ignore the high discount they may at times put on the future, they are concerned with long- and not short-run profits. Their executives are not subject to the same type of short-run political pressures, and a large firm is so organized that there is great incentive for all managers to give a high priority to the growth and survival of the institution itself. Governments can take into consideration a wider public interest than can be expected from firms, and thus cannot give firms a free hand. In other words, governments may use a wider-angle lens through which to view events but the range of the lens is often shorter. This difference can be of great importance in economic development.

On the other hand, as we have seen, international firms may easily overestimate their strength and take a surprisingly short-run view of their own interests, especially in the less-developed countries. It may pay a firm in the short run to take advantage of the veniality of ignorance of the governments of such countries and obtain agreements that contain provisions which are manifestly indefensible from the point of view of the countries' welfare, or to resist reasonable changes in such agreements when conditions have changed. In the longer run, however, such 'exploitation' may give rise to resentment that will lead new or better-informed governments to take severe, and perhaps in their turn, unreasonable, action against the firm. In some countries, of course, a foreign firm may conclude that its time is short in any case and that the best strategy therefore is to take the cash and let the credit go.

But even when they try to consider the interest of their hosts, the managers of private firms can see things only through their own eyes, and in inventing for themselves a conception of the international and public interest, they will

[1] P. 537.

have to contend with the widespread human tendency to find, very sincerely, that there is a fortunate coincidence between their own welfare and that of others. Many American firms, for example, are apparently convinced that virtually untrammelled private enterprise, almost in any circumstances, will bring the best results for all concerned. As we have seen, some have no hesitation in trying to persuade their own government to use its influence with other governments to prevent interference with their activities. Or, to give another example, it can be held that underdeveloped countries are better served when foreign management has full control because of the superior efficiency of foreign managers. From this it would follow that a foreign firm that refused to dilute its foreign management and control would be acting in the best interests of the country itself. But who is to adjudicate between this and the contrary view that the economic development of a country requires that its people should obtain, not only training, but also experience, with the corollary that local management should be given authority in spite of the inefficiency that may result in the short run? Particularly difficult is the conflict between the financial interest of the firm as a whole and the financial needs of governments. Firms rarely find that the public interest is best served by making them pay higher taxes! They may of course often be right in their judgment in given circumstances, but who is to decide?

Examples can be multiplied, for there is nothing new about the problem. Yet, as matters stand, the most progressive and internationally-minded of firms has no alternative than to consider the several points of view and to do the best it can to reconcile conflicting interests according to its own conception of justice, while at the same time safeguarding at least the minimum requirements of its own financial position. This is especially unsatisfactory when a firm is placed in a position in which it has to decide between its own financial interest and that of the countries in which it operates. In practice, however, international firms must bow to the wishes of those governments that are in the strongest position to bring pressure on them.

INTERVENTION BY GOVERNMENTS

If firms are privately owned, they are not, in principle, instruments for the implementation of the economic or political policies of any particular government. This is one of the strongest arguments in favour of private enterprise, particularly internationally, where the interests of government may, for political reasons, give rise to more conflict than the economic interests of private groups. The political acceptability of international firms is likely to depend very much on the host governments' conviction not only that the operations of international firms are in the economic interest of their countries, but also that they are not deliberately manipulated in the political or economic interest of any other country. In this respect, the tendency for the parent governments of international firms—notably the UK and the US, where the majority of the world's largest international firms have their headquarters—to use their political power to force the firm to act in the interest of these countries at the expense of other countries is likely to become increasingly resented. The receiving countries often have legitimate and serious grounds for complaint about foreign interference with their industry or even economy. Interference with the commercial decisions of international firms by their parent governments occurs for both economic and political reasons.

In order to strengthen their own balance of payments, both the US and the UK have for some years put pressures on the international firms based in their respective countries to force the firms to discriminate in the conduct of their financial affairs against the other countries in which they operate. International firms have been asked not only to reduce their rate of investment in foreign countries but also to repatriate more of their earnings arising from foreign operations. Since the foreign income of international firms will have been derived from sales made in foreign countries, the demand from the parent government that it should repatriate its profits instead of reinvesting them abroad in the normal course of its international expansion is a way of putting pressure on the balance of payments of other countries in order to improve that of the parent country. Such pressure is most likely to occur during times of international financial strain and countries may well find themselves subjected to this kind of discriminatory treatment when they can least afford it.[1]

Control over affiliates in a foreign country has also been used to bring economic pressure to achieve political ends. There have been numerous reports of such politically motivated pressures arising from the difficulties between the US and Cuba. For example, an American oil Company in Japan (Caltex) was reported to have refused to supply lubricating oil to a British ship at a Japanese port because the ship was on a time charter to the Russians and was carrying Cuban sugar to Siberia.[2] Clearly if the lubricating oil industry of a country is controlled by foreign firms who discriminate among their customers according to the political quarrels of their home country and regardless of whether the political policies of the host country coincide with those of the home country, the government of a host country has cause for concern.

Strangely, neither the UK nor the US Governments seem to be inhibited by their own discriminatory policies from complaining about discrimination by other countries against their firms. For example, again in Japan, in 1962, the Government was considering measures to deal with the problems of the oil refining industry. Since it was known that the Japanese Government was uneasy about the extent of foreign control of the industry, the US and UK Governments, fearing that some of the measures would discriminate against the affiliates of American or British Companies, made representations to the Japanese Government about their proposed measures. That the representations involved heavy pressure (and were partly successful) seems to be agreed.

There is considerable danger that the developed, or 'advanced', countries may fail to appreciate the implications of the assumptions commonly made by their economists and politicians in judging the policies of the developing countries. To be sure many of these policies are unreasonable and unsound;

[1] I have argued above that the indefinite growth of a foreign firm through the reinvestment of retained earnings may not be to the economic advantage of the host country, and that arrangements for a gradual diminution of the foreign equity may well be desirable. But this is not an argument that can be used to defend the kind of interference from the government of the investing country that we are discussing here, for this involves actions that are unrelated in their purpose, timing, extent or incidence to the needs of the receiving countries.

[2] *Petroleum Press Service*, September 1962, p. 332.

but is it really established that foreign private investment should be unreservedly welcomed as an unmitigated benefit in all circumstances, as is commonly implied by policy makers and others in the developed world? In particular, can one seriously assume that private business will not take advantage of monopoly positions to 'exploit' the countries in which it works?

In enacting its foreign aid programmes the US Congress has always laid great stress on the importance of private enterprise, and AID is specifically enjoined to take account of the 'progress' being made by recipient countries toward the recognition of the importance of private enterprise. Moreover, aid legislation is used to protect US investors overseas from uncompensated expropriation. For example, the 'Hickenlooper Amendment' of January 1962 requires the President to suspend aid under the Foreign Assistance Act to any country nationalizing property owned by US nationals and failing to make prompt, adequate and effective compensation in convertible foreign exchange. It was under this provision of the law that aid was withdrawn from Ceylon after the nationalization of the distribution facilities of the oil Companies. Yet, as we have seen, the Companies were very clearly using their monopoly position in Ceylon to try to maintain prices there while giving way elsewhere.[1] This was, equally clearly, a discriminatory exploitation of the country.[2]

Ceylon is a poor country, yet the only ways she had of 'legally' protecting herself against this type of policy on the part of the Companies were expensive: to build new—and redundant—facilities to displace those of the Companies or to take over the existing ones compulsorily with compensation. She chose the latter, offered compensation, which was deemed too little, and gave no assurance as to when it would be paid. But what else was she to do? The US has, of course, the right to lay down the conditions under which it will give economic assistance to foreign countries. But if the US Government feels it necessary to base its aid on such unconditional support of private business, other governments have grounds for considering carefully the terms on which they can accept either.

The governments of host countries may also attempt to force discriminatory policies on international firms, particularly with respect to pricing and exports, through their political control of the subsidiaries operating within their frontiers. Such actions may well affect the acceptability of the subsidiaries of the international firm in other countries, but for the most part the power of a host government to enforce such policies does not depend on the fact that it is dealing with a subsidiary of an international firm and not with a locally owned firm. For example, in the petroleum industry the governments of the exporting countries in the Middle East have not only attempted to determine the prices at which crude oil should be transferred to subsidiaries abroad by international Companies, but have also at times placed restrictions on exports for political reasons. Nevertheless, if oil production had been in the hands of local firms,

[1] Pp. 229–30 above.

[2] A more extensive analysis of the meaning and significance of 'exploitation' can be found in my essay 'International Economic Relations and the Large International Firm', in *New Orientations in International Relations*, Vol. 1, edited by Peter Lyon and E. F. Penrose (Frank Cass, 1968).

the governments could equally well have fixed prices and enforced restrictions on exports. The reach of governments is the same whether or not the local company is domestic or is foreign-owned.

But when the home government of an international firm enforces restrictions on the firm's management of its financial affairs and those of its subsidiaries throughout the world, effective action is possible only *because* the government is dealing with firms that have direct investments abroad and not with firms producing only in the domestic economy. Any government may take action affecting exports, imports, and foreign exchange transactions to the detriment of trading partners, and international arrangements and codes of conduct to limit such actions have been adopted in various forms. Similar codes should perhaps be adopted with respect to the type of interference with international firms by their home governments that should be permitted. Attempts have been made to establish rules governing the treatment of foreign investors by the governments of host countries, but perhaps any foreign investors' 'charter' should deal with both types of problem.

Here, however, we are primarily concerned with the actions of governments that affect the international acceptability of firms, that is, actions that affect the willingness of all countries to accept investment from them. It is true that the line between the 'local' and the 'international' activities of firms is difficult to draw sharply. The international acceptability of a firm will not normally be affected by the necessity of complying with the policies of host or home governments with respect to labour relations, methods of domestic competition, use of resources and capital markets, domestic investments, etc. It will be affected by government controls over export prices, investment in other countries, international remittances, etc. From the point of view of the host countries, and particularly of the developing countries which have no international firms of their own, so to speak, the issues arise most acutely when the home governments of international firms intervene to force international firms to adopt discriminatory policies in their international relationships.

International firms are well aware not only of the financial cost or other inconvenience to themselves when their home governments place restrictions on their international financial freedom, but also of the potential damage it may do them in other countries. When the American and British Governments imposed such restrictions to protect their respective balance of payments many firms protested, but the complaints were muted, at least in public. In the US the firms may have feared that they would be criticized for putting their private interest above the interest of the domestic economy. In Great Britain, there was much debate over the question whether the economy gained or lost from British investment abroad generally.

CONCLUSION

We have not discussed here the economics of direct private foreign investment from the point of view of either the domestic or the foreign economy apart from the organization and policies of the international firms responsible for it. Nor have we analysed the reasons for the growth of this type of investment, particularly from the US since the war. For the domestic or 'home' economy, direct investment to keep markets is often profitable as a substitute for exports, as a means of acquiring cheaper imports, or as a way of acquiring access to foreign resources, for example raw materials such as oil or even foreign scientists in a research programme. Or it may be undertaken for a variety of other reasons: to get inside tariff walls, to comply with the requirements of foreign governments, to maintain patent rights, to compete more effectively with other foreign or domestic firms in foreign or regional markets, and so on.

The receiving countries may benefit from the enterprise and managerial skills made available from foreign countries, with the consequence that their resources are used more productively than would otherwise have been possible; the act of investment makes foreign exchange available to them which may be used to import foreign equipment and to pay for foreign skills, or simply to pay for increased imports of consumer goods and thus reduce the risk of inflation as investment proceeds. Moreover, certain types of products, some of great importance—for example, pharmaceuticals—are protected by patents, and it may be that only if the foreign patentee is prepared to establish its own manufacturing plant can production be undertaken at all.

Economists have traditionally emphasized the benefits to be gained from an international division of labour and from the flow of resources, particularly capital and skills, from countries in which they are relatively abundant to those in which they are relatively scarce. Nothing in this study should be taken as an attempt to deny the importance of such benefits, for we have not been concerned specifically with this question. Any analysis of the complicated problems involved in a general discussion of the 'welfare' aspects of foreign investment must take into consideration the kinds of changes that may be expected to take place in the economies of both the investing and receiving countries in response to foreign investment or in response to barriers raised against it. For any particular country such an analysis would inevitably involve a large number of judgments, or even guesses, about these responses, but there is a presumption that in general foreign investment contributes significantly to the growth and development of the receiving countries.

In this study, we have been concerned not with these matters, but with the economics of the institution through which direct investment takes place—the international firm—and have attempted to analyse its suitability as a vehicle for the flow of such investment, with special reference to a particular industry. We have found it defective, not so much because of the nature of the institution itself—privately owned and often very large—nor even because of the monopolistic power it often possesses, but because these characteristics, given the international environment in which the firms operate, may seriously distort the international distribution of the benefits obtainable from foreign investment.

An international firm functions within an environment of sovereign countries, each of whose governments has a direct responsibility for the domestic economy. Each depends for its revenue largely on taxation of one kind or another. The differential impact of taxation on productive enterprises would in any circumstances affect the distribution of investment, but with international firms it is not only the distribution of investment among countries that is affected but also the distribution of the income arising from the investment that does take place. An integrated firm always has an incentive to adjust the international transfer prices of its subsidiaries and in other ways to take advantage of the intricacies of corporate organization to maximize its consolidated profit after taxes. Again, governments have always restricted imports, taxed exports, manipulated their foreign exchange markets and financial transactions to the detriment of other countries (and probably more often than not of their own economy as well), but the parent government of an international firm can in a very real sense reach across frontiers and directly intervene in the economy of another country.

Thus, the chief dangers of an international firm as a vehicle of international investment arises from the fact that it is forced to discriminate among countries, because of its legal status and because of the commercial and financial policies it is almost bound to pursue to minimize its tax payments. In consequence the head offices of international firms are inhibited in adopting a genuine international neutrality in their operations. The only fundamental solution lies in the development of a method by which international firms could be incorporated under international law and subject to a single international income tax. With international 'citizenship', they would be freed from the control of a home government; an international income tax would eliminate the significance for them of the fact that different countries impose income taxes at different rates. Formulae would have to be worked out for the distribution among countries of the proceeds of international taxation. The difficulties would be great, but not so great as any attempt to persuade all countries to adopt the same effective rates of corporate taxation.

Such a change in the status of international firms would not affect the economic distortions that arise from the exercise of monopoly power in particular markets, which would require separate consideration, but it would remove or reduce the types of economic discrimination discussed above. It might also pave the way for further 'internationalization' of their attitudes, of the composition of their management, and of their ownership. As a privately owned international organization, a firm might, for example, find it advantageous, both financially and from the point of view of public relations, to take special measures to facilitate the purchase for local currency of the shares of the parent company in the countries in which it has subsidiaries and thus widen internationally the flow of dividends to include all countries in which it makes its profits.

Developments such as these can be expected only in a far distant future, if ever. In the meantime, the large international firm will remain the chief instrument of direct foreign investment. And because such investment is probably profitable to the investing countries and is desired by large numbers of receiving countries, international firms will continue to play an important, and perhaps an increasingly important, part in international commercial relationships. Many of them are already making great efforts to ensure their international acceptability by joining in partnership with the businessmen of the countries in which they operate, or by giving their local subsidiaries a large measure of independence. But such steps often reduce the freedom of action of the international firms and, combined with government regulation, may lead to their elimination as international economic organizations capable of independent international planning.

Chapter XI

THE OIL INDUSTRY IN LATIN AMERICA

by

PETER R. ODELL

As in many other fields of economic endeavour, the attitudes and policies of Latin American countries towards the oil industry—and particularly towards the international firms concerned with it—have provided forerunners to attitudes and policies adopted by developing nations elsewhere in the world in more recent years. The more-than-a-century-old political independence of most of the twenty-odd Latin American nations placed them early in a position from which they could adopt nationally determined attitudes towards the international petroleum Companies. Thus, Argentina, Chile, Bolivia and other countries of Latin America were already reacting to the activities of these firms in the 1920s—through the establishment of state controls over oil and, sometimes, even of state companies with rights and responsibilities for organizing the supply and/or distribution of oil. By 1938, Mexico felt so strongly about the policies of the international firms that it expropriated all their assets—more than a decade before Iran took similar action. And as early as 1943 Venezuela re-negotiated its concessions agreements with the major international Companies and secured, within the framework of the new agreements, provisions favourable to the government of a type not achieved by other major producing nations until the late 1950s and early 1960s. The process of a nationalist reaction to the stimulus offered by the international oil Companies has been under way for a longer period of time in Latin America in comparison with other parts of the developing world.

In most of Europe (including the Soviet Union and most of the East European countries) and much of North America and Japan, coal has provided the basis for economic expansion in general and industrial expansion in particular. Oil has only gradually, and often recently, increased in importance in providing their base load energy requirements. In contrast, Latin America has, by and large, been an oil-consuming continent from the earliest days of its commercial demand for energy. The continent's annual coal consumption has risen little above the 10 million tons[1] it used in 1937, whereas the consumption of oil which then stood at 12 million tons[2] has risen steadily and in 1966 stood

[1] The equivalent of little more than 6 million tons of oil.
[2] UN, *Energy in Latin America*, 1957, p. 27.

at approximately 110 million tons.[1] A continually increasing supply of petroleum fuels has thus been a prerequisite for economic development, and it appears unlikely that the overall contribution of these fuels will fall much below its present level of about 75 per cent in the foreseeable future. This will be so, not only because of the limited availability of alternative fossil fuels (except for local supplies of natural gas which are, in any case, mainly associated with the production of oil), but also in light of the restraints on the expansion of hydro-electric and nuclear power capacity arising from the scarcity of capital in all the Latin American countries—except Venezuela which, however, has large reserves of oil and gas.

This dominant—and fundamental—role of oil in Latin American economic development thus sets up the industry as one of the commanding heights of the nations' economies. As might be expected in a region which derives much of its basic political philosophy from Western European socialist movements, this situation produces in itself an adequate political reason for governmental intervention. The fact that the Companies producing and/or supplying crude oil, owning and operating the refineries, and distributing and selling the petroleum products were foreign—and, in the main, from the US, with which Latin American politico-economic relationships have undergone a long period of strain—further enhanced the difficulties in the way of a continued acceptance of the oil industry as part and parcel of a relatively unhindered private sector.

These well-formulated attitudes to the international petroleum Companies have emerged from the fact that the Companies have been concerned with Latin American oil from the earliest days of the industry—Standard Oil of New Jersey, for example, was in business in Chile as early as 1913 and in Bolivia in 1920, whilst Shell secured an interest in Venezuela in 1913. Over this long period these Companies have achieved a widespread reputation (whether justified or not, it is not the author's intention to judge) of being 'agents of economic imperialism'. This reputation made the international oil Companies susceptible to attack from a wide spectrum of political groups within the Latin American countries, for they thus became involved in the major political issue of nationalism and its associated anti-Americanism of more recent years. The oil companies' involvement in this issue automatically implied a general concensus—amongst parties and groups deeply divided over many other issues—on the need to bring oil under 'effective national control'.[2]

THE CHILEAN CASE

A pragmatic evaluation by Chile's ruling groups of the economic and political realities arising from the presence in the country of the powerful international firms of Shell and Esso (Jersey Standard's subsidiary) formed—and still forms—the background to Chile's establishment and continuing exercise of national sovereignty over its oil sector. This evaluation may not have had the sophistication of the arguments on the nature of the international firm put for-

[1] British Petroleum, *Statistical Review of the World Oil Industry*, 1966.

[2] For a well-documented and detailed appraisal of the development of this political attitude in Latin America see H. S. Klein, 'American Oil Companies in Latin America; the Bolivian Experience', *Inter-American Economic Affairs*, Vol. 18, 1964, pp. 47–72.

ward earlier in this book by Professor Penrose, but it has had the effect of severely curtailing the freedom of action of the international companies within the Chilean economy. It also illustrates clearly the politico-economic philosophy underlying Latin American attitudes towards these Companies and hence is not an inappropriate country with which to start this study of the oil industry in Latin America.

Chilean action in this respect goes back to 1927 when, following some seismic and other geological work, Shell and Esso sought concessions to develop potential oil areas in Tierra del Fuego. Chile chose not to grant these concessions on the grounds that they would constitute an infringement of Chilean sovereignty. The Government was fearful that foreign enterprise in this area, very remote from the political control of Santiago and with its pastoral industries already largely based on British capital, might lead to a separatist tendency. Instead the Congress passed a monopoly law which had the effect of confining the development of the oil industry to the State. Since then a series of legislative measures have confirmed the exploration, production and refining functions of the industry in State hands. Throughout the 40 years since 1927, challenges to this fundamental attitude have been weak and even under the 'businessman's' government of Alessandri (1958–64) a proposal, in 1961, to break the State monopoly by opening up Chile's northernmost provinces to exploitation by foreign private Companies attracted fewer than one-third of the votes in Congress.

State attitudes to the industry have, over this period, gradually become more institutionalized. In the late 1920s and the 1930s a government department initiated a little exploration in the areas for which private concessions had been sought but it was not until 1939, when oil development was made the responsibility of La Corporación de Fomento de la Producción (CORFO)—a state development corporation—that serious attention was given to exploration efforts. CORFO's exploration finally bore fruit in 1945 with the discovery of the Manantiales field in Tierra del Fuego. The complexity of the industry's development led to the formation of La Empresa Nacional de Petróleo (ENAP) in 1950 with a legal monopoly to explore for, develop and refine oil in Chile—under legislation whose monopoly provisions appear to be unchallengeable (except by new legislation). This became clear when Esso made an abortive attempt to establish a joint refinery with ENAP in 1963. Since 1950 ENAP has in general terms and certainly in comparison with other State oil entities in Latin America, been accepted as a technical rather than a political organization in the 'oil world'. It employs drillers under contract, calls in foreign advisers when necessary and has to be self-financing by meeting its investment needs out of profits on sales. In these respects it does, of course, behave in much the same way as a private company might do but the president of ENAP, Fernando Salas, has indicated clearly[1] not only the bases on which State ownership is considered to be advantageous to Chile but also the ways in which this works out in practice.

Geologically, Salas argues, Chile is a country poor in oil opportunities with only very small areas of sedimentary basins compared with Venezuela and Middle Eastern countries. The prospects, therefore, of Chile becoming an oil

[1] In conversation with the author, March 1964.

exporting country are remote and any production can only be for local consumption. If this were to be developed by the international Companies, Salas continues, their Chilean operations would be a very small percentage of their total operations and in times of price weakness their marginal costs in the main producing areas would almost certainly be much below their marginal costs in Chile such that there would be a natural inclination of their part to import rather than to produce locally. In contrast, Salas indicates, with a State monopoly its first consideration in such circumstances is the overall effect of production or imports on the Chilean economy. For a state entity, Salas suggests, it is unlikely that the cost of producing the additional barrel would exceed the price at which oil could be imported. Even if it did, a small discrepancy could still be justified because the saving effected in foreign exchange would mean its availability for use in another sector of the economy where its value might be greater in terms of its potential contribution to national well-being.

Based on this line of argument, ENAP has since 1950 gradually expanded its producing operations in the extreme south of Chile substituting domestic oil for products previously imported from the Caribbean and Peru. Production rose gradually to just over 2 million tons by 1964[1]—sufficient to meet 66 per cent of the country's requirements. Maintenance of import substitution in a market growing at the rate of 6 per cent per annum demands, of course, continuing investment in exploration and development. It was with this in mind that ENAP turned its attention to the desert north of Chile. It did this, however, in the full knowledge that the geological prospects were not particularly attractive—and certainly in the belief that they were not attractive enough to persuade foreign Companies to do the work. From ENAP's point of view, however, the risk was considered to be economically justified—for its need of new crude greatly exceeded the need of any foreign company. In addition, Salas' view was that a State monopoly had a social reason for undertaking the work; namely as a contribution to an area in which other opportunities for economic development do not exist under present conditions.

In this explanation by ENAP's President of his organization's motives and way of working lies the essence of the nationalistic argument against allowing the international Companies to dominate the oil industries of economically weak countries such as Chile. Paradoxically it has had little to do with securing oil products at prices lower than those which would necessarily apply if the international Companies were responsible for the country's oil supply. ENAP, in fact, makes use of the Caribbean quoted price system for valuing and selling its own products—and out of its allegiance to such a system ensures, of course, its own profitability. Prices of products ex-refinery in Chile are based on the 'low of Platts' in the Caribbean plus AFRA freight rates from the Caribbean to Chile plus whatever import duties would have been charged had the products been imported. The inclusion of a tariff component presupposes that a tariff would have been charged—an assumption which is probably only valid if there is domestic production to protect. Had there been no development of Chile's indigenous oil resources, then it seems highly unlikely, for

[1] Since 1964, however, production has declined, and in 1966 domestic crude met only 54 per cent of total requirement.

example, that a 70 per cent tariff on the c.i.f. price of fuel oil could have been sustained in a period when the country was attempting to industrialize. In this case fuel oil imported even at Caribbean posted prices through subsidiaries of the international Companies would have been available at less than the ENAP administered price—to the benefit of the country's economy. The validity of this sort of argument has apparently been somewhat belatedly recognized by ENAP which took a decision in 1963 to price products with a notional tariff component at less than the price permitted by the formula. Thus, by 1965 ex-refinery prices included only 75 per cent of the notional tariff component and there were hopes that this would be further reduced. Even so, most product prices remained high.

But there are still other price complications emerging from Chile's attempt to maintain 'sovereignty' over its oil sector in an oil world dominated by the major international Companies. As already indicated, as indigenous crude production and refinery developments have made locally produced products available, the local marketing companies have been obliged to switch to these away from their traditional—and, no doubt, preferred—Caribbean sources. To this extent imports have been substituted and foreign exchange has thus been saved. This requirement, however, has never applied to certain major oil consumers such as the copper, nitrate and iron-ore mining companies, the international airlines and the steel company. These consumers have not only enjoyed the right to import but also the right to negotiate internationally for their supplies, which would have been brought directly into their own storage facilities. As the world oil market weakened they enjoyed the advantage of falling prices—but, of course, they have to face the tariff barriers designed to protect local production and have thus had their costs increased above the levels that might have been operative in the absence of domestic production. It is also the declared intention to ensure, if local supplies of products increase either from domestic or imported crude, that these import privileges are withdrawn. This will produce the likelihood of somewhat higher product prices for these consumers—among which are Chile's main export earners, whose international competitiveness may thus be adversely affected.

A final complication in relations between the state and the international oil Companies in Chile is introduced in the local marketing sector of the industry, for even in this sector state intervention was considered desirable to end the dominance of Shell and Esso whose former joint control of the market made for oligopoly and excessive profits. Thus, in 1935, with Government assistance, a local Chilean marketing company, COPEC, was formed to offer a degree of competition to the foreign Companies. COPEC, however, was unable to compete with the established marketers and the Government again had to step in to help in 1942. It introduced a quota system which effectively guaranteed 50 per cent of the market to COPEC—and the other half to Shell and Esso. This system, coupled with Government price controls on most products, persists and has achieved its objective of 'Chileanizing' a large part of the distribution system. The 'cost' of it lies in the absence of competition which would otherwise have led to lower prices for consumers and also in the fact that the two international Companies concerned are more or less guaranteed appropriate profits. Such a system has the inherent possibility of excessive distribution and retailing margins—in the very part of the oil business where the international Companies have had the greatest propensity to compete with each other to the advantage of consumers.

Thus, Chilean policy represents a comprehensive attempt by a small oil-consuming nation—and one whose economy in general is relatively weak and unstable—to secure its 'independence' from the international oil industry. It has achieved this political objective quite effectively—but with room for some differences of opinion on its economic effects. The protagonists of ENAP's activities argue that the costs incurred in developing local production and refining and in establishing a marketing system with a Chilean component amount to no more than the total payments that would have had to be made to the international Companies as they sold products out of the Caribbean to a Chilean market held tightly within the framework of their posted price system with the opportunity that it gave to them to earn monopoly rents. It is further argued that the costs of notional tariff protection—designed to provide ENAP with its profits—must be set against the benefits arising from the employment in the oil industry of Chilean resources, particularly labour, which would otherwise have remained unemployed. The claims on neither of these counts have been substantiated, however, and for the moment it is possible only to conclude that even in a world of imperfect competition in oil, Chile may or may not have secured economic advantages from its declared policy of independence from the international oil Companies.

STATE REFINING IN URUGUAY

It is impossible to be any more confident about a net national gain from State intervention in the oil industry in any other Latin American country—though one can have confidence in such a result in certain countries at particular periods. In recent years, for example, Uruguay has achieved such a net gain through its national ownership of the country's oil refining industry—as part of the Administración Nacional de Combustibles, Alcohol y Portland (ANCAP), which was originally formed in 1931. This gain has arisen since 1961 in the period of 'surplus' crude oil seeking outlets in non-integrated refineries. Before that date crude oil moved to Uruguay at near posted prices. However, in negotiations since then, ANCAP has been able to secure its crude oil requirements at prices well below those posted in the producing areas and paying only a freight rate component reflecting the discounts available off intascale rates due to surplus tanker tonnage. In 1961 a four-year agreement was signed between ANCAP and Gulf Oil whereby the latter agreed to deliver the former's requirements to Montevideo at a landed price of \$2·05 per barrel. With a posted price of \$1·59 per barrel for this Kuwait crude and an AFRA rate for freight of \$1·05 per barrel from the Persian Gulf to the River Plate, a transfer between associates of the same international Company at Montevideo would have taken place at \$2·64 per barrel. Uruguay thus saved over 50 cents per barrel—a discount of some 20 per cent. This rate of saving was, moreover, applied to all Uruguay's imports of crude oil under an arrangement whereby the marketing companies in Uruguay (Esso, Shell and Texaco), supplying crude to the ANCAP refinery in relation to their market share, agreed to bring in their supplies at prices equivalent to those obtained by ANCAP under its own long-term contracts. Thus, the foreign exchange saving on oil amounted to about 3 per cent of the country's total import bill over the whole four-year period.

ANCAP was in the market again for crude oil in 1965. By then it found the situation even more advantageous and from a large number of bids finally accepted the offer made by British Petroleum. BP not only agreed to supply ANCAP with Nigerian and Middle Eastern oil for three years at a delivered price of only $2·00 per barrel (thus giving ANCAP an even larger discount than in 1961) but also offered to loan the State company $5 million for refinery expansion at a rate of interest of merely 6 per cent per annum. These financing arrangements were very advantageous to Uruguay, which, given its persistent economic difficulties, would have had very great difficulty in securing ANCAP's capital needs on the open market even at higher rates of interest. They were, in effect, tantamount to an additional discount of some 2 to 3 cents per barrel on the nineteen million barrels of crude oil covered by the contract. Uruguay's success in securing its oil supplies since 1961 at discounts of some 20 to 25 per cent off the posted prices and AFRA rates at which transfers between the international majors and their Uruguayan affiliates would otherwise have taken place is, of course, a function of the availability of 'surplus' crude oil and tankers. This has given Uruguay the bargaining power to buy at or near the supply price of the best placed company. Should either or both of the 'surpluses' disappear then Uruguay will inevitably go back to the pre-1961 situation and find its import prices rising significantly. In the absence of any local production and/or any Uruguayan commitment to explore for and produce oil overseas there is nothing that it can do to safeguard against this possible development: but in this respect it is no worse off than many of the much larger oil-consuming nations of Western Europe.

BRAZIL: 'O PETRÓLEO E NOSSO'

On a very much larger scale Brazil has had the opportunity to pursue a Uruguayan-type policy in that since 1950 almost all refinery expansion[1] has been restricted to the public sector, and in that since 1964 all imports—including those for the private refineries—have been the responsibility of the State oil company—Petróbras (Petróleo Brasileiro).

Until 1964, the private refineries in Brazil were supplied with their crude-oil requirements by the international Companies (Shell, Esso and Gulf) which had distributing and marketing facilities in the country. Transfers thus tended to take place at or near the posted prices ex-Caribbean or the Middle East plus a freight component reflecting AFRA rates. Even as late as 1963 this oil, moving essentially within the framework of the international Companies, was landed in Brazil at an average price per barrel of $2·48: in contrast, Petróbras, responsible for all other imports, was by then paying an average of $2·21 per barrel for its crude-oil supplies. The difference of 27 cents—over 12 per cent—

[1] Since 1950 approximately 300,000 b/d (=15 million tons per annum) of refining capacity has been built in the public sector. Privately-owned refineries have been authorized to build 50,000 b/d of new capacity only.

gives a quantitative indication of the cost[1] of having refineries associated in any way with the international oil groups—in this case through formal supply arrangements rather than by ownership. Thus, in 1963, a Brazilian Congressional Inquiry recommended that Petróbras be given a complete monopoly over all imports. The recommendation came into effect early in 1964 with the intention of effecting savings in the foreign exchange costs of imports through the elimination of the differential cost between 'public' and 'private' oil as noted above. *Petroleum Press Service* expressed the opinion at that time that 'the savings achieved are not likely to be very important' and put forward a defence of the price differential—that it was due to 'the private refineries' needs consisting largely of speciality crudes in relatively small quantities'.[2]

But by 1966 the cost of importing a barrel of oil to the private refineries was only $1·95[3]—a fall of over 50 cents per barrel in only three years. In 1966 crude oil was brought into the privately-owned refineries at an almost identical cost as the crude moving to Petróbras' refineries (average cost per barrel—$1·96). Part of the fall can, without doubt, be attributed to savings arising from economies in operations as the smaller private refineries could, in part, be served within the framework of the larger-scale transport operations required for the State-owned refineries. But most of the price reduction must be attributed to the effective change in status accorded to the private refineries once Petróbras became responsible for their crude-oil supply. Petróbras was able to negotiate unequivocably on their behalf with possible suppliers as a third party and, in effect, the refineries formally achieved independent status. The effect of 'shopping around' for the cheapest supplies is reflected not only in the low delivered cost per barrel of crude oil—which at a $1·95 shows a saving of around 25 per cent off the formally posted prices and assessed freight rates—but also in the proliferation of supply points and of suppliers. In 1962 about 70 per cent of Brazil's oil requirements still came from Venezuela: in 1966 Venezuelan oil accounted for only 27·5 per cent of the total: the USSR supplied 20 per cent; Saudi Arabia over 18 per cent; Iraq almost 18 per cent; Kuwait nearly 11 per cent and Nigeria over 3 per cent. This spread of supply points by 1966 gives a clear demonstration of the battle for 'third-party' business then under way—even amongst the major international Companies, which control supplies from all the sources indicated above with the exception of the USSR. National refineries—of the importance of Brazil's—are obviously a desired market even at the heavily discounted prices required to obtain the business.

But it is not enough to estimate that Brazil is now achieving a gross annual foreign exchange saving of the order of $40 to $50 million (equal to 3–4 per cent of its present annual availability of foreign exchange) as a result of her national refining policy. There is, first of all, the offset arising from the foreign exchange cost of building the refineries in the first place—this can be estimated

[1] The net cost in foreign exchange is, of course, less than this gross figure because the building of state refineries involved the use of foreign currency whereas private refineries could be financed abroad. See below p. 282.

[2] *Petroleum Press Service*, March 1964. p. 95.

[3] The contracts signed for 1966–7 appear to indicate a further fall to an average of $1·75 per barrel.

at $250 million in the absence of domestic refinery equipment manufacturing industry. This cost would not, of course, have arisen directly had the international Companies been permitted to undertake the required expansion. Apart from this consideration, however, one cannot accept that the only other way in which Brazil could have met her oil requirements was by having the international oil groups with marketing interests in Brazil import crude oil for processing in refineries built by themselves and operated as part of their international networks. There are other alternatives—with contrasting foreign exchange implications.

On the one hand, the Companies might, if left to themselves, have built less refinery capacity in Brazil and have continued to bring in refined products at posted prices from the Caribbean. This was the situation in a number of Latin American countries where neither political nor commercial pressure was put on them to build refineries until recent years—as, for example, in Central America until the 1960s.[1] Such a policy might well have been favourable to their internationally integrated profits by ensuring the optimum use of the large resource-orientated export refineries of Venezuela, Trinidad and the Dutch West Indies,[2] but it would also have produced a situation in which Brazil's oil requirements would have cost up to $400 million a year in foreign exchange[3] compared with less than $200 million in 1966, given national refining.

On the other hand, the international Companies might have made such a success of local exploration for and development of indigenous Brazilian oil resources, given the freedom to do so, that the import component in the total oil requirement would have fallen much below its present level of approximately two-thirds—and reduced the foreign exchange cost of oil merely to that incurred in the overseas remission of profits—say, a maximum of some $50 million per year. This alternative possibility of giving the international petroleum Companies the responsibilities for the domestic production of crude oil in developing countries is one which has found much favour with both international and US aid and loan-giving agencies. The view, as Professor Penrose has shown earlier in this book,[4] is predicated on the assumption that highly risky investment—as oil exploration undoubtedly is—is an inappropriate use of scarce national, and particularly public, capital which could be better employed in another sector of the economy where international private funds are not available.

W. J. Levy, in his study made for the World Bank,[5] illustrates this thesis with particular reference to Latin American nations, including Brazil. His analysis of the cost to the Brazilian exchequer—which provided about three-

[1] See P. R. Odell, *An Economic Geography of Oil*, Bell, 1963, pp. 133–9, for an examination of the Central American situation.

[2] See P. R. Odell, 'The Refining Industry of the Caribbean and the Middle East', *Tijdschrift voor Economische en Sociale Geografie*, Vol. 54, 1963, pp. 208–13 for details of the major companies' refining interests and policies in the Caribbean.

[3] An estimate—using 1966 Caribbean product prices—of the foreign exchange cost to Brazil of importing the products that it would have required after allowing for the availability of products manufactured from indigenous crudes.

[4] See Chapter VIII, pp. 238–40.

[5] W. J. Levy, *The Search for Oil in Developing Countries*, 1960.

fifths of Petróbras' total available funds between 1953 and 1959 either directly or indirectly[1]—appears to be a formidable indictment of 'Statism' in this field. However, as already indicated in the case of Chile, state involvement in oil exploration and production can be justified both conceptually and in practice, and what Levy does is to provide evidence of the non-technical and non-commercial character of Petróbras whose essentially political role arises from the impact of the history of state intervention in oil in Brazil. The effect of this cannot, of course, be ignored in an economic appraisal of the country's oil industry.

Ironical though it may now seem, early public interest in oil in Brazil arose from the lack of interest by the international Companies in exploring the country's oil resources—for these Companies had little incentive to spend money on very risky Brazilian exploration in difficult physical conditions when the market could be easily served by the rapidly expanding crude-oil production of Venezuela and the new refineries of Aruba and Curaçao. As a reaction against this attitude by the Companies the Brazilian Government, in 1938, declared petroleum to be a public utility and placed it under the direction of a National Petroleum Council. The Council, however, took little effective action to develop the country's oil resources and by the late 1940s the international Companies—now under pressure from left-wing political forces in Venezuela and therefore anxious to diversify their sources of supply—were showing a keener interest in the possibilities of exploration in Brazil, where the demand for oil was now growing rapidly (at a rate exceeding 12 per cent per annum) under the stimulus of officially sponsored industrialization. They were able to point out with justification how little had been done to initiate oil production by the Brazilian Petroleum Council (production in 1946 was only 200 barrels a day), and a Bill to permit foreign participation in the search for crude oil was introduced into the Congress. It was not passed, however, and the issue became a central one in the Brazilian Presidential election of 1950. Getulio Vargas, President of Brazil from 1930–45, sought re-election under the slogan *O Petróleo e Nosso* ('Petroleum is Ours') and was successful. He immediately proceeded to put through comprehensive oil legislation which not only greatly widened the powers of the National Petroleum Council but also formed Petróleo Brasileiro (Petróbras) with monopoly rights for exploration, production and all new refining developments. Since 1950 there has been no serious political effort to undermine the authority of Petróbras, which has instead been given increasing powers and additional responsibilities.

This attitude persists in spite of Petróbras' failure to find and produce oil in the quantity demanded by the rapidly expanding Brazilian economy and in spite of a series of financial and organizational scandals. Its failures to make effective use of the funds with which it has been endowed have, however, remained in the eyes of most Brazilians less important than its successes in developing the Bahia fields and, more recently, in finding new fields in Sergipe and Alagôas. Its success here has been particularly significant in light of the adverse geological report made in 1960 on the possibilities of new oil developments in Brazil by Mr W. Link—an eminent US oil geologist and adviser to Petróbras from 1954. His report on the prospects for new oil discoveries in

[1] W. J. Levy, Chapter III, p. 15.

Brazil was, in fact, so pessimistic that nationalistic opinion accused him of being in league with the international Companies—which, following his report, were presumed to aim to step in to take over from a finally disillusioned Petróbras! In fact, Link's report seems to have so impressed the Companies in the other direction that they gave thanks that Brazilian nationalism had kept them out of such unpromising territory. Thereafter, pressure to have Brazil 'opened up' for international oil investment evaporated—and had Petróbras not already existed in 1960 it would probably have had to be formed in order to ensure that any search for indigenous oil resources continued. Thus, in light of the short period of only about a decade in which the international oil Companies had much interest in developing Brazilian oil resources it is difficult to envisage a situation in which they could have produced more domestic oil than Petróbras—and hence taken import substitution to a greater degree than has in fact occurred. Whether the Brazilian economic infrastructure or the country's social services would have benefited from investment which in the event went to Petróbras; or whether Brazilian inflation might have been contained to some degree, with consequent overall economic advantage, had Petróbras not had to be financed—these remain unanswered, and perhaps unanswerable, questions. Unless and until they are answered the net costs or benefits to Brazil of its persistent nationalistic policy towards oil exploration and production, as well as to refining, cannot be assessed.

ARGENTINA: THE 'BATTLE' FOR OIL

By way of contrast, the vacillations in Government policy towards oil in Argentina—ranging over time from ultra-nationalism on the one hand to possibly excessive freedom for foreign private Companies on the other—appear almost to have produced the worst of both approaches and, in particular, to have caused Argentina to suffer significant and long-term foreign exchange losses that need not have occurred under a more rational approach to the industry. It was as long ago as 1910 that Argentina first took steps to develop oil production through a State entity with the formation in that year of the *Dirección General de Explotación del Petróleo en Comodoro Rivadavia.*[1] At that time, however, foreign companies—notably Standard Oil—retained their rights to explore for and develop the country's resources. But in 1927 the State intervened on a much greater scale with the formation of a national petroleum authority—Yacimientos Petroliferos Fiscales (YPF) which was given monopoly rights to all future petroleum exploitation and production. For thirty years both YPF and those international Companies which had developed concessions before 1927 were responsible for the expansion of the country's industry. Production, however, was unable to keep pace with the rapidly rising demand occasioned by the country's policy of industrialization from the mid-1930s onwards, and particularly under President Perón from the end of the Second World War. The international Companies were, of course, effectively prevented from any expansion by their inability to work any other than their existing fields, whilst YPF was unable to push production above 5 million tons even by 1957 (compared with almost 2·5 million tons in 1943 and

[1] A. Frondizi: *Politica y Petróleo*, Buenos Aires 1953, p. 50.

3 million tons in 1948) as a result of serious political intervention in its activities. This intervention seriously affected its effectiveness as an organization and inhibited it from raising funds for development work. Rigid price control over petroleum products prevented it from self-financing its operations and the Government generally had other calls on national revenue which were given priority over the requirements of the oil industry.

Argentina's long-standing intensely nationalistic attitude towards the oil industry[1]—and its particular hostility to the international oil companies—was well demonstrated in the political reaction in 1956 to Perón's belated recognition of the need to change the basis of the country's petroleum policy. He opened negotiations with an American oil Company for the possible development of certain areas of more-or-less-proven reserves. The negotiations, however, became known to the public and were used to arouse popular opinion against him, contributing a few months later to the termination of his twelve years old dictatorship. But the country's economy was by 1957 in such a parlous state that some action had to be taken to lighten the economic burden of the nationalistic oil policy. Domestic oil production was now providing less than 33 per cent of total national requirements and the remaining requirement, available in general only at, or very near, internationally posted prices and assessed freight rates, was costing the country about 25 per cent of its annual availability of foreign exchange of about $1,000 million. YPF had embarked on a major period of refinery expansion so that foreign crude—available at a lower foreign exchange cost—could be substituted for more expensive products. This, however, diverted YPF's available capital from the development of crude-oil reserves which it had discovered earlier and thus, even at a time of economic crisis, Argentina's proven oil reserves remained in the ground. The election of Arturo Frondizi to the Presidency in 1958 did not appear to auger well for any fundamental change, for his book *Politica y Petróleo* provided the classic case for Argentina's nationalistic attitude towards oil and, moreover, his election campaign had been based, in part, on a promise to maintain these nationalist attitudes and policies which reserved petroleum development to YPF. However, in spite of intense political opposition, President Frondizi recognized the essential illogicality of the petroleum situation and speedily signed exploration and production contracts with various international and US oil Companies.[2] These quickly moved in to develop the proven areas and also to explore for oil in other parts of the country.

Frondizi gave clear evidence that, as far as oil was concerned, he was switching his policy from political to economic nationalism—a switch which necessitated not only a change of attitude within the Argentine establishment but also a re-evaluation of their position by the international oil Companies already operating in the country. These Companies continued to produce a little oil (about 0·7 million tons in 1956) from their long-established concession areas but, more profitably, were still the main participants in the refining and marketing functions of the country's oil industry. Here they were able to use

[1] See, for example, A. Frondizi, op. cit. and also M. Kaplan, *Economia y Politica del Petróleo Argentino* 1939–56, Buenos Aires, 1960.

[2] See *Petroleum Press Service*, April 1962, pp. 133–5, and December 1963, pp. 448–52, for details of the different kinds of contracts signed.

mainly crude oil and products from their international operations. Price controls on products inside Argentina had not worried them a great deal for prices had been fixed in relation to the high costs of YPF's domestically produced oil and to import costs related to Caribbean posted prices. The absence of generous margins to cover marketing costs—particularly in a situation of continuing inflation—was of minor significance compared with the international profits which accrued to the parent companies from the transfer of crude oil and oil products to Argentine associates at Caribbean posted prices. Frondizi's new policy would, of course, if successful, have eliminated the profits going to the international Companies, for domestic production would automatically be given precedence over imports such that the refining and marketing operations of Shell and Esso—barely profitable in themselves because of price controls—could no longer be justified on the grounds that they enabled profits to be made on crude oil produced elsewhere. This strong possibility was sufficient to persuade Shell and Esso to accept commitments for additional capital expenditure on production in Argentina in spite of the virtual certainty that costs would exceed those in the major producing areas of the world. However, the use of their own domestically produced crude oil gave a greater chance of profitable operations than did the use of domestic oil produced by other Companies and, hence, both Shell and Esso joined the search for Argentine oil in 1959.

At this stage, therefore, Argentina adopted a policy out of the main stream of oil industry developments in Latin America—with the pendulum swinging strongly back in favour of oil development by private foreign Companies including some of the international Companies. There were, of course, political safeguards, in light of the long history of Argentine nationalism over this issue, with Frondizi insisting that this nationalism was not compromised because ownership of the oil remained in national hands with YPF paying 'fees' to the contracting Companies as they made available to YPF the oil which they produced. Frondizi sets out a persuasive defence of this policy in his book *Petróleo y Nación*—published after his fall from power in 1962—a fall due in some part to his petroleum policy, which appears not to have been acceptable to the Argentine nation.

The significance of Frondizi's petroleum policy lies only partly in the fact that it proved to be politically unacceptable in the country. It lies even less in the fact that technically—and, in part, economically—it proved to be a great success. The investment of foreign capital and know-how in the known oil areas of Argentina produced an immediate return in the form of a rapidly rising level of production. Output increased from 5 million tons in 1957 to approximately 15 million tons by 1963, whilst imports, which cost Argentina over $271 million in 1957, were all but eliminated within five years. Moreover, the development of oil produced a bonus in the form of a greatly increased output of natural gas, for which pipelines were constructed to the main consuming centres of the country. The only question mark hanging over the success arises from the nature of the contracts which were signed. In retrospect there is some evidence that they were exceedingly generous, particularly to those Companies which were able to secure production contracts for the proven areas and, virtually without risk, produce oil for which there was both a guaranteed market and a guaranteed price. Although the uncontrolled remission of profits by these Companies constituted a drain on the balance of payments, it was nevertheless short of the drain created by rising imports at an earlier period. The net overall effect of the contractual arrangements on the economy of Argentina, however, remains another unanswered question on state/international oil Company relations in Latin America.

The real significance of the Frondizi contracts lies in the reaction of the international oil and business world to them. In this milieu they were seen as the desirable result of international pressure designed to achieve a 'more sensible' attitude on the part of developing nations towards oil investment and development. W. J. Levy's study for the World Bank had provided the documented evidence needed for challenging the nationalistic policies of Latin American countries towards their oil industries. The insistence of these countries on financing their own oil developments was considered to be entirely illogical, creating balance of payments and monetary difficulties such that the IMF could bring pressure to bear. The World Bank noted that international private capital was available for petroleum developments and thus not only refused to give loans itself for this purpose but took a long second look before giving loans for other purposes to countries which still persisted in using national resources for petroleum exploration. The US Government also refused to make grants or loans to nationalized oil companies[1] and appears to have played its part in trying to persuade developing countries to change their attitudes towards the oil industry. International banking houses, too, commented adversely on nationalistic excesses.

In the light of all this, it seems evident that Frondizi was under considerable external pressure to adjust Argentine oil policies. When he did so the 'international' reaction was tremendous and the contractual arrangements which he agreed were 'hailed' as introducing a new, viable and acceptable form of cooperative enterprise between international oil capital and the developing nations which insisted on maintaining 'sovereignty' over their oil resources. Argentine-type contracts were suggested as possible prototypes for other countries in Latin America—including Chile and Brazil and even for Mexico where, as shown below, oil had been under complete State ownership since 1938. On the other side of the world the possibility of Argentine-type contracts was discussed with reference to Indonesia where Government and Companies were in more-or-less continuous disagreement.

The early and outstanding technical success of the contracts in quickly producing additional oil encouraged the euphoric atmosphere and Argentina—in wider development circles than merely oil—became the nation 'chosen' to demonstrate the ability of international capitalism to secure the growth in wealth of the developing nations of the world. Thus, *Petroleum Press Service* in 1960 ended a survey of the situation in Argentina in the following terms:

'Despite considerable *initial* criticism . . . the present Argentine Government's bold petroleum policy is being vindicated by results . . . the increasing savings in foreign exchange thus being effected will help Argentina to

[1] See P. R. Odell, 'Oil and State in Latin America', *International Affairs*, Vol. 40, 1964, pp. 669–71 for a discussion of US attitudes.

regain the economic prosperity to which, until 15 years ago, she had been accustomed.'[1]

In the four years up to 1962 the 'trend away from nationalism'[2] led to the investment of $400 million of new funds in Argentina, mainly in new industries such as chemicals, motor vehicles and steel. Argentina was obviously on its economic way—the process triggered off by the changed attitude to oil. Unfortunately, however, the Argentine people were not convinced that their economic salvation lay in this direction and a few months later Frondizi was forced out of office. The following year—in March 1963—the new Government of President Illia declared all the contracts signed between YPF andthe petroleum Companies 'null and void'. The Government further indicated its intention to seek legal redress against the oil Companies for the alleged harm which they had done to the country's economy. *Petroleum Press Service* entitled its article on the development 'Argentina Turns Back the Clock'[3]—and indeed it had, for the oil system reverted to State enterprise.

Since 1963 YPF has managed to keep production moving slightly ahead, but the balance which was achieved between domestic production and domestic demand by 1963 has disappeared as the latter has outrun the former—and in both 1965 and 1966 imports of oil ran at about $100 million per year. In June 1967, however, with a new, more right-wing Government in office, a new oil law was passed. This appears to provide an adequate framework which will again enable foreign Companies to participate in Argentina's oil industry which YPF seems inherently unable to develop on a scale large enough to satisfy the country's energy requirements.[4]

NATIONAL OWNERSHIP IN MEXICO

The post-war vicissitudes of the petroleum industry in Argentina have done much to vindicate the point of view of people like Levy that exploration and development should be left to the international oil Companies which, given an adequate opportunity to run a profitable operation, will do their best to produce the goods out of which their profits will arise. These arguments, applied with validity to Argentina, have also been applied to Mexico where the private companies were nationalized in 1938 within the framework of a revolution which the companies were, at that time, quite unwilling to accept. Their unwillingness to re-negotiate their concession arrangements and their largely extra-territorial and statutory positions—so extreme that some Companies

[1] *Petroleum Press Service*, October 1960, p. 380.

[2] Ibid., April 1962, p. 136.

[3] Ibid., December 1963, pp. 448–52.

[4] The new law was passed at the end of June. It goes further than the Frondizi law in providing opportunities for foreign Companies to participate in the oil industry. Although YPF retains its present production areas (370,000 sq. km.), work in them can be contracted out to private companies. In addition over 800,000 sq. km. are made available for new concession agreements between Government and companies. The latter will, if their searches prove successful, have tax obligations only to the Government rather than the complex contractual arrangements with YPF demanded of the companies under Frondizi's legislation. The new legislation thus runs very strongly against the trend of increasingly restrictive attitudes towards the international oil Companies in Latin America.

had what amounted to private armies for protecting their lands—probably made nationalization inevitable.[1]

The nationalization of Mexican oil provides Latin America's—and possibly the world's—outstanding example of action by a poor, undeveloped nation against what in 1938 was termed the international petroleum cartel. Politically it showed that the power of this 'cartel' could be contained by resolute action in which, in the final analysis, the companies would have to acquiesce unless they could persuade the US and Britain to intervene with physical force. Technically, it has demonstrated that a State oil enterprise, responsible for all aspects of oil operations from exploration through to final sales can, in the long run, operate an oil industry. And economically, it has indicated how a nationally owned company can be organized and controlled by the State to meet what are considered to be required national objectives—which in Mexico's case have been increasing supplies of energy at low prices to the consumer.

PEMEX—Petróleos Mexicanos—operating just next door to the militant private enterprise Texan oil industry has, throughout its 30-year history, been subjected to close and hyper-critical observation from spokesmen with an attitude similar to that expressed by the proverbial witness of the dog walking on two legs: it is not surprising that it does not do it well—it is surprising that it does it at all! W. J. Levy—a more objective student of PEMEX than most of his fellow American oilmen—offered documented and constructive criticism of PEMEX's activities in his report to the World Bank in 1960. To set against this study, however, is the official 'success-against-all-odds' story best portrayed in the recent book[2] by A. Bermúdez who was Director-General of PEMEX from 1947 to 1958.

The two main points at issue in an evaluation of PEMEX's activities are, firstly, whether or not it has expanded the industry to an appropriate degree so as to maximize the benefit to Mexico and, secondly, its ability to generate and use funds for the development of the Mexican oil industry. The two issues are, of course, not unrelated for the former is, in part, a function of the latter. In dealing with this latter point first one notes that in recent years PEMEX appears to have enjoyed a status which makes it worthy of support by foreign private and public capital—such financial 'respectability' is confirmed in its A1 credit rating in international money markets and its ability to negotiate long-term loans at favourable rates of interest. Such an attitude appears to confirm its success, but what should also be noted is the length of time which has been required to achieve this success—it cannot be dated earlier than 1961—almost 25 years after the formation of PEMEX. It is a success, moreover, achieved partly as a result of PEMEX's belated domestic victory in gaining the right to charge reasonable prices for its products—that is, with the right to act much more as a commercial enterprise rather than as an instrument of the State, forced to allocate its products at less than average costs. In brief, the growing ability of PEMEX to obtain external (i.e. non-Mexican Government)

[1] See P. R. Odell, op. cit., p. 660, and R. Vernon, *The Dilemma of Mexico's Development*, Harvard University Press, 1963.

[2] A. J. Bermúdez, *The Mexican National Petroleum Industry: A Case Study in Nationalization*, Stanford University Press, 1963.

credit has gone hand in hand with its ability to generate investment funds out of sales income. It now seems to be set on a course which should enable it to finance its developments in a manner identical with that of the international oil Companies—which, as Professor Penrose has demonstrated (see pp. 36 f.), consists mainly of charging consumers prices high enough to generate funds for investment, but supplemented by recourse from time to time to external sources of funds.

The other issue concerns the level of activities and degree of expansion undertaken by PEMEX. In general terms the defenders of PEMEX argue that the organization has done whatever needed to be done to meet the country's required energy supplies—in a period of thirty years when the economy has enjoyed almost continuous growth and a rapidly increasing demand for energy. In general one would have to concede the validity of this claim, for oil and gas have been made available almost as required. For the first decade after 1937, however, this was achieved through the use in the growing domestic market of that part of the country's production which had previously been exported. Exports in 1937 amounted to half of total production—Mexico lost most of these markets, whether she wished to or not, because the 'international oil cartel' instituted what amounted to a virtual embargo on them. But this action, though it harmed Mexico's balance of payments, did at least make productive capacity available to meet growing domestic requirements when PEMEX had neither the funds nor the technical expertise to increase output. This gave PEMEX a breathing space on the exploration and development side while it organized and expanded its national distribution and marketing system—which in the short term had to be based on refining capacity which had been originally located to serve export markets.[1] By 1947 domestic demand had expanded to a level (approximately 6·5 million tons) where it was approaching the country's total productive capacity of about 7 million tons—still only a little higher than in 1937 before nationalization. Thereafter, partly as a result of contracts with foreign firms, production and reserves were quickly increased and by 1952, the latter had been almost doubled and the former increased to 10·5 million tons. The increase in production was accompanied by an expansion of refineries, orientated in part to the major consuming area of Mexico City, and the programme of expansion was extensively augmented in the next 5 years. By 1957 PEMEX had the logistical problem of meeting domestic needs well under control and in the last decade expansion of facilities—from production through refining to transport, storage and distribution—has more than kept abreast of requirements, even at the low prices charged for petroleum products for most of the period.

The charge, then, that PEMEX has failed to satisfy the country's energy requirements appears to lack validity. In fact, the weight of the evidence points in the other direction—that through the existence of PEMEX, and the Government's control over it, the country's energy requirements have been met probably as effectively and certainly at lower direct cost to the consumer than through any possible alternative. This conclusion seems to be the only one that can be reached for it is difficult to see how private foreign Companies

[1] W. J. Levy, Chapter III, p. 41.

would have been able to sell oil products in Mexico at the prices at which PEMEX has had to sell; or that they would have been prepared to meet the requirements for sales at sub-standard prices to bus companies and taxis in Mexico City; or that they would have accepted the social obligations to sell products at all and any locations in the country at other than appropriately higher prices—a situation which would have inhibited the development of Mexico's particularly backward areas. PEMEX has not necessarily liked having to do these things—but it had no option within the framework of Government controls and its role as an instrument of governmental developmental policy. Foreign private Companies would have had more freedom to argue for commercially acceptable pricing policies and they would probably have persuaded the Government to accept their views—particularly as they would have had great bargaining power in light of the significance of petroleum in the country's energy economy.

The offsetting advantages that the international Companies could have offered—such as a higher overall level of efficiency, smarter service stations, superior technical advice to consumers of petroleum products and the availability of higher grade gasoline—would not seem likely to have compensated for these obstacles to development which they would have posed.

In one other important respect, however, one can question the effectiveness of PEMEX in assisting Mexico's economic development. To put it more positively, it seems likely that one aspect of Mexico's petroleum policy as expressed through the activities of PEMEX has hindered the country's development. This is in respect of attitudes and policies towards exports. As already indicated nationalization in 1938 eliminated exports—then running at a rate of over 3 million tons per annum—almost overnight. This immediate loss of foreign exchange earning, amounting to some $25 million per annum, was unavoidable for two reasons: firstly, because the expropriated Companies were in a position to exclude Mexican oil from virtually all world markets both by their direct control over markets and by threats of legal action; and secondly, because PEMEX was not technically in a position to maintain production at a level where significant exports would have been possible given the need to maintain supplies for the domestic markets. A little oil continued to flow mainly to Europe on a barter basis,[1] but this was soon interrupted by the war. By the end of the war domestic consumption had, as already shown, gradually increased almost to Mexico's productive capacity and thus only small exports were possible—in spite of the existence of a post-1945 world-wide buyers' market which could have produced significant foreign exchange earnings had the oil been available.[2] By this time one can certainly say that had the international Companies remained in Mexico they would have created a situation in which the country's oil exports would have earned many tens of millions of dollars a year—for part of the massive investment they put into Venezuela during and immediately after the war would certainly have gone to Mexico's oil development with a view to achieving exports of both crude oil and products.

[1] W. J. Levy, Chapter III, p. 42.

[2] By 1945 compensation terms between the Government and the Companies over the expropriation of the latter's assets had been agreed such that there was no longer a legal threat to PEMEX's exports.

In the later 1950s and the 1960s, as PEMEX has developed the country's oil resources, opportunities for exports expanded—but by this time PEMEX was denied the right to seek customers wherever they might be found. This arose because PEMEX was, in general, forbidden to export crude oil on the grounds that crude oil is an inappropriate export commodity as it can be converted into refined products which are 'worth' several times as much and whose manufacture produces jobs and income within the Mexican economy. But this doctrine of 'inherent worth' ignores the reality of the post-war world trading pattern in oil—a pattern within which there has developed a fairly broad market for non-integrated crude oil but only a very limited one for refined products. As PEMEX had by 1964 built up a reserves-production ratio of 32 years—far exceeding the statutory requirement of 25 years, which is in itself half as high again as it need be[1]—and as it had by then an estimated shut-in productive capacity of over 2·5 million tons per year, Mexico could clearly have embarked on a modest crude-oil export campaign with a good chance of success. An export of 5 million tons per year would have brought in at least $50 million per annum in foreign exchange. It seems unlikely that this intensely nationalistic approach to oil exports—backed only by specious economic arguments—could be to the advantage of Mexico in any way whatsoever. Mexico may well have been in a strong enough economic position to forgo the national economic advantages to be gained from the export of crude—but the attitude certainly hindered the ability of PEMEX to secure revenues required for expansion. The absence of export earnings forced it to resort to a greater degree of overseas borrowing to meet its requirements for imported equipment. PEMEX's unavoidable foreign borrowing then operated to the detriment of Mexico's external financial position. Herein lies one national reaction to the international petroleum situation which appears unworthy of emulation by other developing countries.

STATE INTERVENTION IN VENEZUELA

Ironically, however, the recent petroleum policy of Venezuela—the country to which the international petroleum Companies turned their major attention in the Western Hemisphere when they were expropriated in Mexico—has, at least in part, been based on a theory of an 'intrinsic' value for petroleum rather than a value created by demand in the international market. In Venezuela, however, the Government is attempting to implement its policy through its direction and control of the international Companies which continue to dominate the oil industry of the world's major oil exporting nation. Venezuela is, therefore, of more importance to these Companies than the whole of the rest of Latin America put together. The oil Companies' investment in Venezuela totals some $4,000 million; but in the rest of Latin America only some $1,500 million. Their after-tax profits from Venezuela are running at an annual rate of some $400 million; profits earned by the international oil Companies from the rest of Latin America are difficult to compute because of the absence

[1] Compare US experience where the reserves ratio has not exceeded sixteen years for the last twenty-five years, or Venezuela where an annual production over ten times that of Mexico is sustained on a less-than-sixteen years' reserves ratio.

of information from many wholly-owned subsidiaries of international Companies and because of currency conversion problems, but they are certainly no more than one-fifth of those from Venezuela. Thus, within Latin America the only changes in relationships between Companies and governments that have immediate, rather than merely demonstrative, international significance for the Companies, in terms of both their supply position and their overall profitability, are those that occur within Venezuela.

As Lieuwen has shown[1] the international Companies' early search for and development of oil in Venezuela took place under extremely favourable political circumstances. Early war-time dislocations proved to be only minor setbacks in a period of rapid expansion, and production grew from 20 million tons in 1942 to over 70 million tons in 1947. The 1943 re-negotiations of the concession arrangements and taxation system secured for Venezuela—amongst other advantages—a higher tax per barrel of oil produced, and through the so-called red-line agreement, also obliged the Companies to build large-scale refining capacity in Venezuela.[2] The Companies, however, eager to increase their rate of increase of offtake in response to the buoyancy of war-time demand for Western Hemisphere oil accepted the revised obligations as the price required to secure unlimited access to oil which remained highly profitable within the framework of the Gulf posted price system.[3]

The first potential political threat to the international Companies came in 1948 when Venezuela came under the control of Acción Democratica—a left-wing party long pledged, while in opposition, to the nationalization of the country's oil. The new Government concentrated, however, on increasing the nation's share of the profits from the international Companies' oil operations—this it succeeded in doing, but only at the cost of a virtual cessation of growth in production as the Companies sought their increasing requirements from the growing resources of the Middle East. How far the Acción Democratica Government would or could have gone in restraining the international Companies' freedom to act against what it considered to be the best interests of Venezuela remained, however, unknown, as it was quickly overthrown. Then, for almost another decade the Companies again enjoyed a highly favourable politico-economic environment, during which time they lifted annual output from 75 to 150 million tons at an annual average rate of increase of over 10 per cent—some 25 per cent higher than the rate of increase in world demand. The concept of an equal sharing of total profits was accepted by Government and Companies, and so long as almost all the country's oil was transferred to overseas associates of the producing companies at Caribbean posted prices—thus generating high returns on investment in Venezuela—the formula re-

[1] E. Lieuwen, *Petroleum in Venezuela: a History*, University of California Publications in History, xlvi, 1954.

[2] Formerly Venezuelan oil had been mainly refined in the Dutch islands of Aruba and Curaçao. The companies now accepted a commitment that any future expansion of refining capacity in the Caribbean—as defined by a red line on the map—to run Venezuelan oil must be located in Venezuela. See P. R. Odell, 'The Refining Industries of the Caribbean and the Middle East', op. cit.

[3] In retrospect, however, this 1943 agreement between Venezuela and the international oil Companies can perhaps be viewed as a watershed between 'old' and 'new' types of relationships between Companies and host governments.

mained acceptable. One event in this period, however, turned out to be disadvantageous from the point of view of the international Companies: this was President Perez Jiménez's decision in 1956 to auction off large new concession areas. It was at the time when domestic US companies were eager to diversify into overseas production, in order to secure access to lower-cost oil than in the US and hence improve their overall profitability in the US by using it instead of domestic crude. Many such companies offered large sums in return for the concession rights. The auction of concessions produced on immediate $685 million bonus for Venezuela, but it introduced into the country Companies other than the international majors which had hitherto had the fields to themselves. The desire of these new Companies to produce oil quickly was eventually to help produce downward pressure on prices. This was initially to the dismay of the international Companies and ultimately, through them, to that of the Government, whose revenues under the 50/50 profit-sharing arrangements are related to prices actually realized rather than to posted prices as in the Middle East. This fact caused serious difficulties in oil Companies/Government relationships between 1960 and 1966 when an agreement was finally reached. Under the terms of the 1966 agreement, profit sharing is to be evaluated on the basis of 'tax reference' prices (see below, p. 295). In return the Companies secured an end to the threat of legal action over alleged tax under-payment in the period of price weakness after 1958.

Following the overthrow of Perez Jiménez, a second Acción Democratica Government came to power in 1959 pledged to bring the oil industry in Venezuela under effective national control and to make the physical enclaves in Venezuelan territory of the international oil Companies part of the national socio-economic infra-structure with the possibilities of bringing local multiplier effects into operation. The new President was Romulo Betancourt whose book, *Venezuela: politica y petróleo*, published in 1956, had made the case for State control over oil. The Companies, not surprisingly, were concerned at the prospects, and again reacted defensively by immediately cutting back their exploration and development efforts. However, the mixture of left-wing government and the international oil industry proved to be less inflammable than seemed likely. Changes there had to be—but the changes that have occurred have in general been in the Venezuelan tradition of gradual adjustments rather than revolutionary change *à la* Mexico.

There was no explosion because both Government and Companies—both under some external pressures—recognized that their fundamental interests lay in achieving a *modus vivendi*. The Companies had $4,000 million invested in Venezuela: under the revised concession agreements of 1943 they had the right to exploit their oil concessions until 1983 when the agreements terminated. Only limited further expenditure would be needed to ensure that these concessions continued to produce large quantities of oil for the last twenty years or so of their legal life. This gave the established Companies access to low-cost oil in significant amounts over a period when in much of the rest of the world where lower-cost oil might be available there was great political instability. In such a situation it would be desirable to maintain Venezuela as a major producer if only in the interests of geographical diversification of supplies. Coupled with this commercial motivation—which implied at least the maintenance of existing production levels plus stand-by productive capacity for use when required—the Companies also came under political pressure from the US Government not to take any action which would cause difficulties for the Government of Venezuela, which was seen by the US as an 'appropriate' sort of government between the extremism of Castro in Cuba on the one hand and right-wing military dictatorships on the other. All the large oil Companies operating in Venezuela—except Shell—are American and were persuaded to go along with this policy.

The Betancourt Government, for its part, had little to gain and possibly everything to lose by revolutionary action. Politically, it remained in power with the support of the military establishment. It also recognized that its support by the US rested on an evaluation by the latter that it would not resort to extreme measures. Thus, both internal and external political forces acted as moderating influences. Economically, the Government knew that Venezuelan oil had a rapidly declining comparative advantage in the markets of the world as a result of rapidly increasing supplies of low-cost Middle Eastern crude oil and the erosion of Venezuela's advantage in freight costs to markets.[1] The Government had to recognize that, even apart from the legal problems which would undoubtedly arise over the ownership of the oil, there was no possibility of maintaining the level of exports if the major international Companies operating in the country—and responsible for over 90 per cent of total oil exports—were taken over by the State.

Thus, in the period since 1959 the Venezuelan Government and the international Companies have negotiated from positions from which each could inflict serious damage on the other—but knowing this, both have been prepared to accept changes. Out of this there has emerged some form of equilibrium—with, as yet, insufficient time having elapsed to determine whether it is of the 'stable' or 'unstable' type.

The Government has pushed up the industry's direct costs by giving its support to the oil workers union's efforts to secure higher wages and improved fringe benefits, with the result that the rewards to Venezuelan oil industry employees now appear to have reached levels approaching those of oil-workers in Texas (when contrasting tax obligations are taken into account). It has also reduced, and possibly almost eliminated, the economic rents the Companies were previously earning on their Venezuelan operations. This has been achieved by changes in taxation and royalty arrangements whereby the Companies' share of total industry income has been reduced from over 45 per cent to about 30 per cent. The Government has more recently insisted on changing the pricing basis on which gross industry profits are assessed. As indicated previously, the earlier realized prices basis has been replaced by agreed 'tax reference' prices: the relationship of these to posted prices varies from crude to crude and from product to product but, in general terms, they

[1] Due both to lower freight rates for all tankers and to the physical inability of Venezuelan terminals to handle the largest crude carriers not only because of draft restrictions but also because no single crude stream is large enough to load supertankers without substantial investment in additional storage capacity. By contrast, in the Middle East individual streams are so prolific that the crude can be run from well-head to carrier with relatively little need for storage.

recognize a discount of no more than 10 per cent off Caribbean posted prices for tax calculation purposes. This change seems likely to alter the Government/Company shares of gross taxes by another percentage point or two and should ultimately produce a situation very close to a 75/25 split. There seems little chance that the Government's share of the profits can go beyond this without the Companies being forced to close down their higher cost production and thus reduce the overall level of Venezuelan output. This the Government certainly does not want and, recognizing the Companies' position, is unlikely to take the further legislative action which would make such a reaction inevitable.

The Government has, however, quite deliberately adopted a policy of maximizing its 'take' from existing levels of petroleum production (this 'take' per barrel is now, on average, about $1) in the full knowledge that this would cause the international Companies to switch their interest in increasing production and production capacity to other parts of the world. The rationale for this policy stems from the belief that the oil sector of its economy is too dominant—accounting for over 90 per cent of exports and about 20 per cent of the GNP: but, at the same time, giving employment to only 2 per cent of the labour force in a situation in which a population growing at a rate of about 3·5 per cent per annum demands a large number of new job opportunities. The oil industry, moreover, the Government argues, is in foreign hands with no degree of Venezuelan control over decisions taken in New York or London but which vitally affect the Venezuelan economy.

In light of this interpretation of the petroleum situation, the Government's economic policy is essentially one of diversification away from oil—but with oil revenues in the meantime providing the financial basis for the diversification plan, and hence the need to maximize the revenue from each barrel. However, even assuming the policy of diversification is appropriate, this use of oil revenues to finance diversification could be more effectively achieved by maximizing not the 'take' per barrel but total petroleum revenues. This would involve the encouragement of production and an increase in productive capacity, which in turn are only possible with more favourable tax treatment of the Companies (particularly since most Venezuelan oil is higher cost than that of most overseas competitors) and an eagerness—let alone a willingness—to grant new concessions in areas hitherto unexplored. But Venezuela's policy is strictly one of 'no new concessions'. The explanation for this aspect of government policy takes us back to the observation made earlier on Venezuela's attitude to the size of total oil exports. To increase these, so the argument runs,[1] at a time of chronic price weakness not only reduces the return per barrel but may so exacerbate the world price situation that total returns would also fall. And as petroleum is a wasting asset—whose limitations in Venezuela seem real enough in light of the reserves ratio of only fifteen years—a barrel used today when prices are low will be one barrel less to use tomorrow when prices are higher.

Even ignoring the present-worth-of-a-future-barrel type of argument for questioning the validity of Venezuela's policy, it must also be noted that the policy can only be predicated on an optimistic assessment of future demands

[1] See J. P. Perez Alfonzo: *Petroleum Policy*, Government publication, Caracas, 1963.

for crude oils in competition with other forms of energy. Venezuelan authors tend to look rather rosily to the future demand for energy in general and oil in particular—and, most particularly, to the likely contribution of oil from currently known sources to this future demand.[1] However, assuming even again that these Venezuelan views are valid, the Government has to face up to the immediate practical difficulty that its policy, to be effective, has to secure universal acceptance amongst oil producing and exporting nations—for restrictions on production by any one nation to maintain prices may be offset by any willingness elsewhere to expand without control. It was to overcome this fundamental difficulty that Venezuela sought informal contacts with the Middle Eastern producers 'to understand mutual problems, to directly exchange points of view on the world oil situation and to achieve co-ordinated action in the future'.[2] These contacts later provided the initial basis for the formation of OPEC in 1960. Since then Venezuela has taken every opportunity to convince its fellow exporting countries of the validity of its arguments. Within the Organization itself the 1965 agreement on the following year's production levels represented a success for Venezuela. The failure, however, of most of the other nations to take any steps to persuade the Companies to accept these agreed levels and the deliberate actions on the part of one or two OPEC members to encourage the exploiting Companies to increase their production ahead of the OPEC 'agreed level', indicates that not all other major producing nations see their national interests as identical with those of Venezuela. In fact, Venezuela still stands alone amongst the major petroleum exporters in having pursued a deliberate policy of permanent restraint on the level of output.[3]

LATIN AMERICAN REGIONAL CO-OPERATION ON OIL

But Venezuela's attempt, through international co-ordination by means of OPEC, to achieve what it considers to be its national interests, is not its only venture into international action designed to bring into play government-to-government agreements as a means of restraining the power and activities of the international petroleum Companies. Within Latin America it has recently participated in planning regional co-operation on oil policies. Within this geographical framework, however, the only possible interpretation of Venezuela's position is in terms of 'it against the rest'—with the conflict within Latin America arising from causes similar to those that Professor Penrose has indicated affect relations in general between the major oil exporting nations

[1] See, for example, A. R. Martinez, *Our Gift, Our Oil*, 1966.

[2] Ibid., p. 110.

[3] Venezuela has been particularly hard in recent years towards the production plans of the smaller Companies which secured concessions in 1956–7. Their export prices have been closely vetted and challenged immediately if the discounts were considered too high: in contrast the major Companies' prices have in general only been vetted retroactively. Additionally, regulations designed to prevent the flaring of associated gas and other conservation measures have been vigorously implemented for the smaller companies, thus restricting their production. The international Companies have not suffered the same degree of supervision. A former oil minister, Perez Alfonzo, indicated clearly that he would prefer Venezuela to deal only with the international major Companies. Some of the independents have reacted to this by selling out their interests to the larger companies.

and the developing oil importing nations. The former, of which Venezuela is the only one in Latin America (other net exporters in the continent—Peru, Colombia and Trinidad—sold only 13 million tons abroad in 1966 compared with 165 million tons by Venezuela), want high prices for oil in order to enable them to increase their revenues; whilst the latter, including every other nation of Latin America except Mexico and Bolivia, which are self-sufficient, want their imports at the lowest possible cost in foreign exchange and look rather enviously at the relative stability in oil prices in comparison with the great fluctuations in the price levels of their own primary export commodities. From the late 1950s whatever basis there might formerly have been for Latin American solidarity over oil evaporated as the importers accepted tenders for the lowest-cost crude oil. As already shown in the case of Uruguay and Brazil, the orders for most requirements went to the Middle East, Africa and the Soviet Union. Venezuela was increasingly priced out of the Latin American market—a situation which Shell, for example, tried to bring home to the Venezuelans in 1964 by publicizing its very uncompetitive and unsuccessful bids to sell Venezuelan general-purpose crudes to Petróbras and ANCAP.[1] In 1958 Latin American countries imported 25 million tons—of which over 60 per cent came from Venezuela. In 1966 total imports were 31 million tons, of which Venezuela supplied a little less than 50 per cent.

This deteriorating situation, coupled with fears as to the effects that freer Latin American trade in manufactured goods would have on its high-cost industries, inhibited for some years Venezuela's interest in the moves towards regional integration in Latin America—and from 1959 to 1966 Venezuela remained unattached to the Latin American Free Trade Area (LAFTA). However, in 1961, a 'Conference of State Oil Companies of the Americas' was convened in Venezuela by the Minister of Mines and Hydrocarbons, Dr Perez Alfonzo. This later proved to be the initial step towards the formation of ARPEL[2] in 1965—now with headquarters in Lima. This is a non-governmental Latin American organization with a membership consisting of the State oil entity of each country (or a government representative from a country not having a national company but 'manifesting the desire to constitute one'[3]). The discussions with ARPEL so far have been mainly concerned with possibilities of technical co-operation (for example, on the co-ordination of petrochemical developments and the expansion of petroleum equipment manufacture, etc.), but there have also been discussions on national oil policies and a decision 'to carry out studies to facilitate the development of commercial transactions'.[4] The Venezuelan State oil company, Corporación Venezolano del Petróleo, formed in 1960 and staffed by people close to the Government, has played a full part in these discussions and has almost certainly taken the

[1] Shell have, however, continued to secure a market in Brazil for specially tailored Venezuelan crudes to meet specific requirements. In addition, since 1964, Compania Shell de Venezuela appears to have been given greater freedom of action by its parent companies when tendering for business such as Petrobra's. The new policy has enabled it to secure sales —possibly helped by Brazil's desire not to become entirely dependent on Eastern Hemisphere oil.

[2] Asistencia Recíproca Petrólera Estatal Latino Americano.

[3] A. R. Martinez, p. 184.

[4] Ibid., p. 184.

opportunity of sounding out opinion on the possibilities for a Latin American policy of regional self-sufficiency in oil within the framework of the developing regional trading arrangements.

Venezuela's decision in 1966 to seek membership of LAFTA is probably not unrelated to the nature of these Latin American oil contacts. In theory, at least, oil industry operations in different countries and regions of Latin America are complementary—with regional self-sufficiency for the foreseeable future technically possible. In practice, of course, the substitution of Venezuelan oil for oil from external third parties would necessitate some fundamental rethinking of pricing policies by both Venezuela—particularly its state oil entity which refuses in principle to export at less than posted prices—and the importing nations.[1] But some of the major importers, at least, may be not unwilling to purchase oil from CVP at prices somewhat higher than those prevailing on the open market in return for opportunities for some of their manufactured goods in the rapidly expanding Venezuelan markets.

A Latin-American Petroleum Authority—or some similar regional organization—remains only a possibility but its possible development would certainly be viewed with misgivings by the international oil Companies. At best, from their point of view, such an authority would have the effect of taking their Latin American business out of their globally planned operations and of putting it into a regional 'straitjacket', with consequent adverse effects on their profitability and flexibility. At worst, it would assume control and direction over intra-Latin American trade in crude oil and petroleum products and take over responsibility for determining the patterns of production and refining in the region—a responsibility which implies the power of veto by the Regional Latin American authority over the investment decisions of the international Companies. The latter have learned to live profitably with gradually extending national controls over their activities in Latin America over the last three decades. It is difficult to see, however, how they could remain, or perhaps would wish to remain, as viable entities supplying, refining and distributing petroleum to a regionally controlled and economically unifying Latin America. This added regional dimension to the international oil Companies' long and almost continuous fight against 'Statism' in Latin America[2] may

[1] It should be remembered, however, that even complete protection for Venezuelan oil in Latin America would provide additional outlets for, at most, 10 per cent of its present annual production. Quantitatively for Venezuela, preference in the protected US market remains more important, and the country's diplomacy is concerned with securing for Venezuela the right to sell oil to the US outside the quota system—an advantage currently extended only to Mexico and Canada: the only two countries with land frontiers with the US. An extension to Venezuela would involve a break in the national security principle which availability across a land frontier implies.

[2] This struggle with 'Statism' has not been limited to the countries covered in this study. Additionally, Bolivia has followed policies similar to those of Argentina; Peru has had a State oil company—Empresa Petrolera Fiscal—since 1934 and has a long-standing dispute with Esso over the company's concessions and tax obligations; Colombia has a national company—Ecopetrol—with responsibilities ranging from crude production to distribution of products, and has created a legislative and tax framework which the international Companies operating in Colombia generally consider highly unfavourable to their interests. Even the small Central American countries have intervened to ensure national refinery development and to achieve price control over petroleum products. While Cuba, of course, entirely expropriated the international Companies in 1960. With the exception of this

finally turn the odds too heavily against them—but before this happens the 'developing oil exporter' and the 'developing oil importers' have serious differences to resolve. In the meantime the international Companies will continue to sell oil *in* those Latin American countries where they are still able to do so—and, where they are not, continue to make not unreasonable profits in selling oil *to* most of the others.

Cuban action taken by an avowedly Communist government, the case studies presented in this essay were selected to ensure coverage of the whole range of Latin American responses to the international petroleum Companies.

SELECTED BIBLIOGRAPHY AND WORKS CITED

ADELMAN, M. A., 'Oil Prices in the Long Run (1963–75)', *Journal of Business of the University of Chicago*, April 1964, reprinted in *Publications in Social Science*, Series 4, No. 21 (b).

'Efficiency of Resource Use in Crude Petroleum', *Southern Economic Journal*. October 1964.

'The World Oil Outlook' in *Natural Resources and International Development* (Marion Clawson, ed.). Baltimore, Md., Johns Hopkins Press. 1964.

'Oil Production Costs in Four Areas'. A paper presented at the Annual Meeting of the American Institute of Mining, Metallurgical, and Petroleum Engineers (February 28 – March 2, 1966). *Proceedings of the Council of Economics*.

AL-PACHACHI, NADIM, *Iraqi Oil Policy: August 1954 – December 1957*, Baghdad. Research and Translation Office, Beirut. 1958.

AUSTRALIA, Tariff Board, *Report on the Petroleum Refining Industry*. 1959.

Report on Motor and Aviation Spirits. June 7, 1956.

BERLE, A. A., JR, and G. C. MEANS, *The Modern Corporation and Private Property*. New York, Macmillan. 1933.

BERMÚDEZ, A. J., *The Mexican National Petroleum Industry: A Case Study in Nationalisation*. Stanford University Press. 1963.

BETANCOURT, R., *Venezuela: Politica y Petróleo*. F.C.E., Mexico. 1956.

BYÉ, MAURICE, 'Self-Financed Multiterritorial Units and Their Time Horizon', *International Economic Papers*, No. 8, p. 159. Translated from *Revue d'économie politique*. June 1955.

DASGUPTA, BIBLAPKUMAR, *Oil Prices and the Indian Market, 1886–1964*. Unpublished Ph.D. Dissertation in the University of London. 1965.

DE CHAZEAU, M. G., and A. E. KAHN, *Integration and Competition in the Petroleum Industry*. New Haven, Yale University Press. 1959.

DESCHERT, CHARLES R., *Ente Nazionale Idrocarburi: Profile of a State Corporation*. Leiden, Brill. 1963.

European Coal & Steel Community, *Études sur les perspectives énergétiques à long term de la Communauté européenne*. Luxembourg. December 1962.

FORBES, R. J., *Studies in Early Petroleum History*. Leiden, Brill. 1958.

FORBES, R. J., and D. R. O'BEIRNE, *Technical Development of the Royal Dutch/Shell. 1890–1940*. Leiden, Brill. 1957.

FRANK, HELMUT J., *Crude Oil Prices in the Middle East: A Study in Oligopolistic Price Behavior*. New York, Praeger. 1966.

FRANKEL, P. H., *Essentials of Petroleum*. London, Chapman and Hall. 1940.

'Taxation of Petroleum Products and its Impact on Consumption', Seminar Paper No. 30. United Nations Inter-Regional Seminar on Techniques of Petroleum Development. New York, 23 January – 21 February, 1962.

Oil: the Facts of Life. London, Weidenfeld & Nicolson. 1962.

Mattei: Oil and Power Politics. London, Faber & Faber. 1966.

FRANKEL, P. H., and W. L. NEWTON, 'Recent Developments in the Economics of Petroleum Refining'. Paper presented to the Sixth World Petroleum Congress, 1963.

'Current Economic Trends in Location and Size of Refineries in Europe'. Paper presented to the Fifth World Petroleum Congress, 1959.

FRONDIZI, A., *Politica y Petróleo*. Buenos Aires. 1953.

Petróleo y Nación. Buenos Aires. 1963.

GIBB, GEORGE S., and EVELYN H. KNOWLTON, *The Resurgent Years, 1911–1927, History of Standard Oil Company (New Jersey)*. New York, Harper and Bros. 1956.

GIDDINS, P. H., *Standard Oil Company (Indiana): Oil Pioneer of the Middle West*. New York, Appleton-Century-Croft. 1955.

GERRETSON, F. C., *History of the Royal Dutch* (4 vols. 1953–7). Leiden, Brill. 1957.

HARTSHORN, J. E., *Oil Companies and Governments*. London, Faber & Faber. 2nd. ed. 1967.

HIDY, RALPH W., and MURIEL E. HIDY, *Pioneering in Big Business, 1882–1911, History of Standard Oil Company (New Jersey)*. New York, Harper and Bros., 1955.

HIRST, DAVID, *Oil and Public Opinion in the Middle East*. London, Faber & Faber. 1966.

INDIA, Government of, Ministry of Steel, Mines and Fuel, *Report of the Oil Price Enquiry Committee* (Damle Report). New Delhi, Government of India Press. 1963.

Ministry of Petroleum and Chemicals, Department of Petroleum, *Report of the Working Group on Oil Prices* (Talukdar Report). New Delhi. Government of India Press. 1965.

ISSAWI, CHARLES, and MOHAMMED YEGANEH, *The Economics of Middle Eastern Oil*. New York, Praeger. 1962.

JAMES, MARQUIS, *The Texaco Story, 1902–1952*. Texas Company, 1953.

KAPLAN, M., *Economia y Politica del Petróleo Argentino, 1939–56*. Buenos Aires. 1960.

KLEIN, H. S., 'American Oil Companies in Latin America: The Bolivian Case', *Inter-American Economic Affairs*, Vol. 18, No. 2. 1964.

KUBBAH, A. A. Q., *Libya: Its Oil Industry and Economic System*. Baghdad Arab Petro-Economic Research Center. 1964.

LAMBERTON, D. M., *The Theory of Profit*. Oxford, Blackwell. 1965.

LAUDRAIN, M., *Le prix du Pétrole brut*. Paris. 1958.

LEEMAN, WAYNE, *The Price of Middle East Oil*. Ithaca, N.Y., Cornell University Press. 1962.

LIEUWEN, E., *Petroleum in Venezuela: A History*. Berkeley, University of California Press. 1954.

LENCZOWSKI, GEORGE, *Oil and State in the Middle East*. Ithaca, N.Y., Cornell Univ. Press. 1960.

LEVY, W. J., 'The Past, Present and Likely Future Price Structure for the International Oil Trade', Proceedings of the Third World Petroleum Congress, The Hague, 1951.

LEVY, W. J., Inc. *The Search for Oil in Developing Countries: A Problem of*

Scarce Resources and Its Implications for State and Private Enterprise. Report to the International Bank for Reconstruction and Development. 1960.

LONGHURST, H., *Adventure in Oil.* London, Sidgwick & Jackson. 1959.

LONGRIGG, S. H., *Oil in the Middle East*, Royal Institute of International Affairs, London, Oxford University Press. 2nd. ed. 1961.

MACHLUP, FRITZ, *The Basing-Point System.* Philadelphia, Blakiston. 1949.

MCLEAN, J. G., and R. W. HAIGH, *The Growth of Integrated Oil Companies.* Boston, Harvard University. 1954.

MARRIS, ROBIN, *The Economic Theory of Managerial Capitalism.* London, Macmillan. 1964.

MARTINEZ, A. R., *Our Gift, Our Oil.* Vienna. 1966.

MENDERSHAUSEN, *Dollar Shortage and Oil Surplus in 1949–50.* Essays in International Finance, No. 11. International Finance Section, Department of Economics, Princeton University. 1950.

MIKDASHI, ZUHAYR, *A Financial Analysis of Middle Eastern Oil Concessions: 1901–65.* New York, Praeger. 1966.

MONROE, ELIZABETH, *Britain's Moment in the Middle East, 1914–56.* London, Chatto & Windus. 1963.

MUGHRABY, MUHAMAD A., *Permanent Sovereignty Over Oil Resources: A Study of Middle East Oil Concessions and Legal Change.* Middle East Oil Monographs, No. 5. Beirut, The Middle East Research and Publishing Center. 1966.

ODELL, P. R., 'Oil and State in Latin America', *International Affairs*, Vol. 40, No. 4. October 1964.

An Economic Geography of Oil. London, Bell. 1963.

'The Development of the Middle Eastern and Caribbean Refining Industry, 1939–63'. *Tijdschrift voor Economische en Sociale Geografie*, Vol. 54. October 1963.

ORGANISATION OF PETROLEUM EXPORTING COUNTRIES:

OPEC and the Oil Industry in the Middle East, F. R. Parra, 1962.

Pricing Problems: Further Considerations, no date.

The Development of Petroleum Resources under the Concession System in Non-industrialised Countries, F. R. Parra, 1964.

Elasticity of Demand for Crude Oil; Its Implications for Exporting Countries, Isam K. Kabbani, 1964.

The Oil Industry's Organisation in the Middle East and Some of its Fiscal Consequences, F. R. Parra, 1963.

The Price of Crude Oil: A Rational Approach. Paper presented to the Fourth Arab Petroleum Congress, 1963.

Radical Changes in the International Oil Industry During the Past Decade. Paper presented to the Fourth Arab Petroleum Congress, 1963.

OPEC and the Principle of Negotiation. Paper presented to the Fifth Arab Petroleum Congress, 1965.

Taxation Economics in Crude Production. Paper presented to the Fifth Arab Petroleum Congress, 1965.

From Concessions to Contracts. Paper presented to the Fifth Arab Petroleum Congress, 1965.

Collective Influence in the Recent Trend Towards the Stabilisation of Inter-

national Crude and Product Prices. Paper presented to the Sixth Arab Petroleum Congress, 1967.

PENROSE, EDITH T., *The Theory of the Growth of the Firm*. Oxford, Blackwell. 1959.

'Foreign Investment and the Growth of the Firm', *Economic Journal*, Vol. LXVI. June 1956.

'Vertical Integration with Joint Control of Raw Material Production: Crude Oil in the Middle East', *Journal of Development Studies*, Vol. 1, No. 3. April 1965.

PENROSE, E. F., *The Revolution in International Relations*. London, Frank Cass. 1965.

PEREZ ALFONSO, J. P., *Petroleum Policy*. Venezuelan Government Publication. Caracas. 1963.

POLITICAL and Economic Planning, *A Fuel Policy for Britain*. London, PEP. 1966.

RONDOT, JEAN, *La Compagnie Française des Pétroles: du franc-or au pétrole-franc*. Histoire des Grandes Entreprises – 7. Paris, Plon. 1962.

SCHULMAN, JAMES, *Transfer Pricing in Multinational Business*. (A thesis submitted to the Graduate School of Business Administration, Harvard University, 1966).

SHELL International Petroleum Company Ltd., *Current International Oil Pricing Problems*. London. August 1963.

SHWADRAN, BENJAMIN, *The Middle East, Oil and the Great Powers, 1959*. New York, Council for Middle Eastern Affairs. 2nd. ed. 1959.

STANDARD Oil Company (New Jersey), *A Brief History*. New York, Standard Oil Co. (New Jersey). 1964.

STAUFFER, THOMAS R., *The Erap Agreement—A Study in Marginal Taxation Pricing*. Presented to the Sixth Arab Petroleum Congress, 1966. Also published in *Platts Oilgram*, December 1966.

SWEET-ESCOTT, B. A. C., 'Financing Problems of Integrated Oil Companies', Paper No. 46 (A–1), Fourth Arab Petroleum Congress, November 1963.

SYMONDS, EDWARD, 'Life with a World Oil Surplus', First National City Bank of New York. 1960.

'Oil Prospects and Profits in the Eastern Hemisphere', First National City Bank of New York. 1961.

'Oil Advances in the Eastern Hemisphere', First National City Bank of New York. 1962.

'Financing Oil Expansion in the Development Decade', First National City Bank of New York. 1963.

'Eastern Hemisphere Petroleum—Another Year's Progress Analysed', First National City Bank of New York. 1963.

TUGENDHAT, GEORG, 'An Outsider's View of the Oil Industry', *The Institute of Petroleum Review*, February 1967.

UNITED NATIONS, *Energy in Latin America*. Geneva, 1957.

Economic Commission for Europe, *The Price of Oil in Western Europe*. Geneva, United Nations. 1955.

UNITED STATES Federal Trade Commission, *The International Petroleum Cartel*. U.S. Senate Select Committee on Small Business, Staff Report to the

Federal Trade Commission, 82nd Congress Committee Print No. 6, August 22, 1952.

U.S. House of Representatives, *Current Antitrust Problems*, Hearings Before Antitrust Sub-committee (no. 5) of the Committee on the Judiciary. 84th Congress. 1st Session, Part I. Serial No. 3.

U.S. Senate, Committee on the Judiciary, Sub-committee on Antitrust and Monopoly, *Petroleum, The Antitrust Laws and Government Policies*. Report No. 1147, 85th Congress 1st Session, 1957.

U.S. Senate, Committee on Interior and Insular Affairs, Sub-committee on Public Lands, *Emergency Oil Lift Program and Related Oil Problems*. Pt. II, 85th Congress, 1st Session, 1957.

VERNON, R., *The Dilemma of Mexico's Development*. Harvard University Press. 1963.

VOTAW, DOW, *The Six-Legged Dog: Mattei and ENI—a Study in Power*. Pubn. of the Institute of Business and Economic Research, University of California. Berkeley, University of California Press. 1964.

INDEX